IMPROVING QUALITY THROUGH PLANNED EXPERIMENTATION

McGraw-Hill Series in Industrial Engineering and Management Science

Consulting Editors

Kenneth E. Case, *Department of Industrial Engineering and Management,*
 Oklahoma State University
Philip M. Wolfe, *Department of Industrial and Management Systems Engineering,*
 Arizona State University

IMPROVING QUALITY THROUGH PLANNED EXPERIMENTATION

Ronald D. Moen

Thomas W. Nolan

Lloyd P. Provost

Boston, Massachusetts   Burr Ridge, Illinois
Dubuque, Iowa   Madison, Wisconsin   New York, New York
San Francisco, California   St. Louis, Missouri

This book was set in Times Roman by Publication Services.
The editors were Eric M. Munson and John M. Morriss;
the production supervisor was Annette Mayeski.
The cover was designed by David Romanoff.
Project supervision was done by Publication Services.

McGraw-Hill

A Division of The McGraw·Hill Companies

IMPROVING QUALITY THROUGH PLANNED EXPERIMENTATION

8 9 10 11 12 13 14 BKMBKM 9 9 8 7

ISBN 0-07-042673-2

Library of Congress Cataloging-in-Publication Data

Moen, Ronald D.
    Improving quality though planned experimentation / Ronald D.
  Moen, Thomas W. Nolan, Lloyd P. Provost.
        p.   cm. — (McGraw-Hill series in industrial engineering and
  management science)
    Includes bibliographical references.
    ISBN 0-07-042673-2
    1. Quality control—Statistical methods.   2. Process control—
  Statistical methods.   3. Experimental design.   I. Nolan, Thomas W.
  II. Provost, Lloyd P.   III. Title.   IV. Series.
  TS156.M62   1991
  658.5'62'015195—dc20   90-24317

ABOUT THE AUTHORS

Ronald D. Moen, Thomas W. Nolan, and **Lloyd P. Provost** have worked as a team for almost 20 years. They first met in the early 1970s, when they worked at the U.S. Department of Agriculture. It was there that their interest in planned experimentation began. This was their first opportunity to work with scientists and managers to design experiments.

Their understanding of the needs of experimenters expanded in the early 1980s as they worked with managers, scientists, and engineers engaged in improving the quality of their products and services. This work included providing assistance in the planning of experiments for product design, process design, improvements in manufacturing, start-up of new plants, and research.

In 1984 they founded Associates in Process Improvement, a consulting group that assists organizations to make quality a key business strategy.

Ronald Moen has an M.S. in statistics from the University of Missouri. Thomas Nolan has a Ph.D. in statistics from George Washington University. Lloyd Provost has an M.S. in statistics from the University of Florida.

THIS BOOK IS DEDICATED TO THE MANY STUDENTS AND CLIENTS OVER THE LAST SEVERAL YEARS WHO HAD TO ENDURE VARIOUS STAGES IN THE EVOLUTION OF THIS PRODUCT. UNDER CONTINUAL IMPROVEMENT, A BOOK IS NEVER FINISHED. YOUR FEEDBACK GAVE US THE UPDATED CURRENT KNOWLEDGE NECESSARY FOR IMPROVEMENT.

THANK YOU.

CONTENTS

FOREWORD

This book by Messrs. Ronald D. Moen, Thomas W. Nolan, and Lloyd P. Provost breaks new ground into the problem of prediction based on data from comparisons of two or more methods or treatments, tests of materials, and experiments.

Why does anyone make a comparison of two methods, two treatments, two processes, two materials? Why does anyone carry out a test or an experiment? The answer is to predict; to predict whether one of the methods or materials tested will in the future, under a specified range of conditions, perform better than the other one.

Prediction is the problem, whether we are talking about applied science, research and development, engineering, or management in industry, education, or government.

The question is, what do the data tell us? How do they help us to predict?

Unfortunately, the statistical methods in textbooks and in the classroom do not tell the student that the problem in use of data is prediction. What the student learns is how to calculate a variety of tests (t-test, F-test, chi-square, goodness of fit, etc.) in order to announce that the difference between the two methods or treatments is either significant or not significant. Unfortunately, such calculations are a mere formality. Significance or the lack of it provides no degree of belief— high, moderate, or low— about prediction of performance in the future, which is the only reason to carry out the comparison, test, or experiment in the first place.

Any symmetric function of a set of numbers almost always throws away a large portion of the information in the data. Thus, interchange of any two numbers in the calculation of the mean of a set of numbers, their variance, or their fourth moment does not change the mean, variance, or fourth moment. A statistical test is a symmetric function of the data.

In contrast, interchange of two points in a plot of points may make a big difference in the message that the data are trying to convey for prediction.

The plot of points conserves the information derived from the comparison or experiment. It is for this reason that the methods taught in this book are a major contribution to statistical methods as an aid to engineers, as well as to anyone in industry, education, or government who is trying to understand the meaning of figures derived from comparisons or experiments. The authors are to be commended for their contributions to statistical methods.

W. Edwards Deming
Washington
14 July 1990

PREFACE

This book is about planned experimentation to improve quality. We believe that statistical methods of planned experimentation are powerful aids to managers, engineers and scientists, and technicians. But these methods have been applied in only a small fraction of the circumstances in which they would have been useful. The aim of this book is to provide a system of planned experimentation in such a way that there will be a substantial increase in the number of people who will use these methods.

Our approach to accomplish this aim contained several components. We continually strove to increase our understanding of the needs of managers, engineers, and others with regard to methods of experimentation. We studied and integrated theory and methods of others with our own ideas and experiences. We were especially influenced by studying the works of W. Edwards Deming, George Box, Stuart Hunter, William Hunter, and Genichi Taguchi. Finally we developed a system of experimentation that met the essential needs of experimenters but required a lower level of mathematical and statistical sophistication than was previously necessary.

We learned a great deal from Deming's papers on analytic studies (studies to improve a product or process in the future), including

- Prediction as the aim of an analytic study
- The importance of conducting analytic studies over a wide range of conditions
- The limitations of commonly used statistical methods such as analysis of variance to address the important sources of uncertainty in analytic studies
- The importance of certain graphical methods

Chapters on factorial and fractional factorial designs in *Statistics for Experimenters* by Box, Hunter, and Hunter (John Wiley and Sons, 1978) provided the foundations for our chapters on these subjects. Especially useful to us was their description of factorial designs at two levels as a link of paired comparisons and their system of fractional factorial designs.

Taguchi's contributions to the application of planned experimentation to the design of product and processes have been integrated throughout the book, but especially in Chapter 10. We are in agreement with Taguchi that most experimenters are in need of a small number of design matrices (called orthogonal arrays by Taguchi) that cover most of their applications with only minor adaptations. Our approach to fractional factorial designs blends that approach with the important concepts of confounding included in the book by Box, Hunter, and Hunter.

The spirit of our approach to analysis of data from experiments owes much to Dr. Tukey's methods of exploratory data analysis. We have blended these ideas with Deming's counsel that confirmation of the results of exploratory analysis comes primarily from prediction rather than from the use of formal statistical methods such as confidence intervals. Satisfactory prediction of the results of future studies conducted over a wide range of conditions is the means to increase the degree of belief that the results provide a basis for action.

In 1984 we were asked by Dr. Deming to write a book on planned experimentation from the viewpoint of analytic studies described in several of his papers and books. As notes on earlier drafts became available, they were used in almost a hundred seminars and were improved based on observation of their usefulness to the participants both during and after the seminars.

There are several aspects of this book that, when combined, make it different from others currently available. The book is written to be compatible with Deming's viewpoint of analytic studies. Also, we have presented planned experimentation as a system in the context of a model for improving quality. This system includes the integration of methods of statistical design of experiments with statistical process control. As part of the system we have emphasized the sequential use of experiments and provided guidance on how to build the sequence.

We have included guidance on many of the practical aspects of planning experiments and have provided a form so that these aspects are addressed during the planning phase. Most of the examples in the book are based on our experiences. We have included examples that contain some of the problems often encountered in actual experimentation in a manufacturing plant or research facility. We offer suggestions on what to do in those situations. Another distinctive attribute of the book is the almost exclusive use of graphical methods for analysis of data from experiments.

Since we are aiming at a relatively broad audience, we have sometimes substituted methods that could be learned and used by this broad audience in place of the more traditional methods. Our feedback from experimenters convinced us of the importance of the use of original engineering units in interpreting data from experiments. We have not included designs that would require development of a mathematical model for analysis. That is why designs for mixtures or formulations and central composite designs are only briefly discussed. These are extremely useful designs, but the model-based approach necessary for their analysis would not have been compatible with approaches used throughout the rest of the book. In the case of central composite designs, we have substituted factorial or fractional factorial designs.

We have not included fairly common statistical methods such as standard errors, confidence intervals, and analysis of variance. We recognize that experimenters need to distinguish between variation that is a result of planned changes in the factors and variation that results from other sources, e.g., measurement. Based on the principles of analytic studies in Chapter 3 and the concept of common and special causes in Chapter 2, we have chosen to use graphical methods to help experimenters ascertain how much the planned changes in the factors are contributing to the variation in the data.

ACKNOWLEDGMENTS

We wish to thank Dr. W. Edwards Deming for his continuing encouragement and his work on the distinction between enumerative and analytic studies. We appreciate the comments and suggestions of our associates, Jerry Langley and Kevin Nolan. Most importantly, we wish to thank the experimenters that we had the opportunity to work with and learn from. We appreciate their patience with us as we used earlier, sometimes very rough, drafts of this book.

Ronald D. Moen
Thomas W. Nolan
Lloyd P. Provost

CHAPTER

1

IMPROVEMENT OF QUALITY

1.1 INTRODUCTION

Global competitive pressures are causing organizations to find ways to better meet the needs of their customers, to reduce costs, and to increase productivity. Improvement of quality has developed as a focal point in meeting these objectives. Continuous improvement of quality has become a necessary and integral part of the business strategy of organizations.

Improvement of quality is predicated on change. Imai (1986) describes two kinds of change: gradual and abrupt. Gradual change results from small improvements to the status quo through continuous efforts that involve everyone. Abrupt change comes from innovation—a drastic improvement in the status quo.

The requirements for improvement of quality are a common purpose and knowledge of concepts and methods so that change results in improvement. The overriding goal is continuous improvement in every activity. Getting better and better is more important than whether the current results are good or bad.

Purpose of This Book

The purpose of this book is to provide the philosophy, principles, and methodologies to plan and conduct experiments that lead to improvement of quality. Quality will be improved through understanding current and future needs of the customer, designing the product to meet those needs, and designing the process that results in the product. The better the knowledge of these processes, the more likely planned changes are to result in improvement.

The methods of *planned experimentation* will help people to learn about the many factors that impact the quality of the product or process and to use this knowledge to improve quality and make changes to prevent problems and reduce variation. This planned approach replaces the old philosophy of finding something that works and maximizes learning relative to the resources expended.

The primary reason to carry out an experiment is to provide a basis for action on the product, service, or process to improve its performance in the *future*. Interpreting the results of an experiment is prediction — that a change in a product or process will lead to improvement in the future.

The formulation of a scientific basis for prediction has its beginnings with W. A. Shewhart (1931), who said, "A phenomenon will be said to be controlled when, through the use of past experience, we can predict, at least within limits, how the phenomenon may be expected to vary in the future."

Deming (1950) expands on the idea of improving performance of a product or service in the future by differentiating enumerative and analytic studies. Deming states that most problems in industry are analytic. In the spirit of Shewhart and Deming, this book is concerned with the design of analytic studies to improve quality and with the analysis of data from these studies.

The methodologies presented in this book are integrated into a model for improving quality. This model is designed to increase knowledge of the process, knowledge that in turn leads to improvement of the product or service. The model is fundamental to the improvement of quality and is used throughout the book.

The concepts behind improvement of quality and the model for improving quality are developed in the next two sections of this chapter. Chapter 2 features Shewhart's concept of sources of improvement and control charting. Chapter 3 develops the concepts behind the design and interpretation of analytic studies. Chapter 4 lists the tools and properties of a good experiment.

With these foundations for experimentation in place, subsequent chapters present the methods of planned experimentation. Chapter 10 is devoted to the application of methods to improve the design of a product (new or existing) and the processes to manufacture that product. Chapter 11 shows through two case studies how the model and the methods work together as a system of experimentation.

1.2 IMPROVEMENT OF QUALITY

An organization is composed of people—not only machines, policies, activities, or organization charts. Improvement of quality involves the (external) customers and suppliers as part of the organization of people. How should this expanded organization be viewed?

The Organization as a System

A starting point for improvement is adopting a new view of the organization. Deming (1986) views the organization as a system that includes the goal of improvement of quality in every stage from receipt of incoming materials to the consumer, as well as redesign of products and services for the future. All functions and activities are directed at a *common purpose*. Deming illustrated production as a system by means

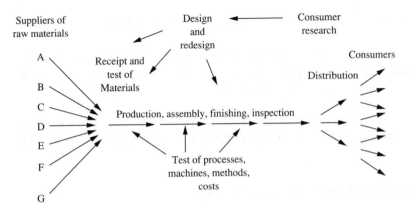

FIGURE 1.1
Deming's view of production as a system

of a flow diagram. This diagram, first used by Dr. Deming in 1950, is reproduced in Figure 1.1. The consumer is the most important part of the production line.

Improvement of quality begins with identifying the future needs of the customer through consumer research. At the design and redesign phase, products or services are designed to better meet those needs. Processes are designed to produce the product or service. These designs and these processes are constantly being improved. The activities for matching products and services to a need are ongoing. The cycle never ends.

Dimensions of Quality

A difficulty in improving quality is translating needs of the customer into measurable characteristics. Garvin (1987) proposed eight dimensions of quality. The following list is an expansion of Garvin's eight dimensions:

Performance	Primary operating characteristics
Features	Secondary operating characteristics, added touches
Time	Time waiting in line, time from concept to production of a new product, time to complete a service
Reliability	Extent of failure-free operation
Durability	Amount of use until replacement is preferable to repair
Uniformity	Low variation among repeated outcomes of a process
Consistency	Match with documentation, advertising, deadlines, or industry standards
Serviceability	Resolution of problems and complaints
Aesthetics	Characteristics that relate to the senses

Personal interface[1]	Characteristics such as punctuality, courtesy, and professionalism
Harmlessness	Characteristics relating to safety, health, or the environment
Perceived quality	Indirect measures or inferences about one or more of the dimensions; reputation

A product or service can rank high on one dimension and low on another. An understanding of the relationships of selected characteristics is basic to any improvement effort.

Basic Activities for Improvement of Quality

Four important activities to improve quality are:

> design of a new product,
> redesign of an existing product,
> design of a new process (including service),
> redesign of an existing process.

These four activities may be carried out within various functions of the organization. Efforts must be coordinated and focused on a common purpose. Barriers between departments should be broken down so that people in research, design, sales, and production can work as a team in performing any of the four basic activities to better match products and services to a need.

The greatest improvement in a product's quality will come during the design of the product and the design of the manufacturing processes; this is *quality by design* (see Chapter 10). The potential for improvement during these phases is many times greater than for stages downstream in the manufacture of the product. The uncertainty of quality improvement at these phases is increased since the results of tests must be extrapolated to predict how the product will perform in the future.

Management must provide the time and resources to allow engineers and people in research to conduct experiments during the generation of technology, development of the product, and development of the process behind the product. The needs of the customer must guide work in all these functions.

Examples of questions that need to be addressed by engineers, R&D personnel, or managers are:

- How do we transform concepts that have the potential to meet customer needs into products?
- How do we select the best concept for meeting customer needs from the many that are in contention?

---

[1] Adapted from Plsek (1987).

- How do we select the vital few parameters for design among the hundreds of choices?
- How do we design a product to work under the wide range of conditions that will be encountered during actual production and use by the customer?
- How do we choose the best operating conditions for a manufacturing process among the hundreds of choices?

Improvement through Learning

The key factor for success in improvement of quality is learning. It is through learning that improvements in products and services are made. The methods of planned experimentation presented in this book will enhance knowledge. This enhanced knowledge will help answer questions such as those given above. It will result in a better design for the product and a better design for the manufacturing processes.

Learning is also enhanced by fostering teamwork. Team activities should be centered around the satisfaction of internal and external customers. Management must provide training for teams in the methods of planned experimentation to increase knowledge in each of the four basic activities for improvement of quality. Only through providing a proper managerial environment, with every person working on improvement of quality to enhance customer satisfaction, will an organization be able to compete in the international marketplace.

Continuous communication and teamwork between customers and suppliers and between managers and workers will be necessary to identify opportunities for improvement. This communication can be started by addressing the following questions:

Supplier to customer: What are some ways in which we could improve our product or service?

Manager to workers: What are some ways in which we could change the system so that you could do your job better?

Customer to supplier: What changes could we make in our system to help you better meet our needs?

A management style must continually encourage these questions and allow people to work in teams to increase their knowledge in order to make improvements or to solve the problems that come to light. Management must view the organization as a system of processes and manage the *linkage* of these processes. Management must remove the obstacles to improvement. Deming (1986) provides both a philosophy and a framework for making continuous improvement of quality a focal point of an organization's business strategy. This philosophy represents a transformation of the style of management present in many organizations today.

Deming's Chain Reaction

Historically, quality control in the manufacturing and service industries has consisted of inspection of the product or service relative to a set of requirements (specifications).

In manufacturing, this function is carried out by the quality control (QC) department. In service industries, the counterpart of the QC department is often called the audit department.

Once the product is inspected, it is sorted into batches of good and bad , and the bad product is reworked or scrapped. Resubmission of a computer run due to errors of input, correction of errors on invoices in accounts payable, and delivery to a passenger's hotel of a bag that was incorrectly routed by an airline are examples of rework in service industries.

There are several well-known inadequacies in this approach to improvement of quality, among them:

- Issues related to quality are not addressed until it is too late; the product or service is already completed.
- Quality is obtained at high cost and with loss of productivity.
- A "firefighting" approach to problem solving is adopted, which results in short-term solutions to immediate problems at the expense of long-term improvement.

A change in the approach to improve quality is needed. The theory needed for this change is provided by Deming (1986, p. 3) and is known as the "quality and productivity chain reaction." Figure 1.2 summarizes this theory, which states that if an organization focuses on the improvement of quality, reduced costs and higher productivity will follow. The organization will then increase market share through the better quality and lower cost. They will stay in business and provide more jobs.

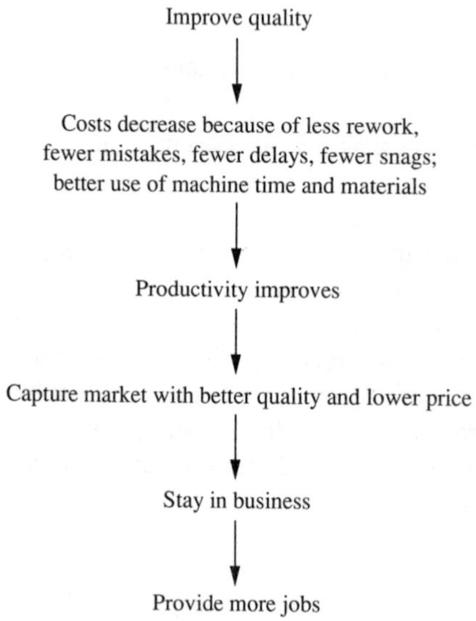

Improve quality

↓

Costs decrease because of less rework,
fewer mistakes, fewer delays, fewer snags;
better use of machine time and materials

↓

Productivity improves

↓

Capture market with better quality and lower price

↓

Stay in business

↓

Provide more jobs

FIGURE 1.2
Deming's quality and productivity chain reaction

This chain reaction for improvement of quality cannot be *initiated* through inspection of outcomes of processes. Instead, attention must be directed toward the process that leads to the product or service. This change in focus from inspection of product to improvement of process is necessary if a higher-quality product at a lower cost is to be achieved. Emphasis on improvement of the process increases the uniformity of product output. Lower cost is achieved by reducing the amount of rework, and the number of mistakes, delays, and snags, and by making better use of machine time and materials.

There are some major differences between inspection of product and improvement of process. The analysis of the process is performed by all members of the organization and thus is a small part of everybody's job rather than the total responsibility of a few. The process is studied, and hence learning takes place, even when no defective products or services are being produced. Quality is increased by the use of new knowledge as a basis for changing the process or the product. Planned experimentation is an important method of obtaining this new knowledge. Since these changes allow tasks to be done better, faster, and easier, decreases in cost accompany the improvements in quality.

Process Model

What is a process? A process is defined (Moen and Nolan, 1987) as follows:

> **Process**: A set of causes and conditions that repeatedly come together to transform inputs into outcomes.

The process model is illustrated in Figure 1.3. The inputs may include people, methods, material, equipment, environment, and information. The outcome is some product or service. There can be several stages to the process. Alternatively, each stage could be viewed as a process.

An example of a process is the design of a product. The stages might be request for a design, preliminary design, review, and approval of design. Inputs include information from marketing as well as engineering knowledge. The outcome is a

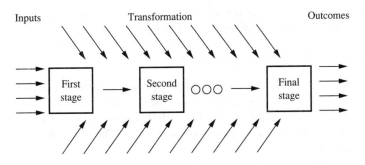

FIGURE 1.3
Process model

written description of a design for a product. Other examples of processes are injection molding, assembly line operation, pressed metal stamping, hiring process, classroom training and education, billing, accounts payable, and managing people.

The next question is, "What is meant by improvement of process?" Process improvement is defined as follows:

> **Process improvement**: The continuous endeavor to learn about the cause system in a process and to use this knowledge to change the process to reduce variation and complexity and to improve customer satisfaction.

Improvement will come through taking action based on a better understanding of the cause system that affects the performance of a process. This improvement includes innovation of processes and products as well as gradual changes.

Basic Sources for Improvement of Quality

For any process, indicators of the performance of the process can be identified and measured. These indicators will be called *quality characteristics*. For manufacturing processes, measures such as length, width, viscosity, color, temperature, line speed, number of accidents, and percent rejected material are examples. Number of errors in billing, number of incorrect transactions in a bank, checkout time in a grocery store, frequency of program restarts in data processing, and actual expenditures are examples of quality characteristics for service processes. Quality characteristics for the organization as a system include absenteeism, turnover, warranty costs, profits, and share of market.

All of these measures will vary over time. Analysis of this variation is used as a basis for action to improve the process. Often, however, this action is inappropriate or counterproductive, because people lack an understanding of the concept of common versus special causes of variation.

A fundamental concept necessary to the study and improvement of processes introduced by Shewhart (1931) is that variation in the outcome of a process is due to two types of causes:

> **Common causes**: Causes that are inherent in the process over time, affect everyone working in the process, and affect all outcomes of the process.
> **Special causes**: Causes that are not present in the process all the time or do not affect everyone, but arise because of specific circumstances.

For example, the attentiveness of 50 people at a presentation is affected by causes common to all of them, such as room temperature and lighting, the speaker's style, and the subject matter. There are other causes that affect attentiveness of certain individuals, such as lack of sleep, family problems, and health. These causes arise because of specific circumstances.

A stable process is defined as follows:

> **Stable process**: A process in which variation in outcomes arises only from common causes.

A stable process is in a state of statistical control; the cause system remains essentially constant over time. This does not mean that there is no variation in the outcome, that the variation is small, or that the outcomes meet customer requirements. A stable process implies only that the variation is predictable within statistically established bounds.

An unstable process is defined as follows:

Unstable process: A process in which variation is a result of both common and special causes.

An unstable process is not necessarily one with large variation. It means that the magnitude of the variation in the outcomes from one time period to the next is unpredictable.

As special causes are identified and removed, a process becomes stable. Deming (1986, p. 340) lists several benefits of a stable process. Some of these are:

- The process has an identity; its performance is predictable.
- Costs are predictable.
- Regularity of output is an important by-product of a stable process. (Then the just-in-time system of delivery of parts follows naturally.)
- Productivity is at a maximum and costs at a minimum under the present system.
- The effect of changes in the process can be measured with greater speed and reliability. In an unstable process it is difficult to separate changes in the process from special causes.

It is vital to know when an adjustment of the process will improve performance. Adjustment of a stable process, that is, one whose output is dominated by common causes, will increase variation.[2] This overadjustment or tampering with a stable system is common in manufacturing processes as well as management processes. Improvement of a stable process is achieved only through a fundamental change in the process that results in the removal of some of the common causes.

One of the first steps in improvement of any process is to learn whether the process is dominated by common causes or special causes. The method to achieve this is the Shewhart control chart (see Chapter 2).

The appropriate people to identify special causes are usually different from those needed to identify common causes. The same is true of those most qualified to remove the two kinds of causes. Removal of common causes is the responsibility of management with the aid of experts in the subject, such as engineers and chemists. Identification of special causes can usually be handled at a local level by those working in the process using control charts. Removal of special causes is usually the responsibility of immediate supervisors.

---

[2] Students of time series analysis will recognize that when the measurements are autocorrelated, adjustments to a stable process using an appropriate feedback controller can reduce variation. However, this case is beyond the scope of this discussion.

Once the process has been brought to a stable state, its capability or performance in the future is predictable. This range of variation is compared to the requirements of the customer to determine if the process is capable of meeting those requirements.

Continued improvement through the reduction of common causes will require a higher level of understanding of the process. The methods of planned experimentation are needed to improve stable processes through identifying major sources of common causes and testing fundamental changes to the process.

Supplier-Customer Relationship for Improvement

An organization as a system can be viewed as a linkage of processes run by internal producers of output and internal customers of this output. The ultimate output of this network is the product or service provided to an external customer. This supplier-customer relationship is illustrated in Figure 1.4.

Each supplier must consider the needs of the customer. Each customer becomes the supplier for subsequent needs. This process repeats until the product or service reaches the final customer.

Traditionally, two feedback loops have provided the basis for action. The *customer feedback loop* comes too late in the process; the product or service has already been completed. The second feedback loop is *inspection*, determining whether the product meets the specifications. Inadequacies of product inspection have already been discussed. Action based on either of these two feedback loops is described by Deming (1985) as "retroactive" management.

The principal basis for action for the supplier is the *process feedback* on defined measures of performance or quality characteristics. The other two loops serve as measurements to give direction and to evaluate progress.

All three loops provide input to an overall model for improving quality. This model is presented in the next section.

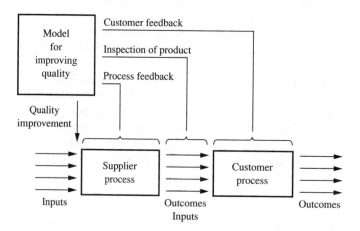

FIGURE 1.4
Supplier-customer relationship for improvement

1.3 MODEL FOR IMPROVING QUALITY

The model for improving quality was introduced by Moen and Nolan (1987). This model involves three major components: the development of a *charter* for the team, a summary of the *current knowledge* of the team, and the use of an *improvement cycle* to increase the team's knowledge and to serve as a basis for taking action. Figure 1.5 provides an overview of this model. A description of each component in the model follows.

Charter of the Team

The first component of the model is development of a charter for the team or individual that will be involved in the improvement of quality. Three reasons for writing a charter are:

> to aid in selection of the team,
> to reduce unwanted variation from the original purpose,
> to help the team choose processes or products for study.

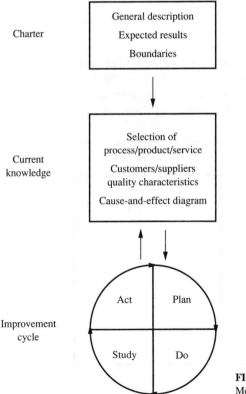

FIGURE 1.5
Model for improving quality

The charter of the team may originate from anyone who has a stake in the performance of the process or product. This sponsor could be a customer, a manager, or an operator. At least a draft of the charter should be completed before final selection of the team.

To prepare for writing a charter, describe the current situation. Recognize the importance of improving the current situation. What data support maintaining the current situation? Who are the customers? Do customers need to be surveyed? The sponsor must ensure that the activity is important and visible enough to get the necessary attention to maintain ongoing priority and commitment.

The next step is to select one of the four basic activities: design of a new product, redesign of an existing product, design of a new process, or redesign of an existing process. List advantages and disadvantages of each option. Costs, resources, timing, and leverage for improvement should be included. People involved in each activity should be solicited for input. A table to aid in the selection is provided in Figure 1.6.

Once the basic activity has been selected, a charter can be written. There is no one correct way to write a charter for a team, but the following format has proved useful in organizing the important content of a charter:

- General description,
- Expected results,
- Boundaries for the activities.

The general description provides an initial orientation for the team. The expected results provide more specific details to support the general description. The boundaries for the activities define the scope of the improvements. The sponsor should assess the impact of the charter on internal and external customers. How does the charter relate

Basic activity	Advantages	Disadvantages
Redesign existing process		
Design new process		
Redesign existing product		
Design new product		

FIGURE 1.6
Table for selection of the basic activity

FIGURE 1.7
Example of a general description for a charter

General description:
- Develop a vinyl flooring that does not require waxing

Expected results
- A floor that will hold its shine
- Assessment of estimated costs and manufacturing feasibility
- Increased sales

Boundaries:
- Appearance (color and pattern) is not degraded
- New methods of installation are not required
- Based on customer research, the price of the flooring
  should not be increased by more than 5% for low grade,
  10% for medium grade, and 20% for high grade

---

to the overall plan of the organization? An example of a team focusing on improving
an existing product is given in Figure 1.7.

Once the charter is written, the team can be selected. The team is composed
of a team leader and other members whose task is to accomplish the charter. Teams
should be small—preferably less than 10—and should comprise the following groups:

- people that will benefit from successful completion of the improvement activity
- people with different responsibilities and different levels within the organization
- people who will contribute to the team and have expressed the desire to be members
- managers who can provide resources and support outside the team's capability
- technical support staff

Current Knowledge

The second component of the model for improvement of quality is a summary of the
current knowledge of the team. If the basic activity selected is to design a new process,
redesign an existing product, or design a new product, a methodology to increase the
current knowledge of these processes is developed in Chapter 10, "Quality by Design."
If the basic activity selected is to redesign an existing process, current knowledge is
summarized through the following steps:

1. List processes related to the charter, and choose the initial process for study.
2. Describe the initial process for study (identify suppliers and inputs, and outcomes
   and their customers).
3. Identify important quality characteristics for selected inputs, for performance of
   the process, and for selected outcomes.

4. Operationally define selected quality characteristics.
5. Develop a flowchart of the process.
6. Develop a cause-and-effect diagram for the important quality characteristics.
7. State some potential improvement cycles for the general plan of improvement.
8. Document other important knowledge of the process and any other related history.
9. After completing the documentation of current knowledge, perform a self-assessment.

For some teams, the current knowledge may be based on well-accepted scientific principles, and there may be widespread agreement as to its validity. For other teams, the current knowledge may consist of widely divergent hunches of what causes problems or variation in the process. As the team learns, the new knowledge should be added to the current knowledge. The methods of planned experimentation will enhance the efficiency of gaining this new knowledge.

This a priori knowledge is essential to planning any change (via an improvement cycle, the third component) to a product or process.

INITIAL PROCESS. The initial process should be selected based on the following:

- the impact on the charter
- the degree of complexity
- the relationship to other processes
- the makeup of the team and necessary resources

Customer feedback and any other information available from customers, inspection of outcomes of the process, and any process feedback from managers and workers involved are most important in selecting the initial process.

DESCRIBE INITIAL PROCESS. The initial process should be described by identifying the process owner, the inputs and suppliers, and the outcomes and customers. Key stages in the process should be listed.

QUALITY CHARACTERISTICS. The next step in the documentation of current knowledge is to identify important quality characteristics or measures of performance of the process. Customers may need to be surveyed to identify their needs. Quality characteristics must be developed that relate to those needs (See Chapter 10). A Pareto analysis may be needed to identify the most important quality characteristics.

Some questions concerning quality characteristics are:

- Has the quality characteristic been stable?
- Has the quality of the test method been studied?

A test method provides a measure of the quality characteristics. What is the "quality" of these measurements? We cannot directly assess the quality of an individual measurement; we can only evaluate a measurement in terms of what we know about the test method. The test method can be characterized in terms of precision, accuracy, bias, sensitivity, robustness, stability, reliability, cost, speed, and simplicity.

The quality characteristic is the window through which we are able to observe processes. If that window does not provide a predictable, consistent view of the process, intelligent decisions about actions to be taken on the process cannot be made.

Examples of quality characteristics for different types of processes are given in Figure 1.8.

FIGURE 1.8
Examples of quality characteristics—by function

Marketing Sales/Service

- Time to process a customer request
- Error in filling out dealer orders
- Overdue accounts

- Customer complaints
- Wrong counts
- Customer satisfaction
- Sales performance
- Slow/missed deliveries

Engineering

- Time to process engineering change
- Number of engineering design changes

- Failure time of product
- Change requests
- Shortage of parts

Manufacturing

- Downtime
- Laboratory precision
- Repair time
- Physical dimensions
- Quality outgoing
- Viscosity of batch process

- Amount of scrap
- Amount of rework
- Level of inventory
- Cost of inspection
- Employee suggestions

Administrative

- Time to process reports
- Errors in accounts receivable
- Cost of inspection
- Incoming calls
- Computer downtime
- Errors in purchase orders
- Idle time of cars

- Telephone usage
- Waiting time
- Transit times
- Time filling orders
- Amount of supplies
- Clerical errors
- Cost of warranty

Management

- Number of accidents
- Time lost by accidents
- Absenteeism
- Turnover of people
- Appraisal of people
- Training and educating people

- Percent of overtime
- Wasted worker hours due to the system
- Variance from budget
- Cost of health care

OPERATIONAL DEFINITIONS. Many of the examples of quality characteristics require operational definitions. What is meant by:

- on-time delivery
- an accident
- machine failure
- due care
- count of people
- speed of light
- on-time performance
- pollution
- percent butterfat
- 50 percent wool blanket
- scratches
- dirt in paint
- good service
- on-time shipment

An operational definition gives communicable meaning to a concept by specifying how the concept is applied within a particular set of circumstances. Therefore, the operational definition will change depending on the application. For example, "clean" will have a different meaning for a private residence than for an operating room in a hospital.

To develop an operational definition, consideration needs to be given to the following:

1. a method of measurement or test
2. a set of criteria for judgment

The concept is thus transformed into an attribute that meets or does not meet the criteria. Now the supplier and customer can understand each other. Operational definitions are necessary for economy and reliability. Without an operational definition, investigations on a problem will be costly and ineffective, almost certain to lead to endless bickering and controversy. Deming (1986, Chapter 9) provides further information on operational definitions.

Since a test method transforms inputs into outcomes, it is a process. This process must be stable; otherwise a method of measurement with predictable performance will not exist. Has the variation of measurement been quantified? The measurement process should be monitored with control charts as a routine part of the process control activities.

FLOWCHART. The next step is to construct a flowchart of the process. A flowchart displays the various stages in the process and, through the use of different types of symbols, demonstrates the flow of a product or service over time. Flowcharts are used for:

defining supplier-customer relationships,

describing the process and making it tangible,

standardizing procedures,

designing a new process or modifying an existing process,

identifying complexity or opportunities for improvement.

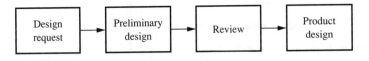

FIGURE 1.9
"Ideal" flowchart of a process to design a product

The most difficult step in drawing a flowchart is deciding how many tasks or operations should be listed and how much detail to include. Usually people err on the side of too much detail or too many tasks included in one flowchart. You might start with the last stage, involving the customer, and work backward, using only enough detail to understand what is happening.

The first flowchart might be an outline of major stages only. An "ideal" flow is from one stage to the next with no complexity. Figure 1.9 is an example of such a process.

At each stage in Figure 1.9, you might list the major obstacles involved in carrying out the tasks of that stage. What are the outcomes at each stage? What is measurable? With the help of people working in the process, chart the actual process in use. Redundant stages or patching of systems may be revealed. Location of major obstacles may be identified. The actual process may look more like Figure 1.10.

Once the team that will be working on the process agree on the flow of the process, the supplier-customer relationships are identified at each stage. What are the customer requirements for that stage? How can these requirements be translated into quality characteristics? What is the mechanism for customer feedback? What are the measures of performance of the outcomes for the supplier's process?

Identification of the fundamental quality characteristics for outcomes of each stage is an important step in documenting the current knowledge. These measures can be identified as "checking points" on the performance of the process as illustrated by the flowchart.

Questions concerning the completed flowchart include:

1. Is this the actual process in use today?
2. Can you identify opportunities for improvement (e.g. complexity, waste, layers of inspection)?
3. Are there obvious improvements to the process that should be made?
4. Are there standards for the process?

CAUSE-AND-EFFECT DIAGRAM. The next step is to develop a cause-and-effect diagram for the important quality characteristics. This diagram organizes all known causes into general categories such as methods, materials, machines, and human, illustrating the common relationships.

The versatility of a cause-and-effect diagram makes it a useful tool for organizing problem-solving efforts in every area of the manufacturing and service industries. The versatility comes from the way in which a cause-and-effect diagram is created. Its

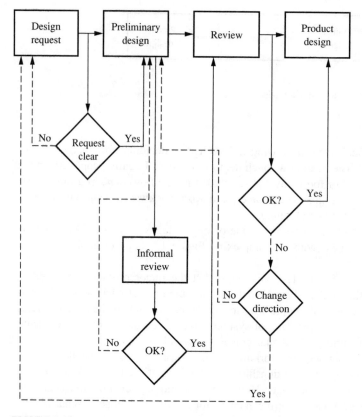

FIGURE 1.10
Actual flowchart of a process to design a product

power comes from the graphic representation of the relationships between problems and their sources.

GENERAL PLAN.   The next step is to develop a general plan for the activities to improve quality. The plan should list some of the early activities for the team's effort and provide guidance to planning in the improvement cycle (the third component of the model).

For example, the activities in the plan may include:

- Standardize the process.
- Study the measurement process.
- Survey customer need.
- Redesign a process.

DOCUMENT KNOWLEDGE.   The next step is a documentation of the history of the process. This documentation is added to as more is learned about the process by using the improvement cycle, the third component in the model for improvement of quality.

SELF-ASSESSMENT. After completing the documentation of the current knowledge, a self-assessment by the team should be done. Are the right people on the team? Does the team still agree on the importance of this effort? Have ground rules and administrative guidelines for the team been established and followed? Is the owner of the process involved in the team's activities?

A worksheet for current knowledge is provided in Section A.1 of the Appendix.

Improvement Cycle

The third component of the model for improvement of quality is the improvement cycle. The improvement cycle is an adaptation of the scientific method. Its application will enhance learning about the product or process. Variations of this cycle have been called the Shewhart cycle, the Deming cycle, and the plan-do-check-act, or PDCA, cycle. Deming himself (1990) calls the cycle "the Shewhart cycle for learning and improvement: the PDSA cycle."

The improvement cycle is used to increase the team's knowledge about the product or process and to provide a systematic way of accomplishing change. Action must be taken to make a change. Will the action result in improved performance of the product or process in the future? What additional knowledge is necessary to take action?

This cycle is a model for learning. A deduction (prediction) based on some theory is made, observation is taken (data collection), a comparison is made of the data to the predicted consequences, and a modification of the theory (learning) is done when the consequences and the data fail to agree. Deming (1985) says, "Experience (by itself) teaches you nothing. You learn by subjecting experience to questions. Questions come from theory."

Knowledge becomes useful when it results in action. Many times a person has a preconceived notion of the course of action and searches for data to support the action. No learning takes place; hence, improvements in quality may not result.

The improvement cycle has four phases. A description of each of these phases and how it is used follows.

PHASE 1: PLAN. The planning phase of the improvement cycle starts with stating the specific objective for the cycle. A set of specific questions to be answered by the data will be necessary input to the plan. The team should predict the answers to these questions using the current knowledge. Does the team agree on these predictions? This book focuses on obtaining data through planning experiments. However, experiments are not the only way to obtain data to answer questions.

Some examples of objectives of a cycle are the following:

- Conduct a survey to understand customer needs.
- Do a Pareto analysis to set priorities.
- Develop control charts to study the stability of the process.
- Develop standards or standardized procedures for the process.

- Conduct an experiment to study the cause-and-effect relationships in the process.
- Conduct a test to evaluate changes to a product or process.
- Run an experiment to choose among competing concepts for a new product or process.

Develop a plan to answer the questions raised by the specific objective of the cycle. Consider the following methods to help answer the questions:

• data collection forms	• frequency plots
• Pareto diagram	• planned experiments
• scatter diagram	• survey methods
• run-order plots	• simulation/modeling
• control charts	• engineering analysis

Questions to ask during the planning phase include:

- Can the plan be carried out on a small scale?
- Have responsibilities for collection and analysis of the data been assigned?
- Has a schedule been developed ?
- Is training needed?
- Have the people outside the team who will be affected by this plan been considered?

PHASE 2: DO.   The second phase begins by carrying out the plan developed in the previous phase. The plan could be a change or a test aimed at improvement. Observations made in carrying out the plan should be documented. Identify the things observed that were not part of the plan. Evaluate the data for changes over time. Document what went wrong during the collection of data. This phase includes control of the quality of the data being obtained.

PHASE 3: STUDY.   Once the data are obtained, they are analyzed using methods considered during the planning phase. Study the results. Compare the results of the analysis of the data to the predictions made from the current knowledge. What have you learned?

The current knowledge is modified if the data contradict certain beliefs about the process. If the data confirm the existing knowledge about the product or process, then the team will have an increased degree of confidence that the current knowledge provides sufficient basis for action.

Compare the analysis of the data to the current knowledge. Do the results of this cycle agree with the predictions made in the planning phase? Under what conditions could the conclusions from this cycle differ? What are the implications of the unplanned observations and problems during the collection of the data?

Summarize the new knowledge gained in this cycle. Revise the current knowledge to reflect this new information (update flowcharts and cause-and-effect diagrams). Will this new knowledge apply elsewhere?

PHASE 4: ACT.   Based on the results of phase 3, the team will decide whether or not to make a change to the product or process. The decision may be to go through the cycle without making a change and to plan a new cycle. Questions to consider include:

- Is the cause system sufficiently understood?
- Has the appropriate action or change been developed or selected?
- Have the changes been tested on a small scale?
- Will the actions or changes improve performance in the future?

Assign responsibilities for implementing and evaluating the changes to the process. List forces in the organization that will help or hinder the changes. Identify the organizations and people who are affected by the changes. Communicate and implement the changes. What should be the objective of the next cycle?

A worksheet for the improvement cycle is provided in Section A.1 of the Appendix.

By repeated use of the improvement cycle, knowledge of the product or process is increased sequentially. The more complete the current knowledge, the better the predictions will be.

Among the important attributes of the improvement cycle are:

- Planning is based on theory.
- The same people that plan a change carry out the change.
- It provides focus and discipline to the team.
- It provides a framework for the application of statistical methods.
- It encourages repeated use of the cycle and enhances the iterative learning process.
- It requires documenting what was learned.

Planned experimentation helps us learn about the cause system in a product or process and use this knowledge to make improvements. Do we need to modify our cause-and-effect theory? Have we increased our knowledge so that we may predict the results of future experiments? What will be the impact downstream, at a later stage in the process? Should we begin a new cycle by testing a new condition? As we increase our knowledge, the focus moves from screening experiments to confirmatory experiments.

Choice of method depends on the objective of the cycle. Figure 1.11 lists a typical sequence of cycles that a team might perform. The order of the cycles will differ depending on the process.

FIGURE 1.11
Potential Improvement Cycles

Improve a process	Improve a product
Cycle 1. Standardize the process.	*Cycle 1*. Survey to determine quality-characteristics.
Cycle 2. Study the measurement system	*Cycle 2*. Test different concepts.
Cycle 3. Identify and remove special causes.	*Cycle 3*. Run a planned experiment on prototypes.
Cycle 4. Remove dominant common causes.	*Cycle 4*. Pilot-test the new product.
Cycle 5. Monitor process.	*Cycle 5*. Survey to determine the acceptance of the new product.

The methods selected for an improvement cycle will enhance the learning process. For example, simple graphical analysis will help ensure that the team's knowledge of the subject matter is included in the analysis. The two most common methods used for the improvement cycle are the control chart and the planned experiment. Control charts will be discussed in the next chapter. The philosophy, principles, and methods of planned experiments will be discussed in the remaining chapters. A worksheet for the improvement cycle is provided in Section A.1 of the Appendix.

1.4   SUMMARY

Improvement of quality results from change—gradual change or abrupt change through innovation. A common purpose and knowledge of methods so that change results in improvement are required.

Improvement of quality of a product or service is accomplished by design or redesign of products, and design or redesign of the processes that produce that product or service. The starting point is to view the organization as a system. Every activity, every job is part of a process and can be improved. Matching of products and services from the system to the needs of the customer are ongoing. Needs of the customer are translated into quality characteristics.

Improvement of quality comes through enhancing people's knowledge. The model for improvement of quality just presented provides a "knowledge-based roadmap" for improvement. The model requires a charter for a team with a common purpose, selection of the basic activity (design or redesign of product, design or redesign of process), defining the current knowledge, and a sequential building on that knowledge to make a change in the product or process using the improvement cycle. Imbedded in the cycle are methods that will enhance the learning process, the most common of which are the control chart and the planned experiment (developed in Chapters 2 through 9).

This model can apply to an overall strategy for the evolution of a new product as well as the improvement of existing product or service in marketing, engineering, manufacturing, administrative, or management areas.

Management must provide an environment that discourages a "firefighting" approach to problem solving, one that requires daily progress reports and leaves little time for planning.

REFERENCES

Deming, W. Edwards (1950): (See Deming 1950 reference end of Chapter 3.)

Deming, W. Edwards (1985): "Four-day Dr. Deming Seminar," Dec. 3–6, 1985, Washington,D.C.

Deming, W. Edwards (1986): *Out of the Crisis*, Massachusetts Institute of Technology, Center for Advanced Engineering Study, Cambridge, Mass.

Deming, W. Edwards (1990): "Four-day Dr. Deming Seminar," Jan. 30–Feb. 2, 1990, Washington, D.C.

Garvin, David A. (1987):"Competing on the Eight Dimensions of Quality," *Harvard Business Review*, vol. 87, no. 6, pp. 101–109, November.

Imai, M. (1986):*Kaizen*, Random House, New York.

Moen, R. D., and T. W. Nolan (1987): "Process Improvement," *Quality Progress*, Vol. 23, no. 5, pp. 62–68, September.

Plsek, Paul E. (1987): "Defining Quality at the Marketing/Development Interface",*Quality Progress*, June 1987.

Shewhart, W. A. (1931): *The Economic Control of Quality of Manufactured Product*, American Society for Quality Control, Milwaukee (Reprinted 1980.)

EXERCISES

1.1 *Questions for group discussion.**
    (a) What is wrong with the practice of relying on inspection to achieve quality, sorting good product from bad product, as standard procedure?
    (b) Under what conditions may it be desirable to sort out good from bad?
    (c) What benefits should customer and supplier expect from a close working relationship?
    (d) What changes would be desirable in your own relationships with your suppliers and your customers?
    (e) Why is continual improvement of quality and of service necessary?
    (f) What improvements have you seen in the past two years in the quality of your company's output or service?
    (g) How have you improved your own work?
    (h) What are some of the handicaps that affect your own work?
    (i) Which handicaps can be corrected locally? Which handicaps require action by management?
    (j) How would management learn about handicaps?
    (k) How does a producer achieve innovation of product or of service?
    (l) Can improvement of operations alone ensure success of a company?
    (m) What is a process?
    (n) What have you done about improvement of a stable system?
    (o) Why are operational definitions desirable for business?

1.2 *Teams and teamwork.*
    (a) List reasons why an organization should use teams.
    (b) Identify what you would like to see as effective team behavior.
    (c) Identify behavior that you would least like to see of people in teams.
    (d) List the obstacles to teamwork in your organization.

---

* Questions from the working group session "Four-day Dr. Deming Seminar," November 1, 1988, Washington, D.C.

1.3 *Supplier/customer relationships*. List major processes in your organization. Identify suppliers and inputs for each process. Identify outcomes and customers for each process. List key stages in each process. What are the quality characteristics for selected inputs and outcomes of each process?

1.4 *Going to work*. Draw a flowchart for tasks involved in going "from bed to work." Draw a cause-and-effect diagram for "late to work."

1.5 *Operational definitions*. Identify concepts in your organization that need operational definitions. Provide those definitions.

1.6 *Process feedback*. Identify (1) process feedback and (2) customer feedback for: (*a*) Engineering process (*b*) Manufacturing process (*c*) Managing process (*d*) Teaching process

1.7 *Improvement cycle.*Discuss how the improvement cycle may have aided your past experimental investigations. Can you think of an experimental investigation that was not iterative?

1.8 *Planned experimentation*. List some examples of planned experimentation that have been carried out in your organization. List some potential applications of planned experimentation in your work area.

CHAPTER
2

CONTROL CHARTS

2.1 INTRODUCTION

Chapter 1 discussed the concept, credited to Walter Shewhart (1931), that variation in a quality characteristic of a process is due to two types of causes:

Common causes of variation: Causes that are inherent in the process over time, affect everyone working in the process, and affect all outcomes of the process.

Special causes of variation: Causes that are not part of the process all the time or do not affect everyone, but arise because of specific circumstances.

A process whose outcomes are affected only by common causes is called a stable process or one that is in a state of statistical control. A *stable process* implies only that the variation is predictable within bounds. A process whose outcomes are affected by both common causes and special causes is called an *unstable process*. For an unstable process, the variation from one time period to the next is unpredictable. As special causes are identified and removed, the process becomes stable.

The **control chart** provides an operational definition of these concepts. The control chart is a statistical tool used to distinguish variation in a process due to common causes and variation due to special causes. Dr. Walter A. Shewhart is credited with developing the control chart. In 1931 Shewhart published *Economic Control of Quality of Manufactured Product*, which discussed the theory and application of control charts.

The control chart method has general applicability throughout an organization. Top managers can use a control chart to study variation in sales, supervisors can use the tool to assign responsibility for improvement of a process, administrative personnel can use it to identify opportunities for improvement, and operators can use a control chart to determine when to adjust a process.

The control chart method is often confused with the use of specifications and tolerances. Figure 2.1 contrasts these two very different approaches to the management of variability.

The control chart consists of three lines and points plotted on a graph. Figure 2.2 illustrates the form of a typical control chart. The control chart is constructed by obtaining measurements of some characteristic of a process. The data are then grouped by time period, location, or other descriptive variable in the process. These sets of data are called *subgroups*. Multiple subgroups, often obtained over time, are required for the control chart.

A descriptive statistic (the result of doing arithmetic on the data), such as the average, range, or percent defective, is computed from the measurements in each subgroup. The statistic is then plotted on the chart, with the horizontal axis representing the subgroup number (often in time sequence) and the vertical axis being a scale for the statistic.

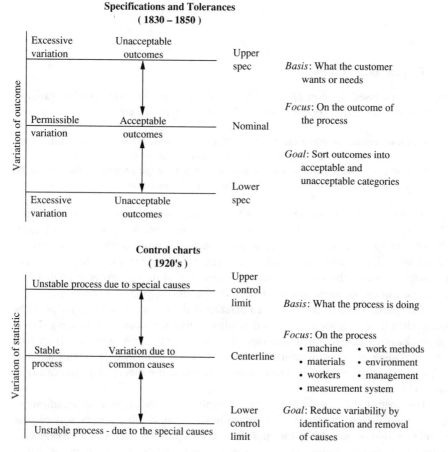

FIGURE 2.1

Two approaches to managing variability

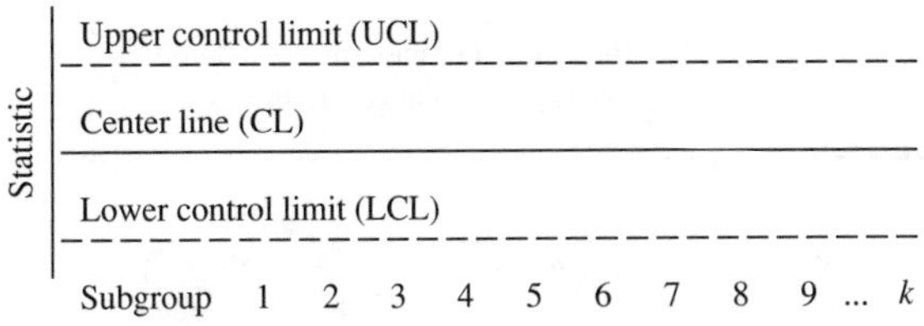

Upper control limit (UCL)	
Center line (CL)	
Lower control limit (LCL)	

Statistic

Subgroup 1 2 3 4 5 6 7 8 9 ... k

FIGURE 2.2

Illustration of the form of a control chart

When 20 to 30 subgroups have been plotted on the chart, the control limits can be computed. The control limits bound the variation of the statistic that is due to common causes. Formulas for control limits have been developed for all common types of control charts. Although the limits are based on statistical theory, Deming (1986, p. 334) states that control limits should not be associated with any calculation of probability.

2.2 SUBGROUPING

The concept of subgrouping is one of the most important components of the control chart method. Shewhart's principle is to organize (classify, stratify, group, etc.) data from the process in a way that ensures the greatest similarity among the data in each subgroup and the greatest difference among the data in different subgroups. The aim of rational subgrouping is to include only common causes of variation within a subgroup, with all special causes of variation occurring between subgroups.

The most common method to obtain rational subgroups is to hold time "constant" within a subgroup. Only data taken at the same time (or in some selected time period) are included in a subgroup. Data from different time periods will be in different subgroups. This use of time as the basis of subgrouping allows the detection of causes of variation related to time.

The statistics from subgroups are usually plotted in order of time. The subgroups can also be ordered by other factors, such as supplier, shift, operator, or part position, to investigate the importance of these factors.

As an example of subgrouping, consider a study planned to reduce late payments. Historical data from the accounting files would be used to study the variation in late payments. What is a good way to subgroup the historical data on late payments? The data could be grouped by billing month, by receiving month, by major account, by product line, or by account manager. Knowledge or theories about the process should be used to develop rational subgroups. Some combination of time (either receiving or billing month) and one or more of the other process variables would be a reasonable way to develop the first control chart.

2.3 INTERPRETATION OF A CONTROL CHART

The control chart provides a basis for taking action to improve a process. A process is considered to be stable when there is a random distribution of the plotted points within the control limits. If there are points outside the limits, or if the distribution

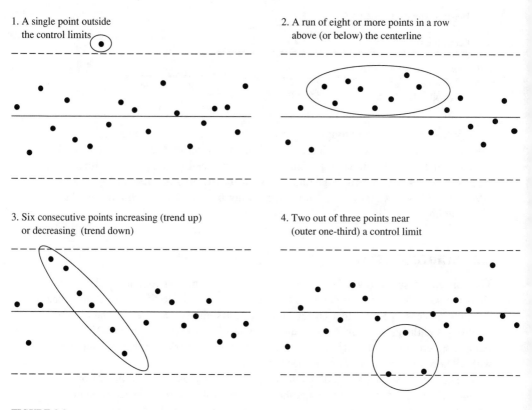

FIGURE 2.3
Rules for determining a special cause on a control chart

of points within the limits is not random, the process is considered to be unstable, and action should be taken to uncover the special causes of variation.

Figure 2.3 lists four rules that are recommended for general use with control charts. Special circumstances may warrant use of some of the additional tests given by Nelson (1984). Deming (1986, p. 319) emphasizes that it is necessary to state in advance what rules to apply.

Control limits for the chart should be established using 20 to 30 subgroups from a period when the process is stable. If it is desirable to extend the control limits, any points affected by special causes should be removed and the control limits recalculated. The limits should only be extended when they are calculated using data without special causes.

Revision of the control limits should be done only when the existing limits are no longer appropriate. This is the case when improvements have been made to the process and the improvements result in special causes on the control chart. Control limits should then be calculated for the new process.

2.4 TYPES OF CONTROL CHARTS

There are many different types of control charts. This section will discuss the various types. Examples will be given for the X-bar and R chart, the P chart, the C chart, and the U chart.

Selection of the control chart to use in a particular application primarily depends on the type of data. The different types of data can be classified into three categories:

1. classification data
2. count data
3. continuous data

The first two types are called *attribute data*; the third type is called *variable data*. For classification data, the quality characteristic is recorded in one of two classes. Examples of classes are conforming units/nonconforming units, go/no go, and good/bad. To obtain count data, the number of incidences of a particular type is recorded, such as number of mistakes, number of accidents, or number of sales leads. For continuous data, a measured numerical value of the quality characteristic is recorded, such as a dimension, a physical attribute, cost, or time.

In general, data should be collected as variable data whenever possible since learning then requires many fewer measurements compared with the use of attribute classifications or counts. The control charts for variable data thus require fewer measurements in each subgroup than the attribute control chart. Typical subgroup sizes for variable charts range from 1 to 10, whereas subgroup sizes for attribute charts range from 30 to 1,000.

Figure 2.4 contains a summary of frequently used charts and the types of data to which they apply.

When variable data can be put in rational subgroups, the X-bar and R chart is the most frequently used. The individual chart (or X chart) is used when the data cannot be organized in subgroups. For the X-bar and R chart, the statistics used are the average of the measurements in each subgroup and the range (largest measurement minus the smallest measurement) in each subgroup. The average is designated as $\overline{X}$ and the range as R. Two control charts are required: the X-bar chart for the averages and the R chart for the ranges. Two to 10 measurements are included in each subgroup, with subgroup sizes of 3 to 6 being most common.

Figure 2.5 shows a worksheet that can be used to calculate limits for an X-bar and R chart. Figure 2.6 shows an example of an X-bar and R chart.

If the subgroup size is variable (i.e., the number of measurements changes from subgroup to subgroup), the standard deviation, another descriptive statistic for variation, is used in place of the range. The X-bar and S chart is used in this case, where S is the symbol for the standard deviation of the subgroup values. For most applications with variable data, the subgroup size can be held constant. The S chart thus has limited use.

In some applications it is not practical to have multiple measurements in a subgroup. For example, there is only one determination of a monthly sales number. In these applications, the single measurement is treated as a subgroup and is plotted

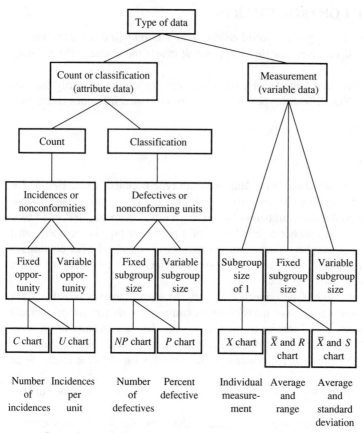

FIGURE 2.4
Selection of a particular type of control chart

on an X chart. An X chart can be developed using the worksheet for the X-bar and R chart (Figure 2.5) with a subgroup size (n) of 1. The average of the moving ranges (the moving range is the range for pairs of consecutive measurements) is used in place of $\overline{R}$.

Attribute Control Charts

When sample units are classified into two categories (conforming and nonconforming units), the P chart (P = percent nonconforming) is appropriate. The P chart

FIGURE 2.5
X-bar and R calculation sheet

X-bar and R Control Chart Calculation Form

Name _____ Date _____

Process _____ Sample description _____

Number of subgroups (k) _____ Between (dates) _____ – _____

Number of samples or measurements per subgroup (n) _____

$$\overline{\overline{X}} = \frac{\Sigma \overline{X}}{k} = \text{———} = \text{———} \qquad\qquad \overline{R} = \frac{\Sigma \overline{R}}{k} = \text{———} = \text{———}$$

$\overline{X}$ Chart	R Chart
UCL $= \overline{\overline{X}} + (A_2 * \overline{R})$	UCL $= D_4 * \overline{R}$
UCL $= + (\quad * \quad)$	UCL $= \quad *$
UCL $= +$	UCL $= \text{———}$
UCL $= \text{———}$	
LCL $= \overline{\overline{X}} - (A_2 * \overline{R})$	LCL $= D_3 * \overline{R}$
LCL $= - (\quad * \quad)$	LCL $= \quad *$
LCL $= -$	LCL $= \text{———}$
LCL $= \text{———}$	

	Factors for Control Limits				Process Capability

n	A_2	D_3	D_4	d_2
*1	2.66	—	3.27	1.128
2	1.88	—	3.27	1.128
3	1.02	—	2.57	1.693
4	0.73	—	2.28	2.059
5	0.58	—	2.11	2.326
6	0.48	—	2.00	2.534
7	0.42	0.08	1.92	2.704
8	0.37	0.14	1.86	2.847
9	0.34	0.18	1.82	2.970
10	0.31	0.22	1.78	3.087

*Use moving range of 2 for determining R

Process Capability

If the process is in statistical control, the standard deviation is:

$\hat{\sigma} = \quad \overline{R} \quad / \quad d_2$

$\hat{\sigma} = \quad\quad /$

$\hat{\sigma} = \text{———}$

The process capability is:

$\overline{\overline{X}} - 3 * \hat{\sigma} \text{ to } \overline{\overline{X}} + 3 * \hat{\sigma}$

$\quad - \quad \text{to} \quad +$

$\text{———} \text{ to } \text{———}$

can be used with either a fixed or a variable subgroup size (so it is never necessary to use the *NP* chart). For an *NP* chart the number of nonconforming units is plotted rather than a percent. The *NP* chart is appropriate only for a fixed subgroup size. Subgroup sizes for *P* charts typically range from 30 to 1,000. Figure 2.7 is a worksheet for calculating limits for a *P* chart. Figure 2.8 shows an example of a *P* chart for both fixed and variable subgroup sizes.

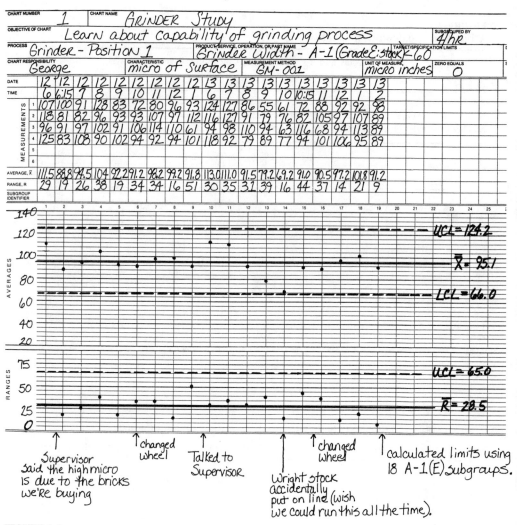

FIGURE 2.6
X-bar and R chart example and discussion

For count data, the *C* chart (number of incidences) is appropriate when the opportunity for occurrence is relatively constant (within 20% of the average) among subgroups. The *U* chart (incidences per unit) is required for count data when the opportunity for occurrence is variable among subgroups. The subgroup size is replaced by the concept of "area of opportunity" for *C* and *U* charts. Figure 2.9 is a worksheet for a *C* chart or *U* chart. Figure 2.10 shows examples of both a *C* chart and a *U* chart.

There are other types of control charts, such as the median control chart and the cumulative sum chart, used for special applications. The seven types of charts shown in Figure 2.4 are the most common Shewhart control charts found in practice.

FIGURE 2.6 *continued*
Discussion of X-Bar and R Chart Example

<hr>

Discussion

This is a control chart for the microfinish (smoothness) of a product from a grinding process. All samples are taken from position 1 on the grinder.

Subgrouping: Four consecutive parts are selected from position 1 each hour to form a subgroup ($n = 4$).

Steps to computer limits: After 19 subgroups (two days), the control limits are calculated. Since subgroup 14 was from a different material supplier, it was not used in the calculations ($k = 18$).

1. The average of the 18 subgroup averages ($\overline{\overline{X}}$) and the average of the ranges ($\overline{R}$) are calculated.

2. The control limits for both the X-bar and R charts are calculated (see Figure 2.5 for formulas and factors):

	X-bar Chart	R chart
Centerline (CL)	96.6	34.6
Upper control limit (UCL)	121.7	98.8
Lower control limit (LCL)	91.5	none

3. The centerlines and control limits are drawn on the chart.

Status of process: Unstable—the special cause for subgroup 14 is due to the material vendor. The process is stable for the 18 subgroups using the A-1 material vendor.

<hr>

FIGURE 2.7
P chart calculation form

---

d = Nonconforming sample units per subgroup
n = Number of sample units per subgroup
k = Number of subgroups
p = Percent nonconforming units = $100 * d/n$

Control limits when subgroup size (n) is constant:

$$\bar{p} = \frac{\Sigma p}{k} = \underline{\hspace{2cm}} = \underline{\hspace{2cm}} \quad \text{(centerline)}$$

$$\hat{\sigma}_p = \sqrt{\frac{\bar{p}*(100 - \bar{p})}{n}} = \sqrt{\frac{\underline{\hspace{0.5cm}}*(100 - \underline{\hspace{0.5cm}})}{\underline{\hspace{0.5cm}}}} = \underline{\hspace{2cm}}$$

UCL = $\bar{p} + (3* \hat{\sigma}_p)$ LCL = $\bar{p} - (3* \hat{\sigma}_p)$

UCL = $\underline{\hspace{0.5cm}} + (3*\underline{\hspace{0.5cm}})$ LCL = $\underline{\hspace{0.5cm}} - (3*\underline{\hspace{0.5cm}})$

UCL = $\underline{\hspace{0.5cm}} + \underline{\hspace{1cm}}$ LCL = $\underline{\hspace{0.5cm}} - \underline{\hspace{1cm}}$

UCL = $\underline{\hspace{1.5cm}}$ LCL = $\underline{\hspace{1.5cm}}$

Control limits when subgroup size (n) is variable:

$$\bar{p} = \frac{\Sigma d}{\Sigma n} * 100 = \underline{\hspace{1.5cm}} * 100 = \underline{\hspace{1.5cm}} \quad \text{(centerline)}$$

$$\hat{\sigma}_p = \frac{\sqrt{\bar{p}(100 - \bar{p})}}{\sqrt{n}} = \frac{\sqrt{\underline{\hspace{0.3cm}}*(100 - \underline{\hspace{0.3cm}})}}{\sqrt{n}} = \frac{\underline{\hspace{1cm}}}{\sqrt{n}}$$

UCL = $\bar{p} + (3* \hat{\sigma}_p)$ LCL = $\bar{p} - (3* \hat{\sigma}_p)$

UCL = $\underline{\hspace{0.5cm}} + (3 * \underline{\hspace{0.5cm}}/\sqrt{n})$ LCL = $\underline{\hspace{0.5cm}} - (3 * \underline{\hspace{0.5cm}}/\sqrt{n})$

UCL = $\underline{\hspace{0.5cm}} + (\underline{\hspace{0.5cm}}/\sqrt{n})$ LCL = $\underline{\hspace{0.5cm}} - (\underline{\hspace{0.5cm}}/\sqrt{n})$

n: $\underline{\hspace{1cm}}$ $\underline{\hspace{1cm}}$ $\underline{\hspace{1cm}}$ $\underline{\hspace{1cm}}$ $\underline{\hspace{1cm}}$

$\sqrt{n}$: $\underline{\hspace{1cm}}$ $\underline{\hspace{1cm}}$ $\underline{\hspace{1cm}}$ $\underline{\hspace{1cm}}$ $\underline{\hspace{1cm}}$

$3 * \hat{\sigma}_p$: $\underline{\hspace{1cm}}$ $\underline{\hspace{1cm}}$ $\underline{\hspace{1cm}}$ $\underline{\hspace{1cm}}$ $\underline{\hspace{1cm}}$

UCL: $\underline{\hspace{1cm}}$ $\underline{\hspace{1cm}}$ $\underline{\hspace{1cm}}$ $\underline{\hspace{1cm}}$ $\underline{\hspace{1cm}}$

LCL: $\underline{\hspace{1cm}}$ $\underline{\hspace{1cm}}$ $\underline{\hspace{1cm}}$ $\underline{\hspace{1cm}}$ $\underline{\hspace{1cm}}$

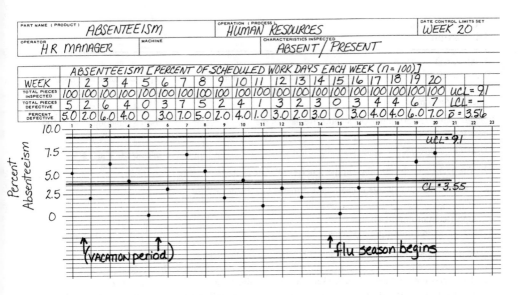

WEEK	1	2	3	4	5	6	7	8	9	10	11	12	13	14	15	16	17	18	19	20		
TOTAL PIECES INSPECTED	100	100	100	100	100	100	100	100	100	100	100	100	100	100	100	100	100	100	100	100	UCL= 9.1	
TOTAL PIECES DEFECTIVE	5	2	6	4	0	3	7	5	2	4	1	3	2	3	0	3	4	4	6	7	LCL= —	
PERCENT DEFECTIVE	5.0	2.0	6.0	4.0	0	3.0	7.0	5.0	2.0	4.0	1.0	3.0	2.0	3.0	0	3.0	4.0	4.0	6.0	7.0	$\bar{p}$ = 3.56	

ABSENTEEISM [PERCENT OF SCHEDULED WORK DAYS EACH WEEK (n = 100)]

PART NAME (PRODUCT): ABSENTEEISM
OPERATION (PROCESS): HUMAN RESOURCES
DATE CONTROL LIMITS SET: WEEK 20
OPERATOR: HR MANAGER
MACHINE:
CHARACTERISTICS INSPECTED: ABSENT / PRESENT

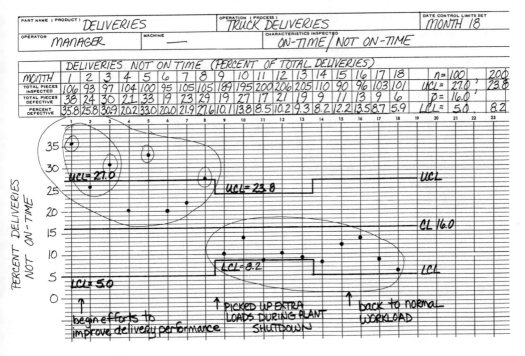

MONTH	1	2	3	4	5	6	7	8	9	10	11	12	13	14	15	16	17	18	n=100	200
TOTAL PIECES INSPECTED	106	93	97	104	100	95	105	105	189	195	200	206	205	110	90	96	103	101	UCL= 27.0	23.8
TOTAL PIECES DEFECTIVE	38	24	30	21	33	19	23	29	19	27	17	21	19	9	11	13	9	6	$\bar{p}$= 16.0	
PERCENT DEFECTIVE	35.8	25.8	30.9	20.2	33.0	20.0	21.9	27.6	10.1	13.8	8.5	10.2	9.3	8.2	12.2	13.5	8.7	5.9	LCL= 5.0	8.2

DELIVERIES NOT ON TIME (PERCENT OF TOTAL DELIVERIES)

PART NAME (PRODUCT): DELIVERIES
OPERATION (PROCESS): TRUCK DELIVERIES
DATE CONTROL LIMITS SET: MONTH 18
OPERATOR: MANAGER
MACHINE: —
CHARACTERISTICS INSPECTED: ON-TIME / NOT ON-TIME

FIGURE 2.8
P chart examples and discussion

FIGURE 2.8 *continued*
Discussion

I. Absenteeism. There are 20 employees in the department, each scheduled to work five days per week. Thus, there are 100 total work days scheduled each week. Each day all 20 employees are each classified as "present" or "absent" (using an operational definition of "absent"). Each week the percent absent (P) is calculated and plotted.

Subgrouping: The 20 employees and the five days of the week are combined to form a subgroup of $n = 100$ each week.

Steps to compute limits: After five months (20 weeks), the control limits are calculated. Thus, $k = 20$.

1. The average of the 20 subgroup percentages ($\overline{P}$) is calculated.

2. The control limits for the P chart are calculated (see Figure 2.7 for formulas):

Centerline (CL)   = 3.55%
Upper control limit (UCL) = 9.1
Lower control limit (LCL) = none

Status of process: Stable—there is a possible trend developing beginning at week 15.

II. Deliveries not on time. A carrier keeps track of delivery performance for each of its shippers. Each month the total number of deliveries and the number of deliveries not arriving on time are recorded. (An operational definition of "on time" was jointly developed with the carrier and shippers.) The percent not on time is calculated and plotted each month.

Subgrouping: All deliveries within a particular month for each shipper are grouped together to form a subgroup. Since the number of deliveries varies, n is variable.

Steps to compute limits: After 18 months, the control limits are calculated. Thus, $k = 18$.

1. The weighted average of the percentages ($\overline{P}$) is calculated by dividing the total number of deliveries not on time by the total number of deliveries.

2. The control limits for the P chart are calculated for $n = 100$ and $n = 200$ (Figure 2.7 contains the formulas):

Centerline (CL)   = 16.0%
Upper control limit (UCL) = 27.0($n = 100$), 23.8($n = 200$)
Lower control limit (LCL) =   5.0($n = 100$),   8.2($n = 200$)

Status of process: Unstable—there are two distinct periods of performance on the chart. An important improvement occurred after the ninth month.

FIGURE 2.9
Control chart calculation sheet for count data

Control Chart Calculation Form (C and U Charts)

- -

C Chart Control Limits (area of opportunity constant)

c = Number of incidences per subgroup
k = Number of subgroups
Note: The subgroup size is defined by the "area of opportunity" for incidences and must be constant.

$$\bar{c} = \frac{\Sigma c}{k} = \underline{\hspace{1cm}} = \underline{\hspace{1cm}} \quad \text{(centerline)}$$

UCL $= \quad \bar{c} + (3* \sqrt{\bar{c}})$ LCL $= \quad \bar{c} - (3* \sqrt{\bar{c}})$

UCL $= \underline{\hspace{0.5cm}} + (3*\underline{\hspace{0.5cm}})$ LCL $= \underline{\hspace{0.5cm}} - (3*\underline{\hspace{0.5cm}})$

UCL $= \underline{\hspace{0.5cm}} + \underline{\hspace{1cm}}$ LCL $= \underline{\hspace{0.5cm}} - \underline{\hspace{1cm}}$

UCL $= \underline{\hspace{1cm}}$ LCL $= \underline{\hspace{1cm}}$

- -

U Chart Control Limits (area of opportunity may vary)

c = Number of incidences per subgroup
n = Number of standard area of opportunities in a subgroup (n may vary)
u = Incidences per standard area of opportunity $= c/n$
k = Number of subgroups
Note: The standard area of opportunity will be defined by the people planning the control chart in units such as worker-hours, miles driven, per 10 invoices, etc.

$$\bar{u} = \frac{\Sigma c}{\Sigma n} = \underline{\hspace{1cm}} = \underline{\hspace{1cm}} \quad \text{(centerline)}$$

UCL $= \quad \bar{u} + (3* \sqrt{\bar{u}}) / \sqrt{n}$ LCL $= \quad \bar{u} - (3* \sqrt{\bar{u}}) / \sqrt{n}$

UCL $= \underline{\hspace{0.5cm}} + (3*\underline{\hspace{0.5cm}}) / \sqrt{n}$ LCL $= \underline{\hspace{0.5cm}} - (3*\underline{\hspace{0.5cm}}) / \sqrt{n}$

UCL $= \underline{\hspace{0.5cm}} + (\underline{\hspace{1cm}} / \sqrt{n})$ LCL $= \underline{\hspace{0.5cm}} - (\underline{\hspace{1cm}} / \sqrt{n})$

n:	___	___	___	___	___
$\sqrt{n}$:	___	___	___	___	___
UCL:	___	___	___	___	___
LCL:	___	___	___	___	___

MISTAKES IN DELIVERIES (C CHART)

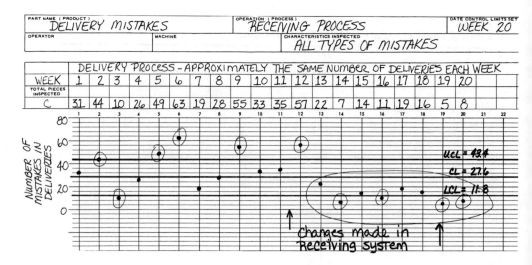

PART NAME (PRODUCT)								OPERATION (PROCESS)											DATE CONTROL LIMITS SET	
DELIVERY MISTAKES								RECEIVING PROCESS											WEEK 20	
OPERATOR					MACHINE			CHARACTERISTICS INSPECTED												
								ALL TYPES OF MISTAKES												

DELIVERY PROCESS – APPROXIMATELY THE SAME NUMBER OF DELIVERIES EACH WEEK

WEEK	1	2	3	4	5	6	7	8	9	10	11	12	13	14	15	16	17	18	19	20	
TOTAL PIECES INSPECTED																					
C	31	44	10	26	49	63	19	28	55	33	35	57	22	7	14	11	19	16	5	8	

UCL = 43.4
CL = 27.6
LCL = 11.8

Changes made in Receiving System

TRAFFIC ACCIDENTS (U CHART)

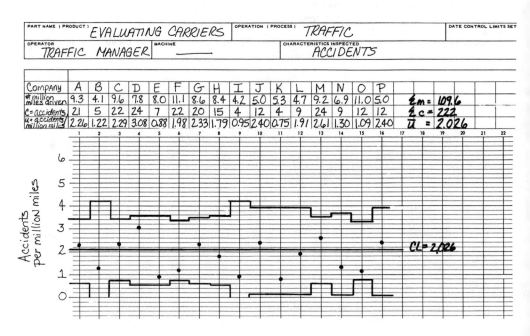

PART NAME (PRODUCT)								OPERATION (PROCESS)									DATE CONTROL LIMITS SET	
EVALUATING CARRIERS								TRAFFIC										
OPERATOR			MACHINE					CHARACTERISTICS INSPECTED										
TRAFFIC MANAGER			———					ACCIDENTS										

Company	A	B	C	D	E	F	G	H	I	J	K	L	M	N	O	P			
# million miles driven	9.3	4.1	9.6	7.8	8.0	11.1	8.6	8.4	4.2	5.0	5.3	4.7	9.2	6.9	11.0	5.0	$\sum m$ =	109.6	
c = accidents	21	5	22	24	7	22	20	15	4	12	4	9	24	9	12	12	$\sum c$ =	222	
u = accidents/ million miles	2.26	1.22	2.29	3.08	0.88	1.98	2.33	1.79	0.95	2.40	0.75	1.91	2.61	1.30	1.09	2.40	$\bar{u}$ =	2.026	

CL = 2.026

FIGURE 2.10
C chart and *U* chart examples and discussion

FIGURE 2.10 *continued*
Discussion of C Chart and U Chart Examples

I. Number of mistakes in deliveries. There are many different types of mistakes that can occur in the delivery/receiving process. The total of all types of mistakes (c) is recorded each week and plotted on a C chart. There are approximately the same number of deliveries (about 500) each week.

Subgrouping: All deliveries during the week are grouped together to form a subgroup.

Steps to compute limits: After 20 weeks, the control limits are calculated ($k = 20$).
1. The average of the 20 subgroup $\bar{c}$ is calculated.
2. The control limits for the C chart are calculated (see Figure 2.9 for formulas):
   Centerline (CL)   27.6
   Upper control limit (UCL)   43.4
   Lower control limit (LCL)   11.8

Status of process: Unstable—there are special causes during the first 12 weeks. Improvement occurred after the 12th week.

II. Traffic accidents. The driving records of the company's 16 major carriers are evaluated each year. The number of miles driven and the number of accidents is recorded for each of the carriers. A U chart is prepared to compare the carriers' records since the number of miles driven varies. The statistic u is computed by dividing the number of accidents by the number of million miles driven.

Subgrouping: A subgroup is constructed from all the deliveries from a given carrier for the past year. The 16 major carriers form 16 subgroups.

Steps to compute limits:
1. The average $\bar{u}$ is calculated by dividing the total number of accidents from all carriers by the total number of million miles driven.
2. The control limits for the U chart are calculated (see Figure 2.9 for formulas):
   Centerline (CL)   2.026
   Upper control limit (UCL)   Different for each carrier
   Lower control limit (LCL)   Different for each carrier

Status of process: Stable—no special causes, no important differences in the carriers' driving records.

2.5 CAPABILITY OF A PROCESS

If a process is found to be stable for a particular quality characteristic, the process capability can be determined. **Process capability** is a prediction of the individual outcomes or measurements of a characteristic of the process.

For variable data, the process capability can be compared to customer specifications to determine if the process is capable of meeting them. The following procedure can be used to determine process capability for variable data:

1. Estimate the standard deviation of a stable process from the centerline ($\bar{R}$) of the range chart using:

$$\hat{\sigma} = \bar{R}/d_2 \qquad (d_2 \text{ is tabulated in Figure 2.5})$$

2. The practical minimum and maximum of the process for the characteristic is calculated using:

$$\text{minimum} = \bar{\bar{X}} - 3\sigma \qquad \text{maximum} = \bar{\bar{X}} + 3\hat{\sigma}$$

3. A process is considered "capable" if the interval from the practical minimum to the practical maximum falls within the customer specifications.

For attribute data, the centerline of the appropriate control chart for a stable process is usually used to express process capability. For a P chart, $100\% - P$ is sometimes presented as the capability. Calculation can be done using the control limit formula to express a range of expected values for a given sample size or area of opportunity.

2.6   PLANNING A CONTROL CHART

The effective use of control charts requires careful planning to develop and maintain the chart. Many attempts to use control charts have not been successful because of lack of planning and preparation. Figure 2.11 shows a planning form that can be used to develop a control chart.

FIGURE 2.11
Form for planning a control chart

---

1. Objective of the chart:

2. Sampling, measurement, and subgrouping
   Variable(s) to be charted:
   Method of measurement:
   Magnitude of measurement variation:
   Point (location) of sampling:
   Strategy for subgrouping:
   Frequency of subgroups:

3. Most likely special causes:

4. Notes required:

   Note   Responsibility

---

5. Reaction plan for out-of-control points (attach copy):

6. Administration:

   Task   Responsibility

   Making measurements
   Recording data on charts
   Computing statistics
   Plotting statistics
   Extending/changing control limits
   Filing

7. Schedule for analysis:

Objective of the Chart

Every control chart should be associated with one or more specific objectives. The objective might be to improve the yield of the process, to identify and remove special causes from a process, or to establish statistical control so that the capability of the process can be determined. The objectives should be summarized on the control chart form. After a period of time, once the objective has been met, the control chart should be discontinued or a new objective developed.

Sampling, Measurement, and Subgrouping

There are a number of measurement and sampling issues that must be resolved prior to beginning a control chart. The most useful control charts will monitor measures of process performance rather than quality characteristics of the outcome of the process. Initially control charts can be used to evaluate characteristics of the outcome that are important to the customer, such as ph, color, weight, yield, or size. Then process variables (such as temperature, reaction time, or voltage) that affect these quality characteristics can be identified. The variables should also be charted as close to the beginning of the process as possible. The type of data for each variable to be charted will determine the type of chart to use.

Information about the variability and stability of the measurement system to be used should be documented. If the variability is not known, an effort to develop that information should be planned.

Important sampling issues for control charts include the point (location) of sampling, the frequency of sampling, and the strategy for subgrouping measurements. The concept of subgrouping (previously discussed) is one of the most important components of the control chart method. After selecting a method of subgrouping, the user of the control chart should be able to state which sources of variation in the process will act to produce variation within subgroups and which sources will affect variation between subgroups. The specific objective of the control chart will often help determine the strategy for subgrouping the data. For example, if the objective is to evaluate differences between raw material suppliers, then only material from a single supplier should be included in data *within* a subgroup.

The frequency of obtaining subgroups should depend on the objective of the chart (i.e., how quickly the objective needs to be obtained). Another consideration in selecting the frequency is how often potential causes act to produce changes in the process. In practice, limitations on the availability of measurements often set the frequency.

Most Likely Special Causes and Notes Required

The documentation of process information and activities is the most important part of many control charts. This documentation includes changes in the process, identification of special causes, investigations of special causes, and other relevant process data. Information from flowcharts and cause-and-effect diagrams should be used to identify particular notes that should be recorded. Responsibility for recording this critical information should be clearly stated.

Reaction Plan for Special Causes

A reaction plan for special causes revealed on the chart should be established. Often a checklist of items to evaluate or a flowchart of the steps to follow is useful. The reaction plan should specify the transfer of responsibility for identification of the special cause if it cannot be done at the local level. As an example, a reaction plan for a control chart in a laboratory to monitor a measurement system might be as follows:

1. Run the quality control standard.
2. Notify operations of a potential problem.
3. Review the log book for any recent instrument changes.
4. Prepare a new QC standard and test it.
5. Replace the column in the instrument.
6. Notify the supervisor and call instrument repair.
7. Document the results of these investigations on the control chart.

Administration of a Control Chart

There are a number of administrative duties required to maintain an effective control chart. Responsibility for timely measurement, recording data, calculating statistics, and plotting the statistics on the chart must be delineated. Proper revision and extension of control limits is an important consideration.

As stated previously, revision of the control limits should be done only when the existing limits are no longer appropriate. In addition to the initial out-of-control points, there are three other circumstances when the original control limits should be recalculated:

1. Control limits should be calculated using 20 to 30 subgroups. However, trial control limits may be calculated when less than 20 subgroups are available. (*Note*: Trial limits should not be calculated with fewer than 12 subgroups.) In this case, the limits should be recalculated when 20 to 30 subgroups become available.
2. If improvements have been made to the process *and* the improvements result in special causes on the control chart, control limits should be calculated for the new process.
3. If the control chart remains out of control for an extended period of time (20 or more subgroups) and approaches to identify and remove the special cause(s) have been exhausted, control limits should be recalculated to determine if the process has stabilized at a different operating level or a different level of variability.

The date the control limits were last calculated should be a part of the ongoing record for the control chart.

The form used to record the data and to plot the control chart is another important consideration. The form should allow for a continuing record and not have to be restarted every day or week. The control chart form should include space to document the important decisions and process information from the planning form. The recorded data should include the time and place and the person making the measurements as well as the results of the measurements.

The scale on the charts should be established to give a clear visual interpretation of the variation in the process. As a guide in selecting the scale, when the control limits are centered on the chart, about half of the scale should be included inside the control limits.

Schedule for Analysis

A schedule for analysis should be established for every active control chart. The frequency of analysis may vary for different levels of management. For example, the quality improvement team may meet to analyze the chart once per week, the department manager may meet with the team once per month to review the chart, and the production vice president might review the chart with the department manager at the end of each quarter.

Figure 2.12 shows an example of a completed planning form for a control chart maintained by an accounting group.

2.7 SUMMARY

The most powerful concept associated with control charts is their universal applicability. Although Shewhart focused his initial work on manufacturing processes, the concepts of common and special causes and of stable and unstable processes have applications in many areas, including administrative and service activities and management and supervision.

Figure 2.13 indicates the roles of workers and managers guided by a control chart to improve a process. All levels of an organization and all departments in the organization should have knowledge of the concepts of common and special causes of variation and the use of control charts to differentiate these causes.

This chapter has presented an overview of control charts including:

construction of a control chart (the concept of subgrouping),

interpretation of a control chart (rules for determining special causes),

types of control charts,

calculation sheets for control limits,

the concept of process capability,

planning a control chart.

The use of control charts in planned experimentation will be presented in later chapters of this book.

FIGURE 2.12
Example of a completed control chart planning form

1. Objective of the chart: To reduce the number of returned invoices that have to be billed again

2. Sampling, measurement, and subgrouping

Variables to be charted: Percent of invoices returned that are not paid

Method of measurement: Accounting supervisor records number of invoices sent each week and number returned unpaid

Magnitude of measurement variation: Complete, accurate counts can be made; totals can be validated

Point (location) of sampling: Master list and returns that cross the supervisor's desk

Strategy for subgrouping: Subgroup will be all invoices mailed in a given week (historically, 35–90 invoices)

Frequency of subgroups: One per week—100% of invoices for that week

3. Most likely special causes: New customers, price changes, computer program updates, new employees in the accounting department

4. Notes required:

Note	Responsibility
Number of new customers each week	Supervisor
New employees	Supervisor
Changes in computer program	Systems

5. Reaction plan for out-of-control points (attach copy):

1. Supervisor will call meeting of department to discuss all special causes.

6. Administration:

Task	Responsibility
Making measurements	Supervisor
Recording data on charts	Supervisor
Computing statistics	Supervisor
Plotting statistics	Supervisor
Extending/changing control limits	Department QI team
Filing	Supervisor

7. Schedule for analysis: QI team review once per month; supervisor sends copies to marketing and customer service group quarterly

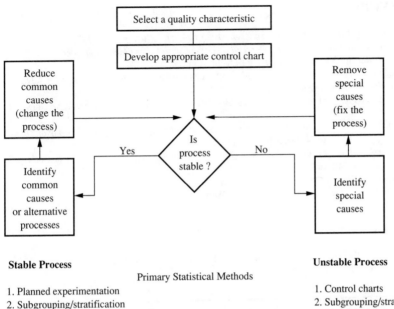

Stable Process   **Unstable Process**

Primary Statistical Methods

1. Planned experimentation   1. Control charts
2. Subgrouping/stratification   2. Subgrouping/stratification
3. Control charts   3. Planned experimentation

Responsibilities for identification

1. Technical experts   1. Workers in the process
2. Supervisors   2. Supervisors
3. Workers in the process   3. Technical experts

Responsibilities for improvement

1. Management   1. Supervisors
2. Technical experts   2. Technical experts
3. Supervisors   3. Management
4. Workers in the process   4. Workers in the process

(Note : Methods and responsibility are ordered by importance.)

FIGURE 2.13
Methods and responsibilities for improvement

REFERENCES

American Society for Quality Control (1985): *Z1.1—1985 Guide for Quality Control Charts*; *Z1.2—1985 Control Chart Method of Analyzing Data*; *Z1.3—1985 Control Chart Method of Controlling Quality During Production*, American Society for Quality Control, Milwaukee.

Deming, W. Edwards (1986): *Out of the Crisis*, Massachusetts Institute of Technology, Center for Advanced Engineering Study, Cambridge.

Grant, Eugene L., and Richard S. Leavenworth (1980): *Statistical Quality Control*, 5th ed., McGraw-Hill, New York.

Nelson, Lloyd S. (1984): "The Shewhart Control Chart—Tests for Special Causes," *Journal of Quality Technology*, vol. 16, no. 4, October 1984, pp. 237–239.

Shewhart, Walter A. (1931): *Economic Control of Quality of Manufactured Product*, American Society for Quality Control, Milwaukee. (Reprinted 1980.)

Shewhart, Walter A. (1939): *Statistical Method from the Viewpoint of Quality Control*, W. E. Deming, ed., Graduate School, Department of Agriculture, Washington.

EXERCISES

2.1. For each of the following situations, list three likely common causes of variation and three likely special causes of variation.

	Process	Quality characteristic
(a)	Service in a commercial bank	Time waiting in line
(b)	Batter in baseball	Number of hits per game
(c)	Plastic extruding process	Shrinkage of parts
(d)	Food service in a restaurant	Temperature of the food
(e)	Delivery of mail	Deliveries to wrong address
(f)	Machining process	Size of a drilled hole
(g)	Teaching a class	Students' scores on a test
(h)	Driving a car	Fuel consumption (mpg)

2.2. Under what conditions should specifications and tolerances be used instead of a control chart?

2.3. Consider a potential control chart for each of the following situations. Describe the variable to be charted, the type of control chart, and the method of obtaining subgroups (including subgroup frequency and subgroup size).

(a) The owner of a new car is interested in understanding the variation of fuel consumption. She uses the car for business and averages about 700 miles per week. The EPA rating on the car is 20 miles per gallon (mpg).

(b) A housekeeper is concerned about the family's high electric bill. She thinks that one important cause is leaving the lights on in the house when they are not being used. To understand this problem, she decided to collect some data. Each day for a month, at 9:00 a.m. and again at 5:00 p.m., she went through the house and counted the number of lights on with no one in the room.

(c) A machine shop is concerned about variation in dimensions of the parts they make. Customer tolerances are ±0.003 in. for most of the critical dimensions. There have been a number of parts returned recently because of the size of the drilled holes. The drilling operation consists of three machines, each with four stations. An operator can drill about 45 parts per hour.

(*d*) A salesman wants to evaluate various marketing strategies. He calls on 20 to 30 prospective buyers each day. Historically, about one-third of his prospects buy his product after a sales call.

(*e*) The typist in a department has to work overtime because of changes to materials he has typed. He wants to collect some data on the extent of this problem to present at the department's monthly quality meeting. He types a total of 15 to 30 reports a day for six department personnel.

(*f*) A runner is interested in progress in improving his times in the mile run. He trains four days a week and has a time trial or a race once per week.

(*g*) The president of an office supply company decided to evaluate the delivery performance of her suppliers. During a period of three months, she evaluated each order from her 30 largest suppliers and recorded the date requested and the actual date of delivery. The number of deliveries from these suppliers ranged from 8 to 24.

2.4. Complete a control chart planning form for the following scenario.

Some of the customers of a specialty chemical have been asking for lower levels of a particular impurity in the product. Although the yield varies widely from batch to batch, the operators felt that attempts to lower the impurity level through temperature and pressure controls resulted in lower yield. Frequent catalyst additions also confused the interpretation of the results.

The chemical process consisted of two parallel, continuous reactors and three sequential distillation towers. The primary feedstock was obtained by pipeline from three suppliers. The impurity was measured by gas chromotography (GC), which had good precision for the levels of interest. The GC was in the unit so that samples could be run as frequently as every 20 minutes. Yield was measured by a material balance, which was done after each shift (every eight hours). The material balance was thought to be accurate to within 1%.

The operators in the unit had recently had SPC training and were eager to use control charts to learn about the process. The marketing manager for the product asked to be kept abreast of progress. He wanted to respond to the customers' requests within two months.

2.5. The personnel director at a trucking company has been keeping track of absenteeism of her 20 employees for about four months. Each week she records the total number of absent days (100 possible). The weekly absenteeism standard for the company is 10%. After looking at the data for the first five months, the director is pleased: her 20 employees' absenteeism has never been over 10% on any week. Develop a control chart for the data given here, and answer the following questions:

(*a*.) Is the company's absenteeism in statistical control?

(*b*.) Should the personnel director be concerned about the absenteeism record?

(*c*.) What does the control chart show that was missed by simple inspection of the data?

week:	1	2	3	4	5	6	7	8	9	10	11	12	13	14	15	16	17	18
days missed:	5	2	3	7	2	4	6	4	2	4	2	3	2	4	5	7	7	8

2.6. The accounting department is instituting process improvement and has been studying causes of delays, rework, and excess overtime. The department's quality control charts indicate that a large number of invoices have to be manually handled (requiring extra phone calls, rerouting documents, and other types of rework) because of mistakes or incomplete information on purchase orders. The head of the accounting department has asked the manager of the purchasing department to investigate this problem.

The purchasing manager decided to select some of the prepared orders for each purchasing agent over a one-week period and review them for completeness and errors. Sixty documents were randomly selected from the work of each of the 20 agents and reviewed. Orders with one or more mistakes were identified. The following data were obtained at the end of the week.

Agent	Number of orders with mistakes	Agent	Number of orders with mistakes
Dan	5	Bart	4
Hank	0	Linda	1
Ann	2	Judy	3
Bill	6	Helen	9
Mary	7	Larry	8
Dave	5	Ron	4
Fred	8	John	5
Sue	4	Mark	0
Chris	0	Emma	6
Tom	5	Tina	8

(a) What can the purchasing manager learn from these results?
(b) Which are responsible for causing the mistakes, special or common causes?
(c) Which agents should be selected for special consideration?
(d) What feedback should the manager give to the head of the accounting department?
(e) Should the purchasing manager continue to collect data? If so, how should he analyze the data?
(f) What other types of data would be useful?

2.7. Approximately 200 deliveries are recorded by the receiving department each week. Each delivery includes accompanying paperwork. There are many different mistakes that can occur with these deliveries, including wrong materials, damage, overage materials, shortages, late arrivals, and errors in the paperwork. Data are collected for a period of 20 weeks to examine the extent of these quality problems.

A clipboard is placed in the delivery area, and all personnel are asked to record the various problems in one of three categories:

critical errors (complete failure of the delivery)
major errors (problems that delay the delivery)
minor errors (errors that do not slow down the delivery)

The following data were obtained from this study.

Week	Critical errors	Major errors	Minor errors
1	4	8	31
2	2	12	44
3	8	6	10
4	3	4	26
5	13	11	49
6	6	7	63
7	7	16	19
8	4	9	28
9	0	5	55
10	5	10	33
11	3	14	35
12	4	8	57
13	7	4	22
14	1	10	7
15	5	13	14
16	2	7	11
17	14	9	19
18	4	15	16
19	7	6	5
20	6	10	8
Total	105	184	552

(a) Prepare appropriate control charts for these data.
(b) What can you learn about the delivery/receiving process?
(c) What are some possible reasons for the special causes?
(d) What other types of data would be useful in improving this process?

2.8. The safety standard for the trucking industry's accident record is no more than two accidents per million miles. The U.S. Department of Transportation gathered accident data for the 14 largest companies for the most recent year to evaluate their safety performance. The data are given below. Prepare an appropriate control chart, and answer the following questions.
(a) Do the accidents come from a stable process?
(b) What does the control chart tell us about the accident records of the 14 trucking companies?
(c) Given the current system, are the companies capable of having no more than two accidents per million miles?

Company	Miles driven (millions)	Accidents
A	9.3	21
B	4.1	5
C	9.6	22
D	7.8	24
E	8.0	17
F	11.1	22
G	8.6	8
H	8.4	15
I	4.2	5
J	5.0	16
K	5.3	6
L	4.7	11
M	9.2	20
N	6.9	8
Total	102.2	200

2.9. A critical contaminant is regularly monitored during continuous production of a chemical. Samples are selected three times (at somewhat regular intervals) during each eight-hour shift. The following data were obtained during the last week.

	Concentration of contaminant (ppm)		
	A shift	**B shift**	**C shift**
Monday	18	17	18
	17	20	18
	17	18	16
Tuesday	19	16	18
	19	17	19
	18	16	19
Wednesday	17	16	20
	17	18	19
	17	19	18
Thursday	17	21	21
	18	22	20
	16	22	21
Friday	24	21	22
	22	21	22
	23	22	22
Saturday	20	20	23
	21	22	22
	23	22	22
Sunday	21	21	22
	20	21	22
	22	21	20

(a) Was the process stable during this week?
(b) What is the capability of the process for this contaminant?

2.10. The materials handling department depends on forecasts from the various outlying plants that receive chemical products. The forecasts are required for planning local production, scheduling transportation equipment, and allocating various resources. Currently, forecasts made one week in advance are used, but this is not enough lead time for many of the activities and resources that must be scheduled.

Forecasts are currently made one, two, three, and four weeks in advance. The one-week forecast is thought to be reasonably reliable. If the two-week forecast were reliable, the materials handling department's job could be done much more efficiently. The following data show the two-week forecasts (stated in thousands of pounds) made by the three largest receiving plants for the past 20 weeks. The actual production usage is also shown.

	Plant A		Plant B		Plant C	
Week	Forecast	Actual	Forecast	Actual	Forecast	Actual
1	350	330	200	210	250	240
2	420	430	220	205	300	300
3	310	300	230	215	130	120
4	340	345	190	200	210	200
5	320	345	200	200	220	215
6	240	245	210	200	210	190
7	200	210	210	205	230	215
8	300	320	190	200	240	215
9	310	330	210	220	160	150
10	320	340	200	195	340	335
11	320	350	180	185	250	245
12	400	385	180	200	340	320
13	400	405	180	240	220	215
14	410	405	220	225	230	235
15	430	440	220	215	320	310
16	330	320	220	220	320	315
17	310	315	210	200	230	215
18	240	240	190	195	160	145
19	210	205	190	185	240	230
20	330	320	200	205	130	120

(a) Is the forecasting process in statistical control for each of the three plants?
(b) What is the current capability of the three forecasting processes?
(c) What recommendations would you make to improve these processes?

2.11. Approximately 40 invoices are prepared by the accounting department each day in one of the product-line divisions. Some of the invoices are returned by customers for corrections and adjustments. Recently, more and more customers have complained to their sales contact about "incorrect invoices." The accounting manager set up a data collection program to study this problem.

All invoices that required adjustment and the reason for adjustment were recorded and grouped by the day the invoice was prepared. Three-day groupings of invoices were studied (approximately 120 invoices). The following data were obtained during the first two months.

	First month			Second month	
Days	Number of invoices	Number requiring adjustment	Days	Number of invoices	Number requiring adjustment
1–3	125	20	1–3	122	21
4–6	120	16	4–6	110	11
7–9	118	9	7–10	125	24
10–12	123	23	11–13	122	20
14–16	130	12	14–17	116	17
17–19	127	14	18–20	117	7
20–22	121	18	22–24	118	23
23–26	114	23	25–27	123	15
27–30	119	19	28–30	128	20
Total	1,097	154		1,081	158

(*a*) Is the process stable?
(*b*) What predictions can be made for future periods?

Process improvement projects were begun in the accounting department. Flowcharts were prepared for the various processes in the invoicing system. A Pareto analysis on the causes of the adjustments was done and used to prioritize the projects. The following data on invoice adjustments were obtained three months after the improvement projects were begun:

	Sixth month			Seventh month	
Days	Number of invoices	Number requiring adjustment	Days	Number of invoices	Number requiring adjustment
1–3	115	3	1–3	126	18
4–6	120	8	4–7	110	13
7–9	128	6	8–9	114	16
10–13	113	4	11–13	122	14
14–16	120	9	14–17	127	3
17–19	117	5	18–20	110	0
21–23	125	13	22–24	112	4
24–26	122	22	25–28	125	8
28–30	129	12	29–31	120	6
Total	1,089	82		1,066	82

(*c*) Have the improvement projects affected the number of required adjustments?
(*d*) What should be the next step to improve the process?

CHAPTER
3

ANALYTIC
STUDIES

3.1 INTRODUCTION

There are two types of statistical studies: *enumerative* and *analytic*. This book is concerned with the design of analytic studies and the analysis of data from these studies. The purpose of this chapter is to describe the nature of analytic studies and to outline some methods for analytic studies. To help the reader understand analytic studies, enumerative studies will also be defined and contrasted with analytic studies. Aspects of enumerative studies will be described only so far as they help clarify the nature of analytic studies. For a thorough treatment of methods for enumerative studies, see Deming (1960).

The terms *enumerative* and *analytic* as applied to statistical studies are due to Deming (1950, 1975). Before defining enumerative studies and analytic studies, we must define the concepts of a *universe* and a *frame*, which are important for enumerative studies.

> The **universe** is the entire group of items (e.g., people, materials, invoices) possessing certain properties of interest.
> A **frame** is a list of identifiable, tangible units, some or all of which belong to the universe and any number of which may be selected and studied.

The frame provides the practical means to access some or all members of the universe. A frame could be a list of manufactured parts, a list of registered voters, a manifest of bales of wool on a ship, or a list of accounts receivable.

Deming (1975) defined enumerative and analytic studies as follows:

> An **enumerative study** is one in which action will be taken on the universe.
> An **analytic study** is one in which action will be taken on a cause-and-effect system to improve performance of a product or a process in the future.

An example of an enumerative study is the United States census, which is carried out every 10 years. The number of representatives in Congress from an area depends on the number of inhabitants in an area, as counted by the last census. Determination of a fair price to pay for an inventory is another enumerative study, as is an exit survey at the polls on election day.

A study of material from three suppliers to decide which supplier should be given a contract is an example of an analytic study. The study is done to determine the supplier whose material will be most advantageous for the plant to use in the future. Use of a control chart to study a process to bring it into a state of statistical control is an analytic study.

A summary of the important aspects of analytic studies is presented in Figure 3.1.

The aim of an analytic study is prediction, prediction that one of several alternatives will be superior to the others in the future. The choice may be among different concepts for a product, among different materials, or among different conditions for operating a process. As part of the analysis of the data from an analytic study, it is generally useful to compare estimates of the performance of the alternatives under the conditions of the study. However, it is important that the experimenters do not lose sight of the fact that the ultimate aim is prediction of performance in the future.

In an analytic study the focus is on the cause-and-effect system. There is no identifiable universe as there is in an enumerative study and, therefore, no frame. As

Aim	Prediction
Focus	Cause-and-effect system
Method of access	Models of the product or process, such as flowcharts and cause-and-effect diagrams.
Major source of uncertainty	Extrapolation to the future
Major source of uncertainty quantifiable ?	No
Environment of the study	Dynamic

FIGURE 3.1
Important aspects of analytic studies.

the purpose of an analytic study is to improve performance in the future, the outcomes of interest have yet to be produced. In some cases the process or product itself may not exist; studies are performed on laboratory or pilot versions of the process or prototypes of a product.

Just as a frame provides access to the universe in an enumerative study, some methods are needed to study the cause-and-effect system and make it tangible in an analytic study. The following methods can be used for this purpose:

- theoretical and empirical models of the product or process
- prototypes of the product
- flowcharts of the process
- quality characteristics of products or processes
- cause-and-effect diagrams outlining the factors and conditions that influence the quality characteristics

These methods and how they fit into a model for improving quality were described in Chapter 1.

A model of the cause-and-effect system in an analytic study will never be complete. All the factors that affect the quality characteristic or outcome are not known. Some of the factors that affect the process are important only during certain periods of time. Changes to the process, such as new equipment or different operating conditions, may change which factors have important effects on the outcomes of the process. For example, temperature may have an effect only when humidity is high. A product may perform differently depending on the condition of use. Thus, the environment in which analytic studies are conducted is dynamic. Methods used to conduct and analyze analytic studies must be suited to this dynamic environment.

Because of the effect of changing conditions, the primary source of uncertainty in an analytic study lies in identifying which variables will have the most influence on future outcomes of the process. There is no statistical theory that allows a quantification of the magnitude of this uncertainty.

There are two types of error in an analytic study:

Type I: Make a change to the process or product that does not result in improvement.

Type II: Fail to change the process or product when a change would have resulted in improvement.

In the environment of an analytic study, the probability of making either error is not known, nor is it usually possible even to know if an error has been made. (Who knows how things would have turned out if another alternative had been selected?) How does one proceed in an analytic study when comforting knowledge such as the magnitude of the uncertainty or the probability of making a wrong decision is not known? The degree of belief that the results of an analytic study provide sufficient basis for action can be increased by the use of methods of design and analysis outlined in the next two sections.

3.2 DESIGN OF ANALYTIC STUDIES

There are important statistical methods that contribute directly to the usefulness of the results of a study as a rational basis for future action. Methods that pertain to the design of analytic studies will be outlined in this section.

Sequential Building of Knowledge

The framework to improve quality was called the "model for improving quality" in Chapter 1, and a description of its key components was given.

To begin an analytic study, the current knowledge about the relevant products and processes is documented by use of models such as flowcharts and cause-and-effect diagrams.

The iterative nature of the model is an important feature for analytic studies. Knowledge of the product or process is increased sequentially. Experiments are designed to answer questions based on a combination of theory from experts in the subject matter and conclusions from analysis of data from past studies. The more complete the current knowledge, the better the prediction will be. Confidence that the current knowledge provides sufficient basis for action is directly related to the ability to predict future performance of the process or results of future studies. Experiments performed to increase the degree of belief, by testing the results of past studies under different conditions, will be referred to as **confirmatory studies**.

Testing over a Wide Range of Conditions

Too often, analytic studies are not conducted over a broad range of conditions. The reasons for this include tight budgets, time constraints, difficulty in analysis of the data, lack of knowledge of how to efficiently include them, and too many possible conditions to consider. Fortunately, there are statistical methods that help to eliminate or reduce these difficulties in planning experiments. The use of blocks (discussed in Chapters 4 and 5) allows the grouping of experimental runs so that a wide variety of conditions can be included in the study and still allow comparisons of different alternatives under uniform conditions. The combination of incomplete blocks with fractional factorial designs, called *outer arrays* by Taguchi (1987), is another way to conduct the study under a wide range of conditions.

The degree of belief in the results of an analytic study is increased as the same conclusions are drawn under a variety of test conditions. If a particular supplier's material proves best over different environmental conditions on different days, one will feel much safer in using the results to select a supplier than if the study were run on a single day under constant environmental conditions. Experimenters might also consider running the study using material from more than one machine setup, batch, shift, or plant.

The concept of testing over a wide range of conditions can be summarized by saying that the degree of belief is increased as results are repeated in subsequent tests. How wide a range is sufficient? How close to field conditions do research conditions have to be? These are questions that only an expert in the subject matter should attempt to answer. They cannot be answered by statistical theory; however, the theory of statistical design of experiments provides tools to increase the degree of belief in the results and hence to increase the expert's potential to make improvements.

Selection of Units
for the Study

In an analytic study, there is no universe from which to draw a sample. However, in designing an analytic study there are decisions to make concerning the conditions under which the process will be run during the study and the outcomes that will be measured for each set of conditions. Deming (1975) makes the point that all analytic studies are conducted on judgment samples. The judgment of the expert in the subject matter determines the conditions to be studied and the measurements to be taken for each set of conditions. It is rare in an analytic study that a random selection of conditions or outcomes is preferred to a judgment selection. The opposite is true in an enumerative study (see Figure 3.2). An example will help to clarify this point. Consider a study to choose one of four different types of instruments for use by operators in the plant.

> *Objective*: Choose an instrument for future use that will provide the best precision when used by any operator.
> *Expected results:*
> **1.** The instrument will have sufficient precision for the application.
> **2.** The instrument will be relatively insensitive to differences among operators.

There are resources sufficient to allow 7 of the 30 operators presently working in the plant to participate in the study. How should the 7 be chosen?

The operators could be randomly selected, and the magnitude of measurement variation under the conditions of the study could be estimated. It is important to realize that the conditions of the study will not be repeated. There will be new operators working under new conditions, training may change, and so on. A judgment selection is more appropriate for a study of this type.

One option is to choose some operators with the most experience and some with the least experience. If a particular instrument performs best when used by both experienced and inexperienced operators, the degree of belief that a good choice of

Type of study	Method of selection	
	Random	*Judgement*
Enumerative	Good	Bad
Analytic	Fair	Good

Figure 3.2
Methods of selection of units for a statistical study

instrument has been made will usually be greater than if the instrument performs best when used by seven randomly selected operators.

The boundaries for the experiment are then defined:

> *Boundaries for the activities:* The operators for the experiment will be selected based on the judgment of the operators' supervisors. The purpose of this selection is to provide a degree of belief that the instrument will be relatively insensitive to the differences among operators selected.

It is often impractical to measure all the outcomes of the process for each set of conditions. However, a random selection is usually not the best method for selecting which outcomes to measure. For example, in a high-volume manufacturing process, a strategy to select units in subgroups is usually preferable so that control charts can be used to interpret the variation in the data.

It is sometimes useful to combine units selected randomly and by judgment. Consider a study of a molding process in which different operating conditions for the process are studied. If the mold has 100 cavities, it may be impractical to study the parts produced in all cavities under each set of conditions. A portion of the experimental units could be selected based on the judgment of an engineer. The engineer may suggest some cavities to be included—possibly those in the corners of the mold, some that are fairly worn out, some near the point of injection, and some far away. Remaining cavities could be selected randomly.

Initially, it is usually best to select conditions or experimental units at the extremes (within practical constraints). Alternatives can be compared at these extremes and further experiments performed as needed. For more on the use of judgment samples, see Deming (1976).

3.3 ANALYSIS OF DATA FROM ANALYTIC STUDIES

Deming (1986, p. 132) states: "Analysis of variance, *t*-tests, confidence intervals, and other statistical techniques taught in the books, however interesting, are inappropriate because they provide no basis for prediction and because they bury the information contained in the order of production."

In an enumerative study, the existence of a distribution for the characteristic of interest is ensured by the existence of a frame. Summary statistics such as a mean and standard deviation can be used to estimate parameters of the distribution. These estimates will have a quantifiable measure of uncertainty if the sample from the frame is chosen using a random number table. This type of analysis is usually an important step in accomplishing the aim of an enumerative study.

In an analytic study the aim is prediction. A distribution useful for even short-term (days or weeks) prediction exists only if the processes related to the study are stable. The concept of a stable process has important implications for analytic studies. A stable process makes it easier to determine the effect that changes in the process have on the outcomes. The stability of a process also provides a rational basis for the experts in the subject matter to make predictions based on the study.

A stable process implies an approximately constant cause system. In making predictions from analytic studies, it is important to consider all available information concerning the stability of the processes involved in the study. This includes the methods of measurement used to obtain the data. In later chapters, emphasis will be placed on plots of the data in order of time to determine the presence and identity of special causes occurring during the study.

The standard error of a statistic or the standard deviation of a process does not address the most important source of uncertainty in an analytic study: factors outside the conditions of the study. These measures of variation may sometimes be useful for assessing the impact that common causes not under consideration in the study have on the results. However, such procedures will not be emphasized in this book. The primary reason for this is that the theoretical assumptions on which they are based are rarely satisfied in an analytic study. This is usually due to circumstances such as lack of stable processes, lack of appropriate randomization due to practical constraints, interaction between the factors under study and the experimental conditions, and interaction between the factors and the experimental units. (See Chapter 4 for definitions of *randomization* and *experimental unit*.)

Often in analytic studies, we are interested in long-term (months or years) predictions or predictions extending far beyond the conditions of the study. Although stability is important in increasing the degree of belief in an analytic study, stable processes in the past do not guarantee a constant cause system in the future. Responsibility for extrapolation still rests with the experts in the subject matter.

Also, in many circumstances the conditions under which the results will be used are very different from the conditions of the study. In these cases, the stability of the processes that generated the data, although extremely desirable, is much less integral in establishing a high degree of belief than are the experimental plan and a knowledge of the subject matter. This is often the case for work in research and development, where a study is commonly conducted on prototypes to learn about how the products will function years later in the field. The same can be said of lab tests on a chemical process intended to foster some insight on how the full-scale process should be designed. In these and many other situations, the experimental plan is the most important statistical tool for increasing the degree of belief. Plans that allow results to be compared over a wide range of conditions are especially useful.

Basic Principles for Analysis

Three basic principles underlie the methods of analysis for analytic studies:

1. The analysis of data, the interpretation of the results, and the actions that are taken as a result of the study will be closely connected to the current knowledge of experts in the relevant subject matter.
2. The conditions of the study will be different from the conditions under which the results will be used. An assessment of the magnitude of this difference and its impact by experts in the subject matter should be an integral part of the interpretation of the results of the experiment.

3. Methods for the analysis of data will be almost exclusively graphical, with minimum aggregation of the data before graphical display. The aim of the graphical display will be to visually partition the data among the sources of variation present in the study.

The first principle relates to the important role that knowledge of the subject matter plays in the analysis and interpretation of data from analytic studies. Part of the model for improvement of quality described in Chapter 1 was the documentation of the relevant knowledge currently held by the experimenters. This knowledge may be based on well-accepted scientific principles or may simply consist of hunches based on past experiences. The improvement cycle is used to build the knowledge necessary for improvement. The first phase in the cycle is *plan*. During the planning phase, predictions of how the results of the study will turn out are made by the experimenters. This is a means of bringing their current knowledge into focus before the data are collected for the study.

In the *do* phase of the cycle, the study is performed and the analysis of the data is begun. The third phase, *study*, ties together the current knowledge and the analysis of the study. This is done by comparing the results of the analysis of the data to the predictions made during the planning phase.

The degree of belief that the study forms a basis for action is directly related to how well the predictions agree with the results of the study. This connection between knowledge of the subject matter and statistical analysis of the data is essential for the proper conduct of an analytic study.

The second principle relates to the difference between the conditions under which an analytic study is run and the conditions under which the results will be used. In some studies, the conditions are not too different. An example of a small difference is a study in which minor changes are made to a stable manufacturing process. Pilot plant studies or medical experimentation on animals to develop drugs or procedures to eventually be used for humans are examples of substantial differences in conditions.

Differences between the conditions of the study and the conditions of use can have substantial effects on the accuracy of the prediction. In most analytic studies, some aspects of the product or the process are changed, and the effects of these changes are observed. Estimation of these effects is an important part of the analysis. However, it should not be assumed that the factors produce the same effects under all conditions. The factors changed in the study may interact with one or more background conditions, such as environmental conditions, time of day, or purity of materials. This interaction can cause the effect of a factor to differ substantially under different conditions. For example, one inspection tool may be the best when used by experienced operators yet perform poorly relative to other alternatives when used by those with little experience.

A wide range of conditions will allow the experimenter to assess the presence of such interactions and the implications for interpretation of results. Graphical methods of analysis will be used to determine the stability of effects over different conditions. During all phases of an analytic study, the experimenters should have firmly in mind the answer to the question: How will the results of this study be used?

The third principle relates to the synthesis of theory and the data presented with the aid of graphical methods. In analytic studies, it is essential to:

- include knowledge of the subject matter in the analysis;
- include those who have process knowledge: operators, technicians, engineers, clerical personnel, scientists, and managers;
- apply this subject matter knowledge creatively to uncover new insights about the relationship between variables in the process.

In later chapters, methods will be described to visually partition the variation in the data into common and special causes of variation as well as variation due to the effect of variables deliberately changed or measured in the experiment. The simplest example of this approach is a control chart. The specific approach to analysis will differ depending on the type of experiment conducted, but the following elements are common to the analysis of all experiments:

1. Plot the data in the order in which the tests were conducted. This is an important means of identifying the presence of special causes of variation in the data.
2. Rearrange this plot to study other potential sources of variation that were included in the study design but not directly related to the aim of the study. Examples of such variables are batches of raw material, measurement, operators, and environmental conditions.
3. Use graphical displays to assess how much of the variation in the data can be explained by factors that were deliberately changed. These displays will differ depending on the type of experiment that was run.
4. Summarize the results of the study with graphical displays.

Statistical methods for design of analytic studies and analysis of data from these studies will be presented in the remainder of this book. The frame of reference for these methods will be the elements of analytic studies presented in this chapter.

REFERENCES

Deming, W. Edwards (1950): *Some Theory of Sampling*, John Wiley & Sons, New York. (Reprinted by Dover Publishing, 1960.)

Deming, W. Edwards (1960): *Sample Design in Business Research*, John Wiley & Sons, New York.

Deming, W. Edwards (1975): "On Probability As a Basis For Action," *The American Statistician*,vol. 29, no. 4, pp.146–152.

Deming, W. Edwards (1976): *Statistical Applications Research*, vol. 23, no. 1, Japanese Union of Scientists and Engineers.

Deming, W. Edwards (1986): *Out of the Crisis*, MIT Press, Cambridge, Mass.

Taguchi, G. (1987): *System of Experimental Design*, Unipub/Krause International Publications, White Plains, N.Y.

EXERCISES

3.1 Consider each of the studies in which you have recently been involved. What was the aim of the study? Was the study an enumerative or an analytic study?

3.2 For any analytic studies identified in Exercise 3.1, assess the difference between the conditions under which the study was conducted and the conditions under which the results were used.

3.3 Planned experimentation is one of the most important statistical tools for analytic studies. Why is planned experimentation not as useful for enumerative studies?

3.4 How would you define "experimental error" for analytic studies?

CHAPTER
4

PRINCIPLES FOR DESIGNING ANALYTIC STUDIES

The first three chapters have discussed the background and role of planned experimentation in the improvement of quality. Before the construction of experimental designs and the analysis of data from planned experimentation is presented, this chapter will cover the language peculiar to experimental design, the properties of a good experimental design, and important statistical tools in planned experimentation. The presentation is designed to give an overall sense of the experimental tools. Specifics in using the tools will come in later chapters.

4.1 DEFINITIONS

An experiment, for the purposes of this book, is an analytic study designed to provide a basis for action. The important aspects of design and analysis of analytic studies have been discussed (Chapter 3). The following important terms associated with planned experimentation are defined for future use in the book:

Response variable: A variable observed or measured in an experiment, sometimes called a dependent variable. The response variable is the outcome of an experiment and is often a quality characteristic or a measure of performance of the process. An experiment will have one or more response variables.

Factor: Sometimes called an independent variable or causal variable, a variable that is deliberately varied or changed in a controlled manner in an experiment to observe its impact on the response variable. The factor can be either qualitative (for example, machine A, B, or C) or quantitative (for example, a temperature of 90°, 100°, or 110°).

Background variable: Sometimes called a noise variable or blocking variable, a variable that potentially can affect a response variable in an experiment but is not of interest as a factor. The objective of the study will differentiate factors and background variables. Typical background variables are lot, time, operator, cavity within a mold, and instrument. Background variables can be controlled in a study by holding the variable constant, by the use of blocks (to be defined), or by measuring the background variable and accounting for the effect in the analysis of the data.

Nuisance variable: An unknown variable that can affect a response variable in an experiment, sometimes called a lurking variable or an extraneous variable. A nuisance variable is a background variable that is unknown at the time the experiment is planned. A nuisance variable will appear in an experiment as noise. The impact of nuisance variables can be minimized by randomization and diagnostic analysis of the response data.

Experimental unit: The smallest division of material in an experiment such that any two units may receive different combinations of factors. Examples of experimental units are parts, batches, one pound of material, an individual person, or a 10-square-foot plot of ground.

Blocks: Groups of experimental units treated similarly in an experimental design. Blocks are usually defined by background variables. The variation of a response variable within a block is expected to be less than the variation within the entire experiment. For example, experimental units produced and tested at one time (a block defined by time) might be expected to vary less than experimental units produced at other time periods.

Level: A given value or specific setting of a quantitative factor or a specific option of a qualitative factor that is included in the experiment. The levels of a factor selected for study in the experiment may be fixed at certain values of interest, or they consist of a random selection from many possible values.

Effect: The change in the response variable that occurs as a factor or background variable is changed from one level to another. The effect must be further described in terms of the context in which it is used (a linear effect, an interaction effect, etc.).

The distinction between response variables, factors, and background variables is especially important in planning an experiment. The factors and background variables can be thought of as causes and the response variables as effects. The objective of the experiment should clarify the distinction between response variables, factors, and background variables.

4.2   PROPERTIES OF A GOOD EXPERIMENTAL DESIGN

Experiments on processes or products are very common in industrial environments. Experiments range from very informal investigations to well-planned studies involving

TABLE 4.1
Types of experiments

Very informal	1.	*Trial-and-error methods* Introduce a change and see what happens
	2.	*Running special lots or batches* Produced under controlled conditions
	3.	*Pilot runs* Set up to produce a desired effect
	4.	*One-factor experiment* Using a control chart to experiment on a process
	5.	*Planned comparison of two methods* Background variables considered in plan
	6.	*Experiment planned with two to four factors* Study separate effects and interactions
	7.	*Experiment with 5 to 20 factors* Screening studies
Very formal	8.	*Comprehensive experimental plan with many phases* Modeling, multiple factor levels, optimization

a series of specific experiments. Table 4.1 (adapted from Western Electric Co., 1956) lists some types of experiments ordered by increasing level of formality. The types of experiments discussed in the chapters of this book begin with number 4, the one-factor experiment.

There are certain properties an experimental design for an analytic study must have to provide an effective basis for action. R. A. Fisher (1947) first expounded principles for sound experimentation, and many others have extended these concepts during the past 40 years. The following five properties are applicable to analytic studies:

1. *Well-defined objective.* The objective of the experiment must be clearly defined, preferably with a statement of planned action based on the results.
2. *Sequential approach.* Experimentation should proceed sequentially, with knowledge gained in previous experiments used to design new experiments.
3. *Partitioning variation.* The experiment should allow the variation in the response variables to be clearly partitioned into components due to factors, due to background variables, and due to nuisance variables.
4. *Degree of belief.* Conclusions from the experiment should be made with an adequate degree of belief. The study should be conducted under a range of conditions to increase the degree of belief.
5. *Simplicity of execution.* The design should be as simple as possible, while still satisfying the first four properties.

Objective

A careful statement of the objective will allow efficient allocation of resources for designing the experiment. If the experiment is part of the plan for an improvement cycle, the objective of the cycle should be stated in such a way that it provides guidance to those designing the experiment. The objective should clarify whether the experiment involves screening a large number of variables to find the most important ones, studying in depth a few variables, or confirming the results of past studies under new conditions.

All interested parties should contribute to the objective before other work on the experiment is begun. For example, an experiment to be designed and conducted in research is often of interest to those in manufacturing, engineering, marketing, or management. Discussion of the objective with people in these groups can often result in a study that is more generally useful.

A statement of the results required for taking action should be considered in the objective when appropriate. Examples of objectives for experiments might be "identify factors that can be used to improve yield by 3%," "determine if the machines differ by more than 10 psi," and "identify the three factors that have the greatest effect on the variation of the parts."

Another consideration in the statement of the objective is to make clear the analytic nature of the experiment and identify the various courses of action that could be followed based on the results of the study. The objective should be helpful in identifying the response variables and appropriate factors for the study.

Sequential Approach

The sequential nature of learning in analytic studies should be considered in planning experiments. The model for improving quality (see Chapter 1) stresses the iterative nature of development of knowledge of the process. Many iterations of the cycle will include the design of an experiment in the planning phase of the cycle. Studies in the initial phase of improvement of quality will be screening studies to focus on the key factors. Screening studies are discussed in later chapters. Chapter 7 discusses fractional factorial designs to examine five or more variables in an initial study. Chapter 8 presents nested designs that can be used to focus on the areas of the process with the most potential for improvement.

As knowledge is gained, experiments will be repeated using new levels of factors previously studied and some new factors. If it is desired to study in depth the relationships between the factors, two to four factors can be studied in a factorial design (Chapter 6). A fractional factorial design can be used for five factors. As new theories are developed, experiments will be confirmatory in order to increase the degree of belief from previous experiments (see Table 4.2).

Partitioning of Variation

Determination of important factors and estimation of the effects of these factors on the response variables is important in any experiment. These decisions and estimates

TABLE 4.2
Strategy for experimentation

Current knowledge	Type of experiments
Little knowledge	Fractional factorials (screening studies)
Some knowledge	Factorial studies (new levels, new factors)
Much knowledge	Confirmatory

should not be confounded by background or nuisance variables. The factors chosen for the experiment are usually those that the experimenter believes will have the greatest effect on the response variable. In many experiments, the variation due to background or nuisance variables will be as great or greater than the variation due to the factors chosen. To help determine if the most significant factors have been studied, the experimental design must allow the variation in the response variable to be partitioned into components due to factors, due to background variables, and due to nuisance variables.

Degree of Belief

The wider the range of conditions included in the experiment, the more generally applicable will be the conclusions from the experiment. The degree of belief in the validity of the conclusions is increased by running the experiment using different machines, different operators, different days, different times of the year, different batches of raw materials, and so on.

The range of conditions selected for the study will ultimately determine the degree of belief in the actions taken as a result of the experiment. The expert in the subject matter must determine what is an "adequate" degree of belief for taking action. This determination will depend on the magnitude of the change in the process being considered and the degree of extrapolation necessary. Changes in existing manufacturing processes may require a completely different degree of belief than changes in the process of development of a new product.

Simplicity of Execution

The simplicity of the experiment should be considered one of the most important properties of a planned study to improve a process. Simplicity is important in the design, the conduct, and the analysis of a planned experiment. An experimental design should be as simple as possible while still satisfying the other properties of a well-planned experiment. Simplicity allows all interested parties to be involved in all aspects of the study. Simplicity also allows the experimenter the flexibility to adjust for changes that are often required during the conduct of the study.

Simplicity requires that all the practical aspects of conducting an experiment be considered. Some important aspects include the degree of difficulty in changing levels of a factor, the ability to control background variables, and the ability to measure important response variables.

4.3   TOOLS FOR EXPERIMENTATION

The five properties discussed in the previous section should always be considered in planning an experiment. R. A. Fisher expounded the use of four tools that can be used to help ensure that an experiment possesses these properties.

1. *Experimental pattern:* The arrangement of factor levels and experimental units in the design.
2. *Planned grouping:* Blocking of experimental units.
3. *Randomization:* The objective assignment of combinations of factor levels to experimental units.
4. *Replication:* Repetition of experiments, experimental units, measurements, treatments, and other components as part of the planned experiment.

Table 4.3 summarizes how these four tools can be used. The use of each of these tools in analytic studies is discussed in the following sections.

Experimental Pattern

The experimental pattern is the schedule for conducting the experiment. Each test to be conducted or measurement to be made is identified in the pattern. The experimental pattern is sometimes called the test plan. The pattern may include consideration of any of the other tools (planned grouping, randomization, and replication).

The selection of an appropriate experimental pattern is the primary tool for assuring the attainment of the objectives of the experiment. The pattern will identify

TABLE 4.3
Using experimental tools to attain the properties of a good experiment

Property	Experimental pattern	Planned grouping	Randomization	Replication
Well-defined objective	X			
Sequential approach	X	X		X
Partitioning variation	X	X	X	X
Degree of belief	X	X		X
Simplicity of execution	X		X	

the factor combinations and factor levels to be included in the study. The experimental pattern selected will also aid in attaining the other properties of a good experiment:

- The degree of belief for the study will be established by the pattern,
- The proper pattern will allow estimation of important effects,
- The properly selected experimental pattern will simplify the analysis of data from the study,
- The costs and other resources required by a study can be controlled through the experimental pattern.

There are many types of experimental patterns. The names of the particular patterns are often used to describe the type of experiment being conducted. One of the most common patterns used in experiments to improve quality is the *factorial design* (Chapter 6). In a factorial design, tests are arranged in a pattern such that multiple factors are studied with each test. Figure 4.1 is an example of a factorial pattern.

Another common experimental pattern is the *nested* or *hierarchical design* (Chapter 8). In a nested design, levels of different factors are studied within a given level of another factor. Figure 4.2 is an example of a nested design.

Incomplete designs have patterns that include only a subset of a full factorial or nested experimental pattern. These designs include the *fractional* factorial design and the *unbalanced* or *partially balanced* nested design.

Composite designs are another class of experimental pattern. These patterns are either combinations of factorial and nested designs or one of the basic experimental patterns augmented with additional tests. Designs for experiments involving mixtures are another important class. Much of the discussion in the remaining chapters will deal

		Pressure 1		*Pressure 2*	
		Temp. 1	*Temp. 2*	*Temp. 1*	*Temp. 2*
Batch 1	Load 1	#6	#2	#8	#15
	Load 2	#11	#13	#4	#7
Batch 2	Load 1	#10	#9	#5	#12
	Load 2	#14	#1	#16	#3

Four-factor experiment   (Factors: pressure, temperature, batch, and load)
Each factor at two levels (pressure 1, pressure 2, etc.)
Test numbers (#1 — #16) shown in a randomized order

FIGURE 4.1
A factorial experimental pattern

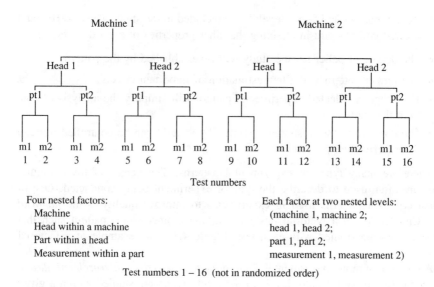

Four nested factors:
  Machine
  Head within a machine
  Part within a head
  Measurement within a part

Each factor at two nested levels:
  (machine 1, machine 2;
  head 1, head 2;
  part 1, part 2;
  measurement 1, measurement 2)

Test numbers 1 – 16 (not in randomized order)

FIGURE 4.2
A nested experimental pattern

with the selection and construction of experimental patterns appropriate for different situations.

Planned Grouping

Planned grouping is an important tool for addressing background variables in an analytic study. There are two important decisions to make concerning background variables:

- How to control the background variables so that the effects of the factors are not distorted by them
- How to use the background variables to establish a wide range of conditions for the study to increase the degree of belief or to aid in designing a robust product or process

There are three basic methods of controlling background variables:

1. Hold them constant in the study.
2. Measure them and adjust for their effects in data analysis.
3. Use planned grouping to set up blocks.

The concept of planned grouping takes advantage of naturally occurring homogeneous experimental units in the conduct of an experiment. When planned grouping is used in developing the experimental pattern, the design is often called a *block design*.

Examples of common background variables used to form blocks are time of day, operator, machine, shift, batch of raw materials, agriculture plot, cavity in a mold, test stand, and season. In studying tire wear for cars during normal use, an individual car is a natural blocking variable. The amount and type of driving for different cars could vary considerably. Different brands (or factor combinations) of tires could be tested on any particular car to minimize the car-to-car variation when comparing the brands. The natural block size is four, so four different brands could be "grouped" on each car. In studying the wear of soles of shoes, a natural block size of two would correspond to the two feet of each individual wearing the shoes.

In an analytic study, planned grouping plays a very important role in establishing the degree of belief for the results of the study. Should the study be conducted on one machine in the plant or on multiple machines? Including different machines will add additional variation but will also increase the degree of belief that the results can be extrapolated to all machines of that type in the plant. By treating the machines as blocks, the experiment can include multiple machines to increase the degree of belief in the results through analysis of interactions between the machines and the factors in the study.

Often, naturally occurring blocks are formed by a combination of background variables. Learning about an important block effect would require additional study to determine the factors or variables that contribute to the effect. Such a background variable has been called a *chunk-type* factor by Ott (1975). For example, if parts for a study were selected from two different lots of available materials, the block effect from the two different lots might be caused by different vendors, or different types of raw materials, different preprocessing equipment, or different storage conditions. The selection of different lots as a chunk variable for the study would potentially allow the effects of any of these factors (plus others unknown) to be exposed in the study. This concept is extremely important in analytic studies. Other common chunk-type block variables are time and location.

Grouping important background variables into blocks is an important part of designing an experiment. If the sizes of parts can possibly alter the effect of certain factors in the study, how can we incorporate part size in the experiment? One way is to form two groups of parts based on their measured sizes. One group would be parts at the low end of the size specification, and the other group would be parts at the high end of the specification. Combinations of factors would be tested using parts from each of these planned groups or blocks.

What if there are five background variables (for example, room temperature, operator, machine, material supplier, and line speed) that need to be considered in the study? Blocks formed by all combinations of these variables may result in too large a study. Two or three blocks could be developed by grouping extreme conditions for each of the background variables. Block 1 could comprise high temperature, operator A, machine 8, supplier X, and low line speed; and block 2 could include low temperature, operator B, machine 1, supplier Z, and high line speed. These blocks would be designed to represent extreme conditions based on the judgment of the expert in the process. If the effect of the block (or the block-factor interaction) were found to be large, the next experiment could be used to determine which background variable or combination of background variables caused the effect.

Figure 4.3 summarizes a one-factor design that creates a chunk variable (four blocks) to incorporate background variables in the design.

Some types of experimental patterns are named for the planned grouping pattern used to accommodate background variables. A *paired-comparison* experiment requires a single factor at two levels and one background variable (which could be a chunk-type variable). Two experimental units must be available for each grouping of the background variable. Treatment differences are evaluated within each block. Examples of background variables appropriate for a paired-comparison design are identical twins, front fenders on a car, and two adjacent plots of land.

The *randomized block design* extends the approach of the paired-comparison design to more than two treatments or factor levels, still using only one background variable (for defining the blocks). Each treatment must occur in each block an equal number of times. An *incomplete block design* relaxes the requirement that each treatment occur an equal number of times in each block.

FIGURE 4.3

Example of planned grouping of background variables using chunk-type blocks

Objective: Run an experiment to compare three material suppliers. Each of the three suppliers will submit four prototypes.

A. Identify background variables in the plant that could affect the response variables of interest:

Background variable	Levels
Machine	#7, #4
Operator	Joe, Susan, George
Gage	G-102, G-322
Saw blade	20 blades available
Time (day-to-day)	Many different days possible.

B. Create four blocks with widely varying conditions based on these background variables:

	Block 1	Block 2	Block 3	Block 4
Machine	#7	#4	#7	#4
Operator	Joe	Susan	George	Joe
Gage	G-102	G-322	G-102	G-322
Saw blade	Blade 1	Blade 2	Blade 3	Blade 4
Time	Day 1	Day 2	Day 3	Day 4

C. Evaluate one prototype from each supplier (A, B, C) in each block (random order within each block):

Test	Block 1	Block 2	Block 3	Block 4
1	B	B	C	A
2	A	C	A	C
3	C	A	B	B

D. Analyze supplier differences within each block; evaluate consistency of differences across blocks.

The *split-plot design* occurs when the different factors or combinations of factors are not assigned to experimental units (possibly within a block) in a random manner (see the following section, on randomization). The subplots or subgroups resulting from the assignment become chunk-type blocks.

Another approach to accommodating background variables is the *outer array* suggested by Taguchi (1979). This concept is discussed further in Chapter 10.

Randomization

Randomization is the use of random numbers to determine the assignment of factor combinations to experimental units or the order of performing some aspect of the study. Whereas the experimental pattern addresses the variables identified as factors in the experiment, and planned grouping is used to accommodate background variables, randomization is a tool that addresses the nuisance variables. Nuisance variables are process variables that affect the response variables but that have not been identified by the experimenter. It is common in conducting studies that many variables that can affect the response variables are not known to the persons conducting the experiment. Typical nuisance variables are environmental effects such as temperature or humidity, drifts in measurement equipment, batch-to-batch variation in raw materials, machine warm-up effects, and position effects in a chamber.

Randomization helps prevent the variation due to nuisance variables from being confused with the variation due to the factors or the background variables. Randomization is often compared to insurance: you only need it when a problem (i.e., a big effect of a nuisance variable) occurs. If important unknown nuisance variables are present during an experiment, randomization can allow their effects to be separated from the effects of factors or background variables. Diagnostic analyses to uncover nuisance variables are also facilitated by randomization. Unlike insurance, when randomization is not done, the experimenter may not be aware of the impact of nuisance variables and might develop incorrect conclusions.

Randomization requires a formal procedure and not just haphazard selection or ordering. Random number or random permutation tables should be used for this purpose. Appendix A.2 contains examples of random number tables and instructions for their use. Mechanical devices that simulate a random process can also be used. Examples of such procedures are flipping a coin, rolling a pair of dice, and pulling numbers (after mixing) from a container.

Randomization is often required at several different levels or phases of an experiment. For example, an experiment to evaluate several factors in an industrial process could require random selection of raw materials stored in a warehouse, random ordering of the combinations of the factors to be evaluated, and random ordering of the tests conducted to obtain the response data on the product during the experiment.

In most situations it is not desirable to completely randomize the conduct of an experiment. In an analytic study, randomization should be restricted as much as possible based on knowledge of background variables. Planned grouping, as previously discussed, is the most common restriction put on randomization. When blocking is employed, the randomization is done within each block rather than over all tests in the design. For example, an experiment might require 20 parts, 5 from each of four cavities of a mold. Randomization would be used to assign combinations of factors to the 5 parts *within* each mold.

In other cases, a random order for the conduct of a test could be prohibitively expensive. For example, changing the levels of one particular factor might require a machine to be shut down for a day, whereas changing all the other factors could be done in a few minutes. Extra care is required in analysis and interpretation in these situations. Some examples will be discussed in later chapters.

The decision of what type and level of randomization to use should be based on the particulars of the experiment. The advantages of randomization should be evaluated against the additional resources required to randomize. Situations in which randomization is particularly important include:

- experiments conducted when the important response variables have not been brought into a state of statistical control,
- experiments that will be conducted by many different operators and technicians,
- experiments in which the variation due to nuisance variables is large relative to the magnitude of effects of important factors,
- formal experiments in which results must be presented to others (such as customers or senior staff) for action to be taken.

A rule of thumb with respect to randomization for analytic studies is, after restricting for important background variables, randomize in all remaining situations unless the constraints to randomization have been objectively considered and found to be prohibitive. Randomization procedures for different types of experimental patterns will be presented in later chapters of this book.

Replication

Replication refers to repeating particular aspects of an experiment. Replication plays an important role in attaining many of the properties of a good experiment. It is the primary tool in analytic studies for studying stability of effects and for increasing the degree of belief in the results. There are many different types of replication, including:

- repeated measurements of experimental units,
- multiple experimental units for each combination of factors,
- partial replication of the experimental pattern,
- complete replication of the experimental pattern.

Each of these types of replication will require different interpretations in the analysis of the results of an experiment. For example, for comparing differences between factor combinations, replications that include all sources of variation typical of a run of either combination are desirable.

Replication plays the primary role in providing a measure of the magnitude of variation in the experiment due to nuisance variables. Replication also aids in minimizing the impact of nuisance variables on factor effects by (1) possibly enabling nuisance variables to be averaged out, and (2) allowing the study of interactions between factors and experimental conditions. Replication, in conjunction with planned grouping, can be used to expand the study to a wide range of conditions.

Since the amount of replication in an experiment directly affects the resources (time, budget, materials, etc.) required for the study, replication is often established by constraints on these resources. Since experiments should be run sequentially for studies to improve quality, the amount of replication for any one experiment is not critical. Usually it is desirable to obtain 5 to 10 comparisons for the different levels of each factor for the initial stages of the study. Another guideline is that the initial experiment should consume about 25% or less of the resources allocated for the study. This will allow the completion of four improvement cycles within the budgetary constraints on the study.

The most important consideration in an analytic study is not the number of experimental units per level of a factor, but the breadth of conditions under which the comparisons can be made. Replication over similar conditions provides little increase in the degree of belief. For example, in a high-speed manufacturing operation, a study based on 5 to 10 experimental units selected from different days' production and made from different material lots would be much preferred to a study based on 50 consecutively produced units.

4.4 FORM FOR DOCUMENTATION OF A PLANNED EXPERIMENT

Figure 4.4 shows a form that can be used to document the planning of experiments. The form is helpful in communicating the experimental plan and in documenting the considerations given to the various tools of experimentation discussed in this chapter. The form can serve as a tool in the planning phase of the improvement cycle. In many studies the form will serve as a summary of more extensive documentation and background information. A master copy of this and other forms used in this book are included in the appendix to this book.

Objective

The first item on the form is a summary statement of the objective of the experiment (see Section 4.2 for a more detailed discussion of the objective). In the statement of the objective, consideration should be given to:

- the level of knowledge about the process under study,
- the actions to be taken as a result of the study,
- the analytic nature of the study,
- a statement of the results required in order to take action.

FIGURE 4.4
Form for documentation of a planned experiment

1. **Objective:**

2. **Background information:**

3. **Experimental variables:**

A. Response variables — Measurement technique

  1.
  2.
  3.

B. Factors under study — Levels

  1.
  2.
  3.
  4.
  5.
  6.
  7.

C. Background variables — Method of control

  1.
  2.
  3.

4. **Replication:**

5. **Methods of randomization:**

6. **Design matrix:** (attach copy)

7. **Data collection forms:** (attach copies)

8. **Planned methods of statistical analysis:**

9. **Estimated cost, schedule, and other resource considerations:**

Background Information

The second item on the form is background information. In analytic studies, it is important to proceed sequentially, building on knowledge gained previously from theory or other improvement cycles. A summary of the current knowledge of the process is a component of the model for improvement of quality presented in Chapter 1. This current knowledge should be used in the design of the current study. The experiment being considered should be put in context with other studies in previous improvement cycles.

Examples of useful background information are summaries of results of previous experiments on the process under study, control charts from the process for the response variables in the study, laboratory prototype experiments, studies on similar products or processes, and studies or control charts on the measurement processes to be used in the experiment.

As part of the summary of the current knowledge of the process, a prediction of the outcome of the study is useful. This prediction can be used during the analysis of the data to help determine whether the study confirms previous beliefs or suggests a need to reconsider those beliefs.

Variables in the Study

The response variables, the factors, and the background variables are next documented. The objective should make clear whether a particular process variable will be a response variable, a factor, or a background variable. In choosing levels for the factors, the levels should be set far enough apart so that their effects will be large relative to the variation caused by the nuisance variables. However, the levels should not be so far apart that the following types of trouble could develop:

- conditions that make the experimental run unsafe
- conditions that cause substantial disruption of a manufacturing facility
- important nonlinearities or discontinuities hidden between the levels
- substantially different cause-and-effect mechanisms under the different conditions in the experiment

The method of control (hold constant, measure, or planned grouping in blocks) for each background variable identified should be included.

Based on the information in the first three areas of the form, an experimental pattern can be developed and the amount and type of replication determined. Any randomization or restrictions on the randomization should be documented, including:

- assignment of treatments to experimental units,
- determination of the order of running the study,
- determination of the order of testing.

Another important part of the documentation of the experiment is the form for recording the data. Careful consideration should be given to developing these forms.

The form should be designed for simplicity of recording, not for analysis. The randomized experimental pattern should be built into the form. The form should also include space for recording any significant events that happened during the study, including things that did not go according to plan. A carefully designed form for recording the data will help communicate the intentions of the planners of the experiment to those running the study.

A brief summary of the planned methods of statistical analysis should next be documented. Finally, some information on cost, schedule, and other necessary resources required by the experiment should be given.

REFERENCES

Fisher, R. A. (1947): *The Design of Experiments*, 4th ed., Oliver and Boyd, Edinburgh.

Ott, Ellis R. (1975): *Process Quality Control*, McGraw-Hill, New York, Chapter 4.

Taguchi, Genichi, and Yu-In Wu (1979): *Introduction to Off-line Quality Control*, Central Japan Quality Control Association, Chapter 5.

Western Electric Co. (1956): *Statistical Quality Control Handbook*, Western Electric Co., Indianapolis, pp. 76–77.

EXERCISES

4.1 Pick a process or product you are familiar with. If you were to plan a study of that process or product, list examples of each of the following:

(*a*) Response variables

(*b*) Factors

(*c*) Background variables

(*d*) Nuisance variables

(*e*) Experimental units

(*f*) Blocks

4.2 How does one distinguish a factor and a background variable in a study?

4.3 What is the relationship between degree of belief and the conditions studied in an experiment?

4.4 What are the four tools for planned experimentation? Give an example of the specific application of each of the tools in a study.

4.5 The manager of a large office wants to compare two different types of computer terminals for future use in the office. The terminal vendors have made three of each type of terminal available for a week on a trial basis. List potential background variables in the office that should be considered in a study of the two types of terminals. Develop three chunk variables to be used as blocks to incorporate these background variables in the study.

4.6 The 14 members of the accounting department will participate in a study. Each person will collect data for a one-week period. Since the study will take more than three months to complete, nuisance variables could affect the outcome. Use each of the three random number tables in Appendix A.2 to develop an order for participation in the study. (*Hint*: To use the 1–8 random permutations, divide the participants into two blocks of seven each.)

4.7 Management has identified some important areas for improvement in a continuous production unit, including yields, costs, efficiencies, scrap, and rework. A number of process changes have been proposed to address these areas, including new mixing procedures, alternative flow control, and changes in the catalyst operation (catalyst type, amount of charge, frequency of catalyst charge). The previous studies of these alternatives have not yielded clear results because of different operating procedures between the shifts. Also, the ambient weather conditions can have a big effect on the performance of the process.

Plan an experiment to improve this process. Complete an experimental design planning form for the study (it is not necessary to complete the design matrix or data collection form for the study). What other important information needs to be determined before planning the study? Use your imagination to fill in these details.

CHAPTER
5

EXPERIMENTS
WITH ONE
FACTOR

This chapter discusses planned experimentation when there is only one factor of interest. The single-factor experiment forms the basis for the multifactor experiments to be presented in later chapters. The principles and procedures introduced in this chapter for developing experimental designs and analyzing data from the experiments will be used in the remainder of the book.

5.1 GENERAL APPROACH TO ONE-FACTOR EXPERIMENTS

In an experiment with only one factor, one or more response variables are measured or observed for different levels of the factor. The levels of the factor may be either qualitative (e.g., machine A, B, or C) or quantitative (e.g., temperature of 60°, 80°, or 100°). One replication of the experiment requires the factor to be tested one time at each level. Additional replications can be done to satisfy the other properties of a good experiment. Each level of the factor is assigned to an experimental unit by a random process.

In an analytic study involving one factor, background variables should always be considered. In the simplest experimental pattern, important background variables are measured or held constant. In block designs the background variables are handled by grouping the experimental units into blocks. Important block designs for one-factor experiments are the paired-comparison design, the randomized block design, and the balanced incomplete block design. Multiple background variables are considered using the concept of chunk variables, introduced in Chapter 4. In these block designs, the factor levels are assigned to the experimental units within each block.

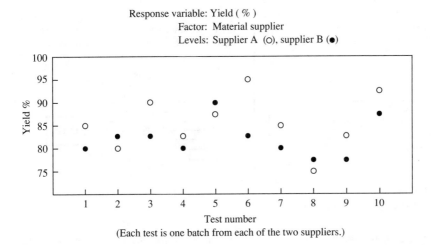

Response variable: Yield (%)
Factor: Material supplier
Levels: Supplier A (○), supplier B (●)

Test number
(Each test is one batch from each of the two suppliers.)

FIGURE 5.1
Example of a run chart for a one-factor experiment

The primary method of analyzing data from a one-factor experiment is the run chart with the factor levels identified. The run chart can also be stratified by factor level. Figure 5.1 is an example of a run chart for a one-factor experiment with two levels of the factor included in the study. The usual steps in the analysis of data from a one-factor design are as follows:

1. Plot a run chart or control chart of the data with the factor levels identified.
2. Reorder the run chart according to factor level.
3. Remove the effects of the background variables and plot the adjusted data by factor level.

Explanation of these tools will be given and clarified in the examples presented in this chapter.

5.2 USING THE CONTROL CHART FOR A ONE-FACTOR EXPERIMENT

Probably the most common use for the one-factor experiment in activities to improve quality is the active use of the control chart. The cause-and-effect diagram (Figure 5.2) is useful in comparing the reactive (passive) and active uses of the control chart.

When a control chart is used in the passive mode, action begins after the effect (i.e., a special cause) has occurred. At that time, a search is done to determine which of the causes or combination of causes produced the effect. Sometimes the search is not successful because of the large number of potential causes (both known and unknown) for the effect.

When the control chart is used in the active mode, changes in the process (the cause variables) are made, and then the effect of these changes on the response variable

Passive Mode

Control chart: Wait until a change in the effect is observed, then determine which causes have changed.

Active Mode

Planned experimentation: Change one or more of the causes and observe the change in the effect.

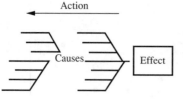
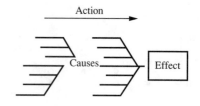

FIGURE 5.2
Cause-and-effect diagram with a control chart

being plotted on the control chart is observed. When only one cause (or factor) is changed, this can be considered a single-factor experimental design. The degree of planning that takes place distinguishes a one-factor experiment from common fiddling with variables in the process. Example 5.1 illustrates the use of a control chart in the active mode.

Example 5.1. *X*-bar and *R* control charts for a critical dimension (measured in thousandths of an inch) had been maintained for six months. The charts had remained in statistical control until recently, when changes in the average occurred for a short period when tools were replaced. A cause-and-effect diagram was prepared to identify the process variables that may have affected the variation in this dimension. One of the variables, clamp pressure, had not been previously studied by any of the operators or engineers on the improvement team. The team decided to study this factor to quantify its effect on the variation in the dimension.

Figure 5.3 shows the form for documentation of a planned experiment developed for this effort. The control chart for the day prior to the study and the three days of the study is shown in Figure 5.4. No special causes that could not be attributed to the clamp pressure changes were noted during the experiment. Increasing the clamp pressure appeared to reduce the variation in the dimension. The averages of the ranges were computed for each day of the experiment and are summarized in Figure 5.5.

Based on this experiment, it was decided to set the clamp pressure at 80 pounds. The control chart and maintenance records will be evaluated during the next week to determine the impact of this change. If the expected reduction in variation is observed, another study will be planned to evaluate the clamp pressure at settings above 80 pounds.

This study is typical of a one-factor experiment by use of a control chart. Some comments on the design of the study:

- Each clamp pressure level was evaluated on only one day (shift), so the effect of clamp pressure is confounded with days (i.e., the observed effect could be either clamp pressure or day-to-day differences, or both).
- Variation from shift to shift has not been observed for the past six weeks, so it was reasonable to include this confounding in the design of the study, as frequent changing of clamp pressure is difficult.

FIGURE 5.3
Documentation form for Example 5.1

1. **Objective:**
Determine the effect of clamp pressure on the variation of the critical dimension.

2. **Background information:**
The variation of the process (range chart) has remained in control for the past six weeks. Previously the variation was reduced by modifying the gage used for the measurement. Currently the variability of the gage is less than 5% of the total variation. No prior testing of clamp pressure has been done. The gage on the clamp pressure adjustment reads from 1 to 1000 pounds, and the current setting is 30 pounds.

3. **Experimental variables:**

   A.

Response variables	Measurement technique
1. Dimension variation	Range of four measurements (in thousandths of an inch) using the modified gage

   B.

Factors under study	Levels
1. Clamp pressure	20, 40 and 80 pounds

   C.

Background variables	Method of control
1. Thirty-five other variables that might affect the range had been identified on the cause-and-effect diagram	None of these variables will be changed during the study

4. **Replication:**
The clamp pressure will be changed at the beginning of the shift and remain at each level for the eight-hour shift. Subgroups will be selected once an hour during the shift for each level. With three levels, the experiment can be completed in three days.

5. **Methods of randomization:**
The order for setting the three levels was determined by selecting the numbers 1 (low level), 2, and 3 (high level) from a table of random numbers.

6. **Design matrix:**

	Day 2	Day 3	Day 4
Level	80	20	40

7. **Data collection forms:**
Data will be recorded and plotted on the current control chart.

8. **Planned methods of statistical analysis:**
Interpretation of the current control chart

9. **Estimated cost, schedule, and other resource considerations:**
Maintenance will be required to set the clamp pressure each day.

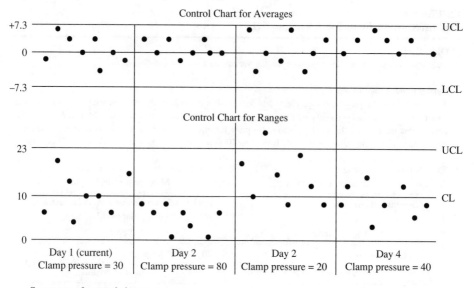

Summary of control charts:

Day 1 (prior to experiment): both average and range charts are in control; control limits are based on data from previous month.

Day 2 (clamp pressure changed to 80 lb.): The average remained in control; all of the eight ranges were below the centerline, indicating a reduction in the variation. Since no other changes in the casual variables were detected, the reduction is probably due to the increased clamp pressure. The average of the eight ranges was 5.0 (thousandths of an inch).

Day 3 (clamp pressure changed to 20 lb.): The variation increased with the range on the third hour above the UCL. This increase is probably due to the low clamp pressure. Average range = 16.0.

Day 4 (clamp pressure changed to 40 lb.): Both charts are in control. Average range = 8.0.

FIGURE 5.4
Control chart for Example 5.1

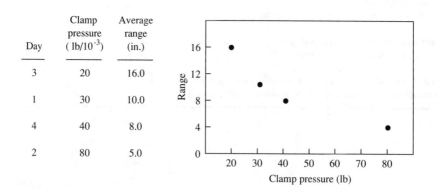

Day	Clamp pressure (lb/10^{-3})	Average range (in.)
3	20	16.0
1	30	10.0
4	40	8.0
2	80	5.0

FIGURE 5.5
Summary of results of experiment for Example 5.1

Some comments on the analysis of the data from the study:

- A special cause in the range chart was indicated two times (a run of eight points on day 2, a point beyond the UCL and two of three points near the UCL on day 3) when the level of the factor was changed. The control chart provided the analysis required to detect this effect of the factor, clamp pressure.
- The summary plot of the average range versus the clamp pressure is important since the levels of the factor are quantitative and can be ordered. The degree of belief that greater clamp pressure will reduce the variation is increased by the consistency in this plot.
- The next cycle is a confirmation of the expected reduced variation with the clamp pressure at 80 pounds. Another cycle may be planned to study the effects of increasing the clamp pressure even more.

There are two important differences between this one-factor experiment and the typical use of the control chart:

1. The active nature of the study. Experimenters made changes in the factor rather than waiting for changes in the response variable.
2. The amount of planning that was done prior to making the changes in the factor levels. The form in Figure 5.3 was the guide used for the planning.

5.3 EXAMPLE OF A ONE-FACTOR DESIGN

Almost all experiments for analytic studies will include background variables, as discussed in Chapter 3. The following example of a one-factor experiment is given to illustrate the importance of run charts in planned experiments. These charts should be developed while the study is in progress to take advantage of information as it is obtained. The run chart will be used to do an initial partitioning of the variation in the response variables into variation due to the factors, variation due to background variables, and variation due to nuisance variables (either common or special causes). The examples in the remainder of the chapter cover the introduction of background variables into one-factor experiments.

Example 5.2. Alternative control methods were being evaluated for a continuous chemical process. These included the current control methods, an approach using a series of manually operated control valves, and an automated flow-metered system. A study was outlined to evaluate the alternatives. Changing the control methods required that the unit be shut down for a significant period of time. The key measures of process performance that could be affected by the control methods were the sulfate content, the percent of water, and the amount of a critical ingredient (X). The target levels for these variables were $\leq 1\%$ sulfate, $\leq 3\%$ water, and ingredient $X = 15\%$.

Figure 5.6 shows the documentation form summarizing the planned study. Table 5.1 lists the data obtained from the study.

FIGURE 5.6
Documentation form for Example 5.2

1. **Objective:**
Compare alternative control techniques in a continuous process for the following quality characteristics: percent sulfate, percent water, and percent ingredient X.

2. **Background information:**
The current method was recently upgraded, and a number of other special causes have been occurring, so little background information is available. Tests are not usually conducted at the point of interest in this study, but the variation of the characteristics of interest is thought to be small, based on sampling and analysis.

3. **Experimental variables:**

   A.

Response variables	Measurement technique	Target level
1. sulfate (%)	Titration	$\leq 1.0\%$
2. water (%)	Titration	$\leq 3.0\%$
3. Ingredient X (%)	Gas chromatograph	$15\% \pm 3\%$

   B.

Factors under study	Levels
1. Control method	Current, valve, meter

   C.

Background variables	Method of control
1. None identified	

4. **Replication:**
A decision on methodology is needed within about three weeks, so tests will be conducted on each method for 5 days (a total of 15 days).

5. **Methods of randomization:**
Because of the extreme expense required to change the control method, tests cannot be randomized. The current method will be tested for five days, then the valve for five days, then the meter for five days.

6. **Design matrix:** (attach copy)
(See Table 5.1.)

7. **Data collection forms:**
Normal laboratory worksheets with "special study" noted at top.

8. **Planned methods of statistical analysis:**
run charts of data; possibly run charts of adjusted data and statistical summaries of response variables if appropriate.

9. **Estimated cost, schedule, and other resource considerations:**
The study can be completed during a 15-day period with one composite sample per day tested for each parameter.

TABLE 5.1
Test results for Example 5.2

Control method	Date	Water (%)	Sulfate (%)	Ingredient X (%)
Current	7/2	2.2	0.5	15.7
	7/3	2.0	0.2	17.2
	7/4	2.4	0.6	15.1
	7/5	2.7	0.3	16.5
	7/6	2.9	0.4	14.8
Valve	7/7	3.3	0.2	11.8
	7/8	3.2	0.5	12.6
	7/9	3.4	0.4	13.4
	7/10	3.6	0.3	19.4
	7/11	3.5	0.5	12.3
Meter	7/12	3.7	0.4	18.8
	7/13	3.8	0.5	17.6
	7/14	4.0	0.3	19.0
	7/15	4.2	0.5	13.0
	7/16	4.1	0.6	18.2

The first step in analysis is a run chart for each of the response variables. Figure 5.7 contains run charts for the three response variables in this study. The following conclusions were made from these plots:

Percent water: The variation in the data can almost all be attributed to a linear trend. The identification of the nuisance variable causing this trend is the most important next step. The effect of the factor is small relative to this nuisance variable effect.

Percent sulfate: The majority of the variation is due to nuisance variables. The process appears stable throughout the study. No factor effects or special causes are obvious.

Ingredient X: The factor, control methodology, has a major effect on ingredient X. Two special causes (July 10 and 15) are also important. Other nuisance variables contribute a small amount of variation. The current control method resulted in a proportion slightly above the target of 15%. The valve showed lower results, and the meter had higher results. The special causes resulted in one of the valve measures being high and one of the meter measures being low.

These conclusions are summarized in Table 5.2. The follow-up studies for this example should focus on the special causes observed in the study. A decision on choice of control methodology may depend on the outcome of this further investigation.

In this design, the factor levels were not randomly assigned to the experimental units (days). If important special causes due to nuisance variables are present in the data, as in this example, the lack of randomization can limit learning about the factors in the study. The run charts play a critical role in detecting these special causes by distinguishing their effects from those due to one of the factor levels.

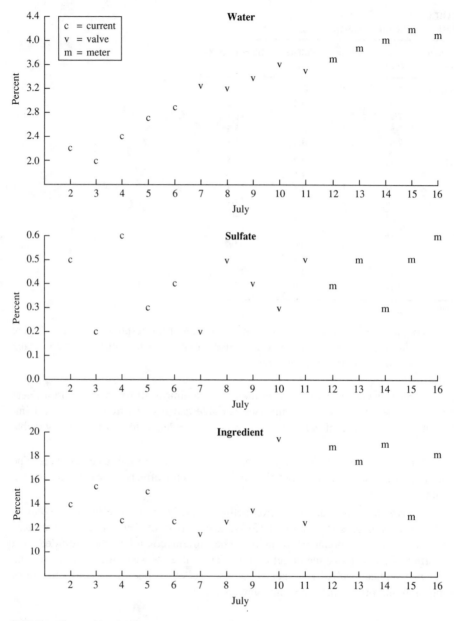

FIGURE 5.7
Run chart of data for Example 5.2

Sometimes the special causes are not obvious by inspection of the run charts. The trends or unusual data values may be hidden by the effects of the factors in the study. Additional analysis may be useful in cases where this might be expected. Each data point in the study can be adjusted by subtracting the factor effect. This analysis is equivalent to "residual analysis" (Box, Hunter, and Hunter, 1978, p. 183) except that

TABLE 5.2
Summary of important effects in Example 5.2

	Response variable		
Source of variation	Percent water	Percent sulfate	Percent ingredient X
Factor	Small	Small	Important
Nuisance variable:			
Common causes	Small	Important	Small
Special causes	Important	None	Important

the overall mean is added to the residual to return to the original engineering units. For this example, the adjusted data values are computed by subtracting the average value for each factor level from the data points obtained with that factor level, and then adding the overall average. Table 5.3 shows the adjusted data calculated for this example. Figure 5.8 shows run charts of these adjusted data.

The special cause (trend) in the water data is no longer obvious in the adjusted data run chart. Since the factor levels were not randomly assigned to the days, the trend is somewhat confounded with the factor effect. Thus, when the factor effect is

TABLE 5.3
Example 5.2 data adjusted for factor levels

Control method	Date	Water (%)		Sulfate (%)		Ingredient X (%)	
		Test result	Adjusted value	Test result	Adjusted value	Test result	Adjusted value
Current	7/2	2.2	3.03	0.5	0.51	15.7	15.53
	7/3	2.0	2.83	0.2	0.21	17.2	17.03
	7/4	2.4	3.23	0.6	0.61	15.1	14.93
	7/5	2.7	3.53	0.3	0.31	16.5	16.33
	7/6	2.9	3.73	0.4	0.41	14.8	14.63
		$\overline{X} = 2.44$		$\overline{X} = 0.40$		$\overline{X} = 15.86$	
Valve	7/7	3.3	3.17	0.2	0.23	11.8	13.59
	7/8	3.2	3.07	0.5	0.53	12.6	14.39
	7/9	3.4	3.27	0.4	0.43	13.4	15.19
	7/10	3.6	3.47	0.3	0.33	19.4	21.19
	7/11	3.5	3.37	0.5	0.53	12.3	14.09
		$\overline{X} = 3.40$		$\overline{X} = 0.38$		$\overline{X} = 13.90$	
Meter	7/12	3.7	3.01	0.4	0.35	18.8	17.17
	7/13	3.8	3.11	0.5	0.45	17.6	15.97
	7/14	4.0	3.31	0.3	0.25	19.0	17.37
	7/15	4.2	3.51	0.5	0.45	13.0	11.37
	7/16	4.1	3.41	0.6	0.55	18.2	16.57
		$\overline{X} = 3.96$		$\overline{X} = 0.46$		$\overline{X} = 17.32$	
		$\overline{\overline{X}} = 3.27$		$\overline{\overline{X}} = 0.41$		$\overline{\overline{X}} = 15.69$	

Adjusted value $= X - \overline{X} + \overline{\overline{X}}$, where X is the data value, $\overline{X}$ is the average for each level of the factor (for each response variable), and $\overline{\overline{X}}$ is the overall average. *Example* (7/2, water): Adjusted value $= 2.2 - 2.44 + 3.27 = 3.03$.

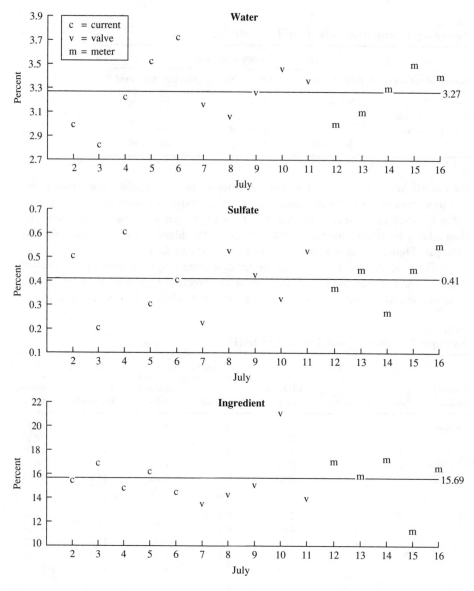

FIGURE 5.8
Run chart of adjusted data for Example 5.2

removed in adjusting the data, some of the trend effect is also eliminated. If the factor levels had been randomly assigned, the trend effect of the special cause would still be obvious in the adjusted data run chart.

The adjusted data for sulfate are very similar to the original data. This is because the factor had no important effect on the data.

The two special causes in the ingredient X data stand out more clearly in the adjusted data run chart than in the original data plot. If it is questionable whether or not these two points should be treated as special causes, a control chart for individuals could be developed for the adjusted data to aid in making this determination.

Run charts of the adjusted data do not need to be developed for every study. If the interpretation of the run chart of the original data is not clear, calculating the adjusted data and plotting them in run order may be useful.

There were no background variables identified in this study. Background variables should be included in analytic studies to evaluate the factors of interest at widely varying conditions in the process. This is important for increasing the degree of belief in the conclusions of the study. The technique of blocking can be used to introduce one or more background variables in an experiment. The next three sections discuss background variables in analytic studies.

5.4 PAIRED-COMPARISON EXPERIMENTS

The simplest type of blocking design is the paired-comparison experiment. This type of design requires a single factor at two levels and one or more background variables. The background variables must be such that two experimental units can be obtained from each grouping of background variables. An example of this is a study to compare the wear properties of two brands of shoes by having different people wear the shoes for a month and then measuring the wear. An important background variable here would be the variation in use of the shoes by the different people in the experiment. A paired-comparison design would assign one shoe of each brand to each person in the experiment. The difference in wear between the two brands could then be evaluated in terms of each person.

Other examples of pairs of units appropriate for a paired-comparison experiment are identical twins, front fenders of a car, two adjacent plots of land, two groupings of raw material from the same lot, and two time periods on the same day. The following example illustrates the design and analysis of a paired-comparison experiment.

> **Example 5.3.** One of the operators in the machining section has suggested an alternative setup procedure designed to increase tool life. The current procedure results in a typical tool life of one-half of a shift. The operator designed an experiment to investigate his suggestion. The single factor to be studied was the setup procedure. The factor was at two levels—the current procedure and the suggested alternative. There was one response variable—the wear rate of the tool.
>
> An important background variable is the quality of the blanks of steel prior to machining. The blanks are purchased in lots representing a single heat at the foundry, with each lot large enough for one shift's production in the machining section. The wear rate of the tool has varied significantly for different lots in the past.
>
> A paired-comparison design was used for the experiment. Figure 5.9 shows the documentation form for the study. Each shift (and, thus, each lot of blanks) was designated a block. The two setup procedures were paired on each shift, and each was used for a four-hour period. The assignment of a particular setup method to the first or second four-hour period was done by flipping a coin. The experiment was run for 10 days. Four parts were selected every 15 minutes, and a critical dimension was measured. Control charts for the average and range were used to summarize the 16 subgroups for each

FIGURE 5.9
Documentation form for Example 5.3

1. **Objective:**
Investigate the effect of an alternative tool setup procedure on tool life. A 10% reduction (0.0004 inch per hour) in wear rate would warrant changing to the new procedure.

2. **Background information:**
Using the current procedure, tool life has averaged about one-half of a shift (wear rate of 0.004 inch per hour). The particular lot of steel significantly affects wear; the standard deviation calculated from the range of tools used on the same batch was 0.00025 inch per hour.

   3. **Experimental variables:**

A.	Response variables	Measurement technique
1. 2.	Wear rate	Slope of straight line fit to data for each tool tested

B.	Factors under study	Levels
1.	Setup procedure	Current (O), new (N)
2.		

C.	Background variables	Method of control
1.	Steel lot	Blocking on steel lot.
2.		

4. **Replication:**
Ten days were available for the study; this would result in 10 comparisons of the two setup procedures.

5. **Methods of randomization:**
Flip a coin to determine setup procedure for first four-hour period (heads: current procedure; tails: new procedure).

6. **Design matrix:** (attach copy)
(See Table 5.4)

7. **Data collection forms:** (attach copy)
Use standard X-bar and R charts for basic data. Read slope from X-bar chart.

8. **Planned methods of statistical analysis:**
Run charts of individual wear rates; run chart with setup types grouped; run chart of adjusted wear rates.

9. **Estimated cost, schedule, and other resource considerations:**
Study can be conducted during production runs during next 10 day shifts of operation.

---

four-hour run. A straight line was fit to the 16 averages, and the slope of this line was used as a measure of tool wear. The results in Table 5.4 were obtained.

The first step in the analysis of a paired-comparison experiment is the preparation and study of a run chart of the response data. A run chart of the wear rates in Table 5.4 is shown in Figure 5.10. The variation in the tool wear rates can be partitioned by studying the run chart:

TABLE 5.4
Results of paired-comparison experiment (Example 5.3)

Wear rates (in. $\times 10^{-3}$/hr.) for each tool on each day

Day (lot)	First four hours		Second four hours		Average for day
	Setup	Tool wear	Setup	Tool wear	
1	New	2.0	Old	3.0	2.50
2	New	4.2	Old	5.3	4.75
3	Old	1.8	New	1.1	1.45
4	New	3.6	Old	4.4	4.00
5	Old	4.1	New	3.0	3.55
6	Old	2.8	New	1.9	2.35
7	Old	3.1	New	2.0	2.55
8	New	2.6	Old	4.0	3.30
9	New	3.2	Old	3.9	3.55
10	New	4.0	Old	5.2	4.60

Overall average $\left(\overline{\overline{X}}\right) = 3.26$

Data adjusted for block (day and steel lot) effects

Day (lot)	Old setup					New setup				
	Wear	− Daily average	+ $\overline{\overline{X}}$	=	Adjusted Wear	Wear	− Daily average	+ $\overline{\overline{X}}$	=	Adjusted Wear
1	3.0	− 2.50	+ 3.26	=	3.76	2.0	− 2.50	+ 3.26	=	2.76
2	5.3	− 4.75	+ 3.26	=	3.81	4.2	− 4.75	+ 3.26	=	2.71
3	1.8	− 1.45	+ 3.26	=	3.61	1.1	− 1.45	+ 3.26	=	2.91
4	4.4	− 4.00	+ 3.26	=	3.66	3.6	− 4.00	+ 3.26	=	2.86
5	4.1	− 3.55	+ 3.26	=	3.81	3.0	− 3.55	+ 3.26	=	2.71
6	2.8	− 2.35	+ 3.26	=	3.71	1.9	− 2.35	+ 3.26	=	2.81
7	3.1	− 2.55	+ 3.26	=	3.81	2.0	− 2.55	+ 3.26	=	2.71
8	4.0	− 3.30	+ 3.26	=	3.96	2.6	− 3.30	+ 3.26	=	2.56
9	3.9	− 3.55	+ 3.26	=	3.61	3.2	− 3.55	+ 3.26	=	2.91
10	5.2	− 4.60	+ 3.26	=	3.86	4.0	− 4.60	+ 3.26	=	2.66
	Average, old setups =				3.76	Average, new setups =				2.76

1. The largest component of the variation is due to the background variable (day/steel lot).
2. The effects of the factor and nuisance variables are somewhat difficult to discern because of the variation caused by the background variable. The wear rates for the new procedure are consistently less than for the old procedure.

The next step in the analysis is to reorder the run chart to group the levels of the factor together. This chart, also shown in Figure 5.10, focuses the analysis on the factor of interest. The wear rates for the new setup procedure tend to be less than those for the old procedure, but the data are still dominated by the effect of the background variables.

In a paired-comparison design, it is possible to remove the effect of the background variable but still evaluate the factor of interest at each level of the background

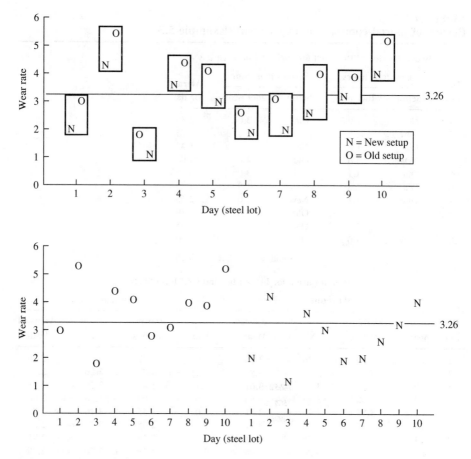

FIGURE 5.10
Run charts of individual wear rates

variable. This is easily done by computing the average of the response variable for each block and then subtracting the appropriate block average from the original data. The overall average of all the data is then added back to keep the data in the original units. These calculations are shown in Table 5.4. The adjusted wear rates contain the effect of the factor of interest (setup procedure), but the effect of the background variable (day/steel lot) has been removed. A run chart of the adjusted wear rates is shown in Figure 5.11.

The run chart of the adjusted wear rates clearly shows the difference between the two setup procedures. The adjusted wear rates for the old procedure are consistently around 3.75, whereas the adjusted wear rates for the new procedure are consistently around 2.75—about one-thousandth of an inch per hour lower. On each of the 10 days (and, thus, for each lot of steel), the new procedure had a lower wear rate than the old procedure.

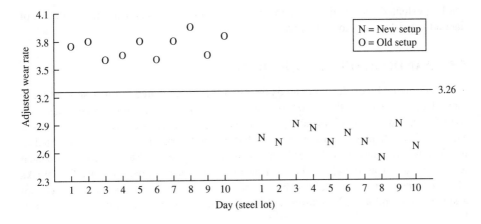

FIGURE 5.11
Run chart of adjusted wear rates

The contribution to the variation of nuisance variables can also be seen in Figure 5.11. The variation of the 10 tests within each setup procedure can be attributed to nuisance variables. This contribution to the variation is small relative to the contributions of the background variable and the factor.

Since the new procedure outperformed the old procedure in each of the 10 comparisons, the process engineer involved in the study would have a high degree of belief that the new setup procedure would result in lower wear rates in the future. The engineer should assess the magnitude of the lot-to-lot variation in the study relative to previous experience. The wear rates in the study ranged from 1.8 to 5.3 for the old procedure, so a wide range of conditions have been included. After implementing the new setup procedure, the next improvement cycle should focus on this lot-to-lot variation.

This example is similar to most paired-comparison experiments for an analytic study. Using this design, it is possible to include background variables that have large effects on the response variable. This allows the comparisons of the factor of interest to be made at a wide range of conditions that might be expected in the future. The steps in the analysis of a paired-comparison design are as follows:

1. Plot a run chart of the original data in order of test or measurement.
2. Reorder the run chart by grouping the levels of the factor.
3. If the effect of the background variable(s) is large, adjust the data to remove the effect. This is done by computing the average of each data pair (each block), subtracting that average from each of the two data points, and then adding the overall average to return the data to the original units.
4. Plot a run chart of the adjusted data, grouped by factor level.

By adjusting the data using the block averages, the effect of the background variable is prevented from interfering with the evaluation of the factor. The levels of the factors can be evaluated for stability over the wide range of conditions using the run chart of data adjusted for the block averages.

5.5 RANDOMIZED BLOCK DESIGNS

The randomized block design extends the approach of the paired-comparison design to more than two levels of the factor. Traditionally, the randomized block design has been used for one background variable, but through the concept of chunk variables (see Figure 4.3 in Chapter 4), any number of background variables can be incorporated in the design. The number of factor levels is limited only by the block "size" since each level must occur an equal number of times (at least once) in each block. The blocks of experimental units are defined by the background variables. The factor levels are assigned in a random order within each block, with a different randomization for each block.

The concept of chunk variables arranged in blocks is very important in analytic studies. The degree of belief concerning the effect of a factor can be greatly increased by varying many of the background variables and evaluating the factor under these widely varying conditions. By organizing these background variables in blocks and then evaluating each level of the factor in each block, the comparisons of the factor levels can be made under uniform conditions. The following study is an example of a randomized block design for a one-factor experiment.

> **Example 5.4.** A program was undertaken to improve the yield of a batch chemical process. Control charts had been continually out of control due to variability of the feedstock. Four batches of the chemical could be made from each tankcar of feedstock at a rate of about one batch per day. A major project was conducted with the feedstock supplier to identify the causes that affected yield from shipment to shipment.
>
> The plant chemist made a suggestion to improve yield through changes in one of the catalysts used in the batch process. Three alternative catalysts were identified, each of which could replace the current catalyst. The production manager asked that a test be run during the next month to evaluate the potential of the alternative catalysts. A documentation form for the study is shown in Figure 5.12.
>
> A randomized block design was used for the study. The single factor to be studied was the catalyst, and the factor had four levels (catalysts X, Y, Z, and the current catalyst, C). An important background variable was the feedstock shipment, so the experimental units were grouped into blocks of four production units from each feedstock shipment. Since the study had to be completed within a month, six feedstock shipments were incorporated into the experimental design. Thus, the experiment would require 24 days to complete.
>
> Each of the four catalyst types was assigned to one of the four batches scheduled for each of the six feedstock shipments in a random order (using a table of random permutations). Yield was calculated for each batch using a material balance method. The results of the study are summarized in Table 5.5.

Figure 5.13 shows run charts of the yields from each individual batch. From the first run chart, the effect of the feedstock shipment (the blocks) is obvious. Differences between the catalyst types can be observed in the second run chart, where the batches

FIGURE 5.12
Documentation form for Example 5.4

1. **Objective:**
Choose catalyst (there are three new alternatives available) to give maximum yield. Any improvement over the current catalyst is worthwhile; a practical minimum difference to change type of catalyst is 0.5%.

2. **Background information:**
The new chemical process is still undergoing major improvements. Feedstock variability has a big effect on yield and is being studied by a joint team including the supplier. Yields during the last 30 batches have ranged from 75% to 95%, with big swings with each new feedstock shipment.

3. **Experimental variables:**

A. | Response variables | Measurement technique |
|---|---|
| 1. Yield (%) | Material balance |

B. | Factors under study | Levels |
|---|---|
| 1. Catalyst | New (X, Y, Z), current (C) |

C. | Background variables | Method of control |
|---|---|
| 1. Feedstock shipment | Block defined by each shipment |

4. **Replication:**
There is enough supply of the experimental catalysts to make six batches from each. The study should be completed within a month. Six replications (blocks) of each of the four catalyst types are planned.

5. **Methods of randomization:**
Table of random permutations of 4 was used to assign the four catalyst types to the four batches made from each feedstock shipment.

6. **Design matrix:** (attach copy)
(See Table 5.5)

7. **Data collection forms:** (attach copy)
Batch cards include material balance calculation of yield.

8. **Planned methods of statistical analysis:**
Run charts of yields; run chart of yields adjusted for background variables.

9. **Estimated cost, schedule, and other resource considerations:**
Initial study can be completed during next 24 days of operation; possible losses or gains in yield from the different catalysts.

---

of a particular catalyst type are grouped together. The highest yield in each block was for the batch using catalyst type Y. Catalysts X and Z had yields close to the current catalyst in each block.

A clear picture of the effect of catalyst type on yield is difficult from the run charts of yield because of the effect of the background variable (feedstock shipment). The effect of the background variable can be removed from the yields by adjusting the data. The approach is similar to the adjusting procedure used with the paired-

TABLE 5.5
Randomized block design to evaluate four catalysts

Results of study: Percent yield (random order)

Block (feedstock shipment)

Catalyst	1	2	3	4	5	6
X	87(4)	79(1)	82(2)	89(4)	83(1)	78(2)
Y	93(1)	84(4)	89(4)	96(2)	86(3)	87(1)
Z	88(3)	80(2)	84(1)	91(3)	83(2)	82(3)
C	88(2)	77(3)	83(3)	90(1)	82(4)	79(4)
Block average	89.0	80.0	84.5	91.5	83.5	81.5

Overall average = 85.0

Yield adjusted for background variables (feedstock shipment)
(Yield − Block average + Overall average)

Block (feedstock shipment)

Catalyst	1	2	3	4	5	6	Average
X	83.0	84.0	82.5	82.5	84.5	81.5	83.0
Y	89.0	89.0	89.5	89.5	87.5	90.5	89.2
Z	84.0	85.0	84.5	84.5	84.5	85.5	84.7
C	84.0	82.0	83.5	83.5	83.5	82.5	83.2

Overall average 85.0

comparison design in the previous section: the average of each block is computed, the block average is subtracted from each individual yield, and the overall average is then added. This calculation removes the effect of the background variable but leaves the adjusted data in the original units (percent yield, in this case). The yields adjusted using this procedure are shown in Table 5.5. The adjusted yields are shown on the third run chart in Figure 5.13, with the data grouped by catalyst type.

The three run charts in Figure 5.13 can be used to partition the variation in the yields:

1. The first two plots (run chart of the yields ordered by time and by catalyst type) indicate that most of the variation is attributable to the background variable, feedstock shipment. The average yields for the five shipments in the study ranged from 80% to 91.5%.

2. The third chart (a run chart of the adjusted yields) indicates that the factor, catalyst type, is the next biggest contributor to the variation in the yields. The difference in yield within a feedstock shipment ranges up to 9%. Catalyst Y consistently was 6% higher than the current catalyst. Catalyst Z had a higher yield than the current catalyst in five of the six shipments, averaging 1.5% higher. Catalyst X was similar to the current catalyst.

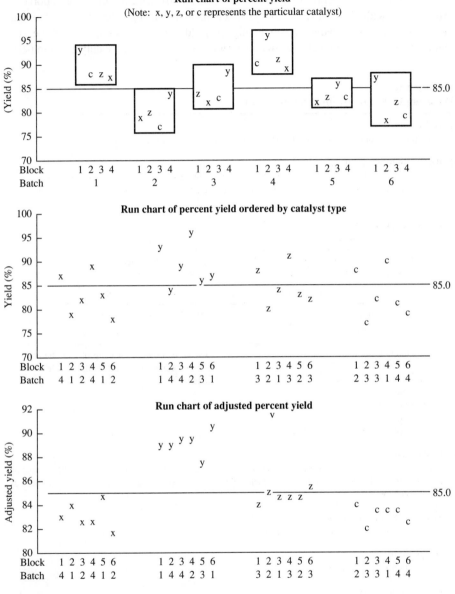

FIGURE 5.13
Run charts for randomized block study

3. The effect of nuisance variables is best seen from the third chart in Figure 5.13. The adjusted yields for each of the three alternative catalysts are clustered tightly, with a range of about 3%. There are no indications of special causes (trends or individual points) in any of the plots.

From the combination of the data from this study with his previous laboratory experiments with the catalysts, the plant chemist had a high degree of belief that the use of catalyst Y would result in higher yields than continuing use of the current catalyst. A critical result was the consistency of factor effect (catalyst differences)

FIGURE 5.14
Documentation form for Example 5.5

1. **Objective:**
   Determine an application rate for glue to minimize loose labels without causing runs or smears.

2. **Background information:**
   About 7% of the labels during the last month have come loose. The operators have changed the application rate a number of times (from 10 to 20 oz/hour). The rate has been turned down when runs occurred on the cans.

3. **Experimental variables:**

   A.

Response variables	Measurement technique
1. Percent cans with loose labels	100% visual inspection
2. Percent cans with smears or runs	100% visual inspection

   B.

Factors under study	Levels
1. Application rate	10, 12, 14, 16, 18 oz/hour

   C.

Background variables	Method of control
1. Humidity in plant	Create two blocks made up of extreme conditions for each background variable (two chunks)
2. Temperature in plant	
3. Label machine speed	
4. Can size	
5. Batch of labels	
6. Batch of cans	
7. Label machine pressure setting	

4. **Replication:**
   Each of the five tests will be run for one hour (1,000 cans). Cans will be run for one-half hour in between tests to let the new factor level reach stability. The tests will then be repeated on another day (the second chunk block).

5. **Methods of randomization:**
   Table of random permutations was used to determine the order of the five test on each day.

6. **Design matrix:**
   (See Table 5.6 for summary.)

7. **Data collection forms:**
   Operator checksheets used to record occurrence of labels and runs.

8. **Planned methods of statistical analysis:**
   Run charts of percent labels loose and runs; response plot for each response variable and block.

9. **Estimated cost, schedule, and other resource considerations:**
   Test can be conducted during production runs. Inspector needed for two days to monitor test and coordinate data collection. Possible higher rework costs due to loose labels and runs.

across all of the blocks. But even with catalyst Y, the batch yields were as low as 84%. The next step in this study would be to begin using catalyst Y, monitor the yields with a control chart, and focus future experiments on the feedstock variation.

The next example is another study based on a randomized block design. In this study, a chunk variable is used to define the block. The example also illustrates an interaction between a factor and a background variable.

Example 5.5. Loose labels on cans had been a problem on and off since the new labeling machine was installed. The application rate of the glue was thought to be the control mechanism for eliminating the problem loose labels. The label operator would turn up the application rate when loose labels were observed. If runs of glue were observed on the labeled cans, the supervisor would reduce the application rate. During the last month, 7% of the cans required rework because of loose labels. The production manager asked the quality team in the label operation to conduct a study to determine the best application rate.

Figure 5.14 shows the documentation form for this study. The application rate would be studied at five levels (10, 12, 14, 16, and 18 ounces/hour). The team felt this

TABLE 5.6
Design matrix and test results for Example 5.5

Development of two chunk blocks

Background variable	Block 1	Block 2
Plant humidity	Low	High
Plant temperature	Cool	Warm
Label machine speed	Low (800/hr)	High (1200/hr)
Can size	Small	Large
Label batch	Batch 803	Batch 615
Can batch	Batch 4,120	Batch 4,125
Pressure setting	Low	High

Design matrix and test results

Application rate (oz/hr)	Block	Run Order	Percent of labels loose	Percent of cans with runs
10	1	3	9.0	0.0
12	1	1	5.4	0.2
14	1	4	1.0	0.4
16	1	5	0.9	1.5
18	1	2	1.0	3.0
10	2	5	10.3	0.0
12	2	1	8.0	0.1
14	2	3	6.7	0.4
16	2	2	7.0	1.3
18	2	4	6.8	2.8

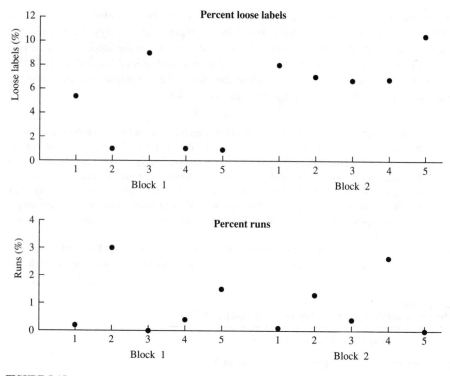

FIGURE 5.15
Run chart for Example 5.15

range covered the application rates that had been tried in the past. Because of the erratic results that had been experienced, the team decided to test each of these levels under different conditions that occurred in the plant. Seven background variables that could affect the labeling were identified, and two chunk-type blocks were defined for running the test. Table 5.6 shows the design matrix and the test results.

Figure 5.15 shows a run chart for the two response variables. During the study, the percentage of loose labels ranged from 1% to 10%. The percentage of cans with runs ranged from 0% to 3%. No outliers or other special causes are apparent on the run charts. Plots of the responses adjusted for the block effect could be prepared but since application rate is a continuous variable, response plots better display the effect of the factor. In a response plot, results for a response variable are plotted on the vertical axis and the factor levels on the horizontal axis.

Figure 5.16 shows response plots for each of the response variables in this study. The results for each block can be compared for stability using these plots. The response plot for loose labels indicates a different effect of the application rate between the two blocks. For the first block, the percent of loose labels decreased as application rate increased. Results at 14 oz/hr and above were near 1% loose labels.

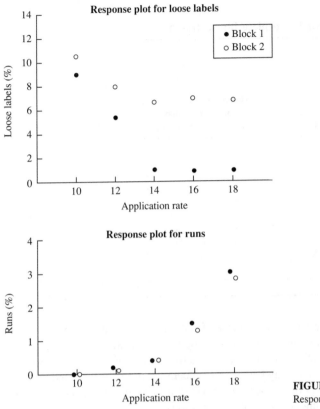

FIGURE 5.16
Response plots for Example 5.5

For the other block, the percentage of loose labels decreased as the application rate changed from 10 to 12 oz/hr, but they remained steady at about 7% for the remaining levels. This indicates that the effect of the application rate is dependent on some of the background variables in the plant. Increasing the application rate above 12 oz/hr does not reduce the loose labels under the conditions in block 2.

The response plots for runs are almost identical for the two blocks. In both cases, the presence of runs was less than 0.5% until the application rate was increased above 14 oz/hr. So, after testing under widely varying conditions, there is a high degree of belief that the runs can be kept below 0.5% if the application rate is maintained at 14 oz/hr or below.

The next cycle in this study should focus on the conditions in block 2 that mitigated the effect of glue application rate on the percent of loose labels. Under the conditions in block 1, an application rate of 14 oz/hr would keep both loose labels and runs at less than 1%. One approach would be to develop a control chart for the loose labels with the background variables noted on the chart. Another would be to set the application rate at 14 oz/hr and study the background variables using a factorial experimental design (see Chapter 6).

TABLE 5.7

Experimental pattern for incomplete block design

Five factor levels (A, B, C, D, E)
10 blocks
Block size = 3

Block	A	B	C	D	E		Block	Factor level		
		Experimental pattern display **Factor level**						**Alternative experimental** **pattern display**		
1	X	X	X				1	A	B	C
2	X	X		X			2	A	B	D
3	X	X			X		3	A	B	E
4	X		X	X			4	A	C	D
5	X		X		X		5	A	C	E
6	X			X	X		6	A	D	E
7		X	X	X			7	B	C	D
8		X	X		X		8	B	C	E
9		X		X	X		9	B	D	E
10			X	X	X		10	C	D	E

This example illustrates the importance of stability of factor effects in an analytic study. The past confusion in understanding the effect of application rate on the percent of loose labels was probably due to the interactive effect of one or more of the background variables. If this study had been conducted with the background variables held constant, the results would not have provided a basis for changing the process to improve performance under future plant conditions.

5.6   INCOMPLETE BLOCK DESIGNS

Suppose that in Example 5.4 only three batches could be made from each feedstock shipment. How should the four catalyst types be tested? An incomplete block design can be used for this situation. An incomplete block design is an experimental pattern that groups experimental units in blocks (to accommodate one or more background variables), where the number of experimental units in each block is less than the number of factor levels.

A special class of such designs is the balanced incomplete block design, which has the following characteristics:

1. Each block is the same size (has the same number of experimental units).
2. Each factor level occurs the same number of times in the design.
3. Any two factor levels occur together in the same block an equal number of times.

Table 5.7 shows two different ways of displaying the experimental pattern for an incomplete block design. Table 5.8 gives some examples of balanced incomplete block designs for experiments with three, four, five, and six levels of the factor or groups to be tested. The design is characterized by the number of levels of the factor,

TABLE 5.8
Some balanced incomplete block designs

				Block number									
t	*b*	*k*	*r*	1	2	3	4	5	6	7	8	9	10
3	2	3	2	A B	A C	B C							
4	2	6	3	A B	A C	A D	B C	B D	C D				
4	3	4	3	A B C	A B D	A C D	B C D						
5	2	10	4	A B	A C	A D	A E	B C	B D	B E	C D	C D	D E
5	3	10	6	A B C	A B D	A B E	A C D	A C E	A D E	B C D	B C E	B D E	C D E
5	4	5	4	A B C D	A B C E	A C D E	A B D E	B C D E					
6	3	10	5	A B E	A B F	A C D	A C F	A D E	B C D	B C E	B D F	C E F	D E F
6	4	15	10	A B C D	A B C E	A B C F	A B D E	A B D F	A B E F	A C D E	A C D F		
				A C E F	A D E F	B C D E	B C D F	B C E F	B D E F	C D E F			
6	5	6	5	A B C D E	A B C D F	A B C E F	A B D E F	A C D E F	B C D E F				

t = Number of factor levels or combinations
b = Block size (number of experimental units per block)
k = Number of blocks required for balanced design
r = Number of replications of each factor level required

A, B, C, D, E, and F represent factor levels or factor combinations.

the size of the block, the number of blocks required to achieve the balance, and the amount of replication of each level of the factor in the design. Balanced designs do not exist for some combinations of block size and number of levels.

The concept of a balanced design is not critical to the analysis or interpretation of incomplete block designs presented here. In general, the more balance in the design, the easier the interpretation and the higher degree of belief in the results. Cochran and Cox (1957) give a good presentation of partially balanced incomplete block designs.

The analysis of data from incomplete block designs is similar to that for the complete block design. A run chart of the data is first prepared. Next the data is adjusted to remove the effect of the background variable. The adjusted data is then analyzed as in a single-factor design.

The following example illustrates the design and analysis of an incomplete block design.

Example 5.6. An evaluation of gaging instruments was proposed in order to select the particular instrument with the best repeatability for future plant use. (Repeatability is measured by the standard deviation of repeated measurements of the same part.) The gages were used by over 100 different operators in the plant, and the level of skill among different operators was known to have an impact on repeatability.

Six different instruments were available for testing. The test protocol for an instrument, established by the quality control department, requires each operator to measure four different parts seven times each. These data are evaluated for stability, and then a pooled standard deviation is calculated and used as a measure of repeatability. This protocol requires about one hour per instrument. Since the instrument selected is for general plant use, it is desirable to use a number of different operators in the study. The production manager said that operators could be made available for the study for up to three hours each.

This study is a single-factor (instrument) experiment with an important background variable (operator). The response variable is the calculated standard deviation from each test. There are six levels of the factor (instruments A, B, C, D, E, and F), but the blocks defined by the background variable are only of size three (maximum three hours per operator). Since a randomized block design cannot be used, a balanced incomplete block design is considered. From Table 5.8, such a design is available for six factor levels and a block size of three experimental units. The experimental pattern requires that each instrument be tested five times ($r = 5$) and that 10 operators be used in the study ($k = 10$).

Figure 5.17 shows the completed documentation form for this study. Table 5.9 shows the experimental pattern and the response variable for the completed experiment.

The response variable (the standard deviation of the repeat measurements) is plotted in run order in Figure 5.18. No special causes are obvious in this plot. Some differences among blocks (operators) are present, but there is a wide variation of results for some of the operators. Operator 10 appears to have more variation than the other operators. The second run chart in Figure 5.18 shows the data (the standard deviations) grouped by instrument type. Instrument C appears to have good precision (a low standard deviation) each time it appears. The differences among operators make it difficult to evaluate the other instruments.

FIGURE 5.17
Documentation form for Example 5.6

1. **Objective:**
Determine which of six available gaging instruments has the best repeatability.

2. **Background information:**
Operators in the plant have widely different abilities in using the gaging instruments. Each operator can test three instruments. The standard deviation of the current instrument is about 1.0 unit.

3. **Experimental variables:**

A.	Response variables	Measurement technique
	1. Standard deviation of repeated readings	Calculated using QC department protocol
B.	Factors under study	Levels
	1. Gaging instruments	Instruments A, B, C, D, E, and F (letters randomly assigned to six different brands)
C.	Background variables	Method of control
	1. Operators	10 operators selected by the production manager, each operator considered a block

4. **Replication:**
Each operator can test three instruments, 10 operators are selected for the study.

5. **Methods of randomization:**
Operators assigned numbers 1–10 in order of particapation in the study. Random permutation table used to assign operator to particular block of instruments. Test order for each operator randomized using permutation table.

6. **Design matrix:**
Balanced incomplete block desing with 10 blocks, three tests per block. (see Table 5.9.)

7. **Data collection forms:**
QC gage test forms used during study

8. **Planned methods of statistical analysis:**
Run charts of original and adjusted data.

9. **Estimated cost, schedule, and other resource considerations:**
Schedule developed to complete study in one day, with 10 operators each contributing three hours.

The next step in the analysis is to remove the effect of the operators so that the variation in the factor (instruments) can be studied. This is done in a similar manner to the randomized block design, by subtracting the average for each block (operator) from the original data and then adding the overall average. Since the experimental pattern is balanced (each instrument is evaluated with every other instrument by the same operator twice in the study), evaluation of the adjusted data

TABLE 5.9
Incomplete block design for Example 5.6

Design matrix and results of study:
Standard deviation of repeat measurements
(run order within the block in parentheses)

Operator	A	B	C	D	E	F	Operator average
1	1.1(2)		0.7(1)			0.9(3)	0.90
2		1.3(3)		1.4(1)		1.2(2)	1.30
3	0.6(1)	0.5(3)			0.9(2)		0.67
4		0.9(1)	0.4(3)	1.2(2)			0.83
5				0.8(3)	1.6(1)	1.2(2)	1.20
6	1.6(2)	1.4(3)				1.5(1)	1.50
7	1.0(2)			0.8(1)	1.5(3)		1.10
8			0.3(1)		1.0(2)	0.6(3)	0.63
9		1.2(2)	0.6(1)		1.4(3)		1.07
10	1.8(3)		1.2(1)	1.7(2)			1.57
Average	1.22	1.06	0.64	1.18	1.28	1.08	$\overline{\overline{X}} = 1.077$

Adusted standard deviations
(Standard deviation - Operator average + Overall average)

Operator	A	B	C	D	E	F
1	1.28		0.88			1.08
2		1.08		1.18		0.98
3	1.01	0.91			1.31	
4		1.15	0.65	1.45		
5				0.68	1.48	1.08
6	1.18	0.98				1.08
7	0.98			0.78	1.48	
8			0.75		1.45	1.05
9		1.21	0.61		1.41	
10	1.31		0.71	1.21		
Adjusted average	1.152	1.066	0.720	1.060	1.426	1.054	$\overline{\overline{X}} = 1.080$

to compare instruments is valid. Table 5.9 shows the adjusted standard deviations. Figure 5.19 shows the adjusted data plotted, grouped by instrument type.

From Figure 5.19 it is much clearer that instrument C performed better in the study than the other instruments. Instrument E was consistently high, whereas the effectiveness of instrument D depended on the operator. Instrument C should be considered for selection. A possible next cycle would be to have the five operators that did not test this instrument in the study try out instrument C. Obtain comments from

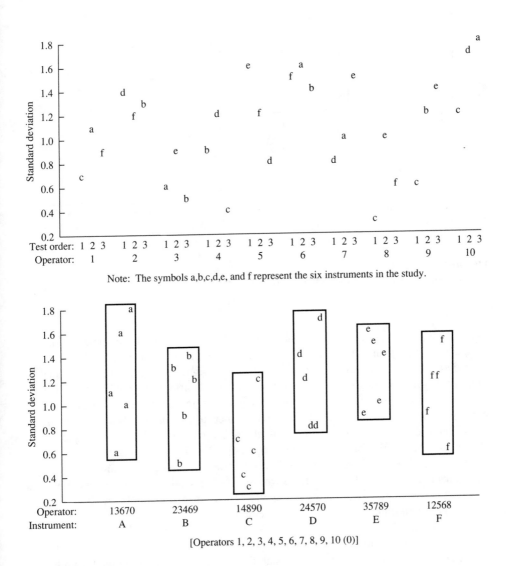

Note: The symbols a,b,c,d,e, and f represent the six instruments in the study.

[Operators 1, 2, 3, 4, 5, 6, 7, 8, 9, 10 (0)]

FIGURE 5.18
Run charts for Example 5.6

these operators on ease of use, general applicability in the plant, and other features before making a decision to purchase instrument C.

Future study should focus on the variability of the operators in using the instruments. The differences in precision among the different operators were at least as great as differences among the instruments in the study. Perhaps a flowchart of the measurement process and a short training session would be appropriate. Operator 10 may require special help.

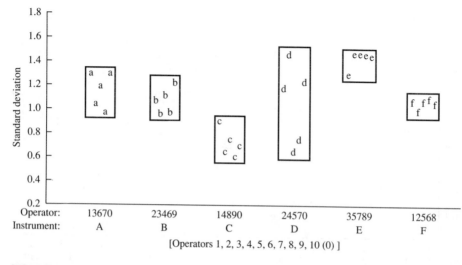

FIGURE 5.19
Adjusted standard deviations for each instrument type, Example 5.6

   The analysis used in this example required that a balanced incomplete block design be used. If an unbalanced block design had been used, the data adjustment procedure would not be valid. With an unbalanced design, there would be an unequal number of differences among the instruments. Paired differences could be used to remove the block effect for unbalanced designs. The response for each instrument could be subtracted from the response of every other instrument in the block to obtain the differences. The results of these calculations for the data in this example are shown in Table 5.10. For example, both operators 3 and 6 tested instrument A and instrument B. The two differences in column B, row A are calculated using the results from operators 3 and 6. For operator 3, the standard deviation for instrument A was 0.6, and the standard deviation for B was 0.5. Thus, the difference B−A (0.5 − 0.6 = −0.1) is determined. Similarly, for operator 6, the B−A difference (1.4 − 1.6 = −0.2) is calculated.

   A plot of these differences grouped by instrument is shown in Figure 5.20. The effect of the factor, instrument, and the stability of the effect can be studied from this plot of differences. An instrument with average precision will tend to have differences centered around 0.0. An instrument with poor precision will have differences greater than zero, and an instrument with good precision will have negative differences. All the differences for instrument C are negative, indicating that instrument C had better precision than all of the other instruments each time it was tested. Instrument E had all positive differences, indicating that it performed poorly (large standard deviation) in all cases. As for the other instruments, B and D were average, A tended toward poor precision, and F tended to have good precision.

TABLE 5.10
Using differences between instruments tested by the
same operator (column − row) to remove block effect

	A	B	C	D	E	F
A		−.1	−.4	−.2	.3	−.2
		−.2	−.6	−.1	.5	−.1
B	.1		−.5	.1	.4	−.1
	.2		−.6	−.3	.2	.1
C	.4	.5		.8	.7	.2
	.6	.6		.5	.8	.3
D	.2	−.1	−.8		.4	−.2
	.1	.3	−.5		.7	−.4
E	−.3	−.4	−.7	−.4		−.8
	−.5	−.2	−.8	−.7		−.4
F	.2	.1	−.2	.2	.8	
	.1	−.1	−.3	.4	.4	
Average	.21	.04	−.54	−.03	.52	−.16
	A	B	C	D	E	F

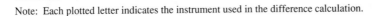

Note: Each plotted letter indicates the instrument used in the difference calculation.

FIGURE 5.20
Plot of differences by instrument type

Incomplete block designs can be used for experiments with more than one factor. Chunk-type blocks can be developed to include multiple background variables. Depending on the size of the blocks created, either a randomized block design or an incomplete block design will be appropriate for studying the factors.

5.7 SUMMARY

A number of important concepts were illustrated in the examples in this chapter:

- The use of the planning form to consider and document the tools for experimentation introduced in Chapter 4
- The use of the run chart as a key tool for the analysis of data from analytic studies
- The use of blocking to incorporate background variables
- The use of grouping and stratification on the run chart to highlight the factor levels under different conditions defined by the background variables
- Adjusting data to remove the effect of background variables to facilitate comparison of the factor levels
- The importance of stability in interpreting the data from an analytic study

Each of these concepts will be expanded upon in the following chapters, when more than one factor is incorporated in a study.

REFERENCES

Box, G., W. Hunter, and J. S. Hunter (1978): *Statistics For Experimenters*, John Wiley & Sons, New York, Chapter 7.

Cochran, W. G., and G. M. Cox (1957): *Experimental Design*, John Wiley & Sons, New York, Chapters 9 and 11.

EXERCISES

5.1 Describe how a control chart can be used to conduct a one-factor experiment on a process. List the key steps involved in conducting such a study. How is this different from using a control chart to control a process?

5.2 Design an experiment with one factor for a process you are familiar with. Complete a planning documentation form, including objective, background information, experimental variables, replication, randomization, design matrix, data collection form, planned analysis, and resource requirements.

5.3 Describe some situations that would make a paired-comparison design appropriate for studying one factor.

5.4 Why are randomized block designs important in analytic studies? Describe how blocking could be used in a study to evaluate three alternative teaching methods in a school.

5.5 The production manager wants to evaluate four new types of drill bits for consideration as the plant standard. There are three machines that could use the drill bits, each with a different operator and each run on two shifts. Historically, there has been variability of drill bit performance among machines and operators.

What type of design is appropriate for this study? Describe the experimental pattern, the blocking, the replication, and the randomization for the study.

5.6 In the one-factor example on control methods for a chemical process (Example 5.2), two other response variables were also measured. The results for purity (%) and a contaminant (ppm) were the following:

Date:	2	3	4	5	6	7	8	9	10	11	12	13	14	15	16
Factor:	c	c	c	c	c	v	v	v	v	v	m	m	m	m	m
Purity:	72	69	67	67	64	62	63	66	68	67	70	70	73	74	76
Contaminant:	33	28	30	26	31	43	31	48	44	46	35	31	33	29	32

(a) Prepare run charts for these data. Partition the variability of the two response variables among the factor and the nuisance variables. What conclusions can be made about the factor for each response variable?

(b) Calculate adjusted data for each response variable, and plot the adjusted data on a run chart. What additional information can be learned from a study of the adjusted data?

5.7 During the paired-comparison experiment evaluating tool wear (Example 5.3), parts were selected from each lot to evaluate the variability of a specified dimension. The quality control manager was concerned that the new setup procedure might affect the variability of the initial parts produced. A subgroup of five parts was selected during the first half-hour from each lot for each of the setup procedures. The dimension (in thousandths of an inch from nominal) was measured for each part, and the range of the five parts was calculated. The following results (range, in thousandths of an inch) were obtained:

Lot:	1	2	3	4	5	6	7	8	9	10
New setup:	5.3	0.2	2.1	4.6	3.0	1.2	5.8	4.0	8.4	3.0
Old setup:	6.4	1.3	2.0	6.2	3.8	3.0	8.5	5.8	8.0	5.3

Prepare run charts of the ranges and the adjusted ranges. Partition the variation in the ranges among the factor, background variable, and the nuisance variables.

5.8 In the study to choose a catalyst (Example 5.4), another response variable (purity) was also evaluated for each of the catalyst types. A composite sample was taken from each batch produced and analyzed for purity (%). The following data were obtained:

	Purity of sample for each feedstock shipment					
Catalyst	1	2	3	4	5	6
X	99.1	99.8	99.5	98.4	99.6	99.7
Y	98.0	99.2	98.3	97.5	98.4	98.9
Z	97.0	98.3	97.9	96.8	97.7	98.3
C	97.9	99.1	98.6	97.2	98.5	99.0

(a) Prepare run charts for the purity data (see Example 5.4 for run order of catalyst within each shipment).

(b) Calculate the purity adjusted for feedstock shipment and prepare a run chart for the adjusted values.

(c) What are the conclusions from this study for purity? Summarize the importance of the factor, background variable, and nuisance variables. Which catalyst would be expected to give the highest purity in future shipments? How strong is the degree of belief in this conclusion?

(d) Combine the conclusions from Example 5.4 on yield and the results on purity to develop recommendations for future catalyst use. The cost of each catalyst ($ /lb) is: current, $12; catalyst X, $10; catalyst Y, $16; and Catalyst Z, $12. What additional information is needed to develop a cost-effective recommendation for catalyst use?

5.9 The accounts payable quality improvement team was studying the process of paying invoices. They wanted to improve the efficiency and accuracy of payments. During the development of flowcharts, the team found that there was variability in the process among the three clerks and at different time periods. In their attempts to standardize the process, four alternative procedures were developed. The team decided to study the four alternatives and choose the process with the highest efficiency (number of invoices paid) and best accuracy (lowest number of errors) as the standard.

The team was concerned that the evaluation could be affected by variation in the payment process from week to week. This variation included number of invoices received, end-of-month payments, and other peculiarities related to different time periods.

A balanced incomplete block design was chosen for the test. A block was defined by the week in which the test was done. There were four levels of the factor, "process alternative," to study. Each of the clerks could use a different alternative during a time period. The design with four factor levels, a block size of three, and four blocks was selected (see Table 5.8). The three process alternatives within a week were randomly assigned to one of the clerks. The experimental pattern was replicated by running the study for eight weeks. The following data were obtained.

	Clerk 1			Clerk 2			Clerk 3		
Week	Process	Number paid	Number of errors	Process	Number paid	Number of errors	Process	Number paid	Number of errors
1	C	32	9	B	38	5	A	30	10
2	A	42	6	B	51	5	D	57	12
3	D	43	7	A	37	9	C	34	3
4	B	70	9	D	74	20	C	55	8
5	B	41	6	C	31	5	A	26	15
6	B	51	3	A	43	8	D	60	10
7	A	35	10	D	45	9	C	33	4
8	C	60	9	D	69	14	B	68	13

(a) Analyze each of the response variables for this study. Prepare run charts, calculate adjusted values, and construct run charts of the differences. Label the factor level (the process alternative) on the run charts. Group the adjusted values by factor level for evaluation of each process alternative.

(b) Which alternative process can be expected to have the highest efficiency? Which alternative will give the least number of errors? What kind of study should be completed next by the team?

CHAPTER

6

EXPERIMENTS WITH MORE THAN ONE FACTOR

In this chapter, designs to study the effects of multiple factors on a response variable are considered. These designs will provide the foundation for studying and improving complex processes and products. Because of the complexity of most processes, several factors are usually studied in an experiment. A common approach to experimentation when there is more than one factor is to study one at a time. A reason often cited in support of this approach is that if more than one factor is changed, the experimenter will not be able to determine which factor was responsible for the change in the response.

There are two major deficiencies with studying one factor at a time. The first is that there are often interactions between the factors under study. An interaction means that the effect a factor has on the response may depend on the levels of some other factors. Figure 6.1 contains response plots illustrating various degrees of interaction. In Figure 6.1a, it is seen that the change in the response as factor 1 is changed is the same regardless of whether factor 2 is at the low or high level (i.e. the slopes of the line are the same). In Figures 6.1b, c, and d, the effect of changing factor 1 from a low to a high level differs depending on the level of factor 2.

The second deficiency in studying one factor at a time is inefficiency. As each factor is studied in turn, the data previously collected to study other factors are set aside and new data are collected. Each set of data supplies information on only one factor.

Factorial designs provide an alternative to studying one factor at a time. Factorial designs allow the study of interactions between factors. The structure of factorial designs allows all the data from the experiment to be used to study each factor. This results in a significant increase in efficiency over studying the factors one at a time.

115

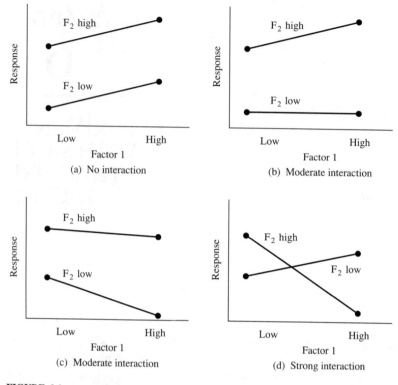

FIGURE 6.1
Examples of interactions

6.1 INTRODUCTION TO FACTORIAL DESIGNS

To set up a factorial design, the investigator determines the factors to be studied and the levels for each of the factors. A full factorial design consists of all possible combinations of the factors and levels. For example, if three factors are to be studied, the first at two levels, the second at three levels, and the third at five levels, a $2 \times 3 \times 5$ factorial design is used. This design requires 30 ($2 \times 3 \times 5$) tests for one replication of the experimental pattern.

In this chapter, designs are considered for experiments in which each factor is studied at only two levels. This class of designs will be referred to as 2^k factorial designs. For example, a 2^3 factorial design would require $2 \times 2 \times 2$ or eight tests. Factorial designs with more than two levels will be considered in Chapter 9.

There are several reasons for emphasizing 2^k designs:

- They are easy to use and the data analysis can be performed by graphical methods. This allows all interested parties to participate in the experimentation.

- Relatively few runs are required. A factorial design for four factors at two levels each requires 16 runs. If each of the factors is studied at three levels, 81 runs are required. For four levels, 256 runs are required.
- The 2^k designs have been found to meet the majority of the experimental needs of those engaged in the improvement of quality.
- The 2^k factorial designs are easy to use in sequential experimentation, so that even complex systems with many variables can be studied in depth using these relatively simple designs.
- When a large number of factors are studied, fractions of the 2^k designs can be used to keep the experiment at a reasonable size.

There are various ways to display the different combinations that make up a factorial design. When factors are studied at two levels, a common convention is to designate the low level of the factor as "$-$" (minus) and the high level as "$+$" (plus). When the factor is qualitative, such as type of material, the "$-$" and "$+$" labels can be assigned arbitrarily. Using this convention, Figures 6.2, 6.3, and 6.4 show examples of displays of 2^2, 2^3, and 2^4 factorial designs.

Figure 6.2 contains a simple listing of the combinations of factors in a form called the design matrix. This display of a factorial design is used primarily for documentation of the tests to be made in the experiment and as a basic form for the collection of data. Also, as described later in this section, this format is useful for estimating the effects of the factors.

	Factor	
Test	1	2
1	$-$	$-$
2	$+$	$-$
3	$-$	$+$
4	$+$	$+$

(a) 2^2 design

	Factor		
Test	1	2	3
1	$-$	$-$	$-$
2	$+$	$-$	$-$
3	$-$	$+$	$-$
4	$+$	$+$	$-$
5	$-$	$-$	$+$
6	$+$	$-$	$+$
7	$-$	$+$	$+$
8	$+$	$+$	$+$

(b) 2^3 design

	Factor			
Test	1	2	3	4
1	$-$	$-$	$-$	$-$
2	$+$	$-$	$-$	$-$
3	$-$	$+$	$-$	$-$
4	$+$	$+$	$-$	$-$
5	$-$	$-$	$+$	$-$
6	$+$	$-$	$+$	$-$
7	$-$	$+$	$+$	$-$
8	$+$	$+$	$+$	$-$
9	$-$	$-$	$-$	$+$
10	$+$	$-$	$-$	$+$
11	$-$	$+$	$-$	$+$
12	$+$	$+$	$-$	$+$
13	$-$	$-$	$+$	$+$
14	$+$	$-$	$+$	$+$
15	$-$	$+$	$+$	$+$
16	$+$	$+$	$+$	$+$

(c) 2^4 design

FIGURE 6.2
Design matrix display

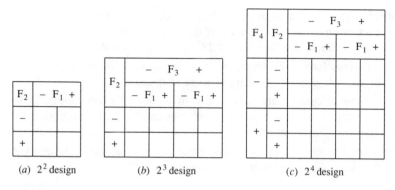

FIGURE 6.3
Tabular display of factorial designs

Figure 6.3 contains a tabular display. Each of the small squares (called cells) in the table corresponds to a specific set of combinations of the factors. The display of the data obtained from the experiment in such a table aids analysis. For example, consider the tabular display for a 2^3 design in Figure 6.3b. By comparing the data

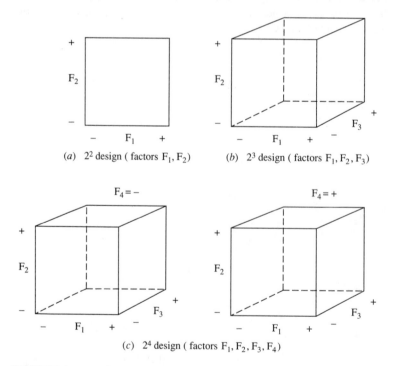

FIGURE 6.4
Geometric display of a factorial design

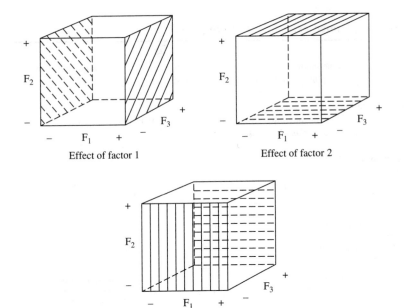

FIGURE 6.5
Comparison of data on a cube

in the four cells on the left side of the table to the data in the four cells on the right side, the effect of factor 3 can be studied.

Figure 6.4 depicts a geometric display. Each corner of the square or cube corresponds to a different set of combinations of the factors. The corners correspond to the cells in the tabular display. The value of the response variable obtained from each test is written at the appropriate corner. The geometric display, like the tabular display, is helpful for analysis.

For the 2^3 design, factor 1 can be studied by comparing the data on the left side of the cube to the data on the right side. Factor 2 is studied by comparing the bottom of the cube to the top, and factor 3 is studied by comparing front to back. Figure 6.5 illustrates the sides of the cube that are used for each comparison. It can be clearly seen that the structure of factorial designs allows all the data from the experiment to be used to study each factor.

Example 6.1   A 2^2 design: Manufacture of plasticizer.   As was the case for the one factor design, careful planning of the experiment is important in order to maximize the amount of information obtained for the expended resources. To aid the planning of experiments with multiple factors, the planning form introduced in Chapter 4 will be used. Figure 6.6 contains the form used to summarize an experiment performed to improve the process of manufacturing a certain type of plasticizer.

Normally the product has been made in batches, and the reaction was allowed to continue until a certain viscosity was obtained. It was desired that this reaction take from

FIGURE 6.6
Documentation of planned experiment, Example 6.1

1. **Objective:**
Find a combination of the amount of ingredient X and the reaction temperature to slow down the reaction time to 7–9 hours in a batch process for the manufacture of plasticizer.

2. **Background information:**
The reaction in plant production has often been proceeding too fast, resulting in batches that were difficult to control and of unacceptably high viscosity.

3. **Experimental variables:**

A.

Response variables	Measurement technique
Reaction time to reach desired viscosity	Viscometer and clock

B.

Factors under study	Levels	
1. Percentage of ingredient X	42% (−)	48% (+)
2. Temperature (°C)	175 (−)	190 (+)

C.

Background variables	Method of control
1. Lab experiment	Confirmation on plant batches
2. Rate of heat-up	Temperature programmer
3. Operator	One lab technician
4. Blend variation in X	Blocking

4. **Replication:**
Two replications of the experimental pattern, resulting in eight batches. Each replication used a different blend of ingredient X.

5. **Methods of randomization:**
Randomize the order of the four runs within each blend of X using a table of random permutations.

6. **Design matrix:**
(See Table 6.1.)

7. **Data collection forms:**
(See Table 6.2)

8. **Planned methods of statistical analysis:**
- Run chart
- Analysis of the square
- Summary of effects if appropriate (response plot)

TABLE 6.1
Design matrix for the plasticizer experiment

X	T
−	−
+	−
−	+
+	+

Percentage of X		Temperature (°C)	
−	+	−	+
42	48	175	195

seven to nine hours. The plant was experiencing other problems, which led to the reaction proceeding too fast. This resulted in batches that were difficult to control and product of unacceptable viscosity. The purpose of the experiment was to find a combination of the percentage of a key ingredient X and the reaction temperature that would result in a reaction that proceeded at the desired rate.

The response variable was the reaction time necessary to reach the desired viscosity. There were two factors under study, the percentage of ingredient X and the reaction temperature. The levels for the ingredient were chosen at 42% and 48%. The temperature levels were chosen at 175°C and 195°C.

Several background variables were considered. The experiment was initially conducted in the lab, so confirmation of the results in the plant would be necessary. The rate of heat-up was controlled by a temperature programmer. One lab operator was used. The difference in blends of X was handled by completing the experimental pattern using only one blend of X.

In analytic studies, it is important to run studies sequentially over a variety of conditions to build up the degree of belief. For this experiment, two replications of the experimental pattern were run. An adequate amount of ingredient X was obtained and

TABLE 6.2
Data collection form

Batch ID	Percentage of X	Temperature	Reaction time
	Replication 1	**(blend 1 of X)**	
1	42	195	5:5
2	42	175	9.0
3	48	175	9.0
4	48	195	1.5
	Replication 2	**(blend 2 of X)**	
5	42	195	6.5
6	48	195	1.0
7	42	175	9.5
8	48	175	8.0

mixed so that the four combinations of the two factors needed for one replication of a 2^2 pattern could be performed using a homogeneous blend of ingredient X. The second replication of the pattern was performed using a blend of ingredient X from a different shipment.

The order of the four runs in each replication was randomized separately using a table of random permutations.

As with the one-factor design, the first step in the analysis of data from a factorial design is a run chart of the data. This chart appears in Figure 6.7.

No obvious time trends or outlying values are seen, so analysis of the effects of the factors can be performed. The data from a factorial design can be analyzed as a series of paired comparisons, one factor at a time. The comparisons are carried out under various conditions of the other factors, which is very desirable for increasing the degree of belief in the results. This concept is displayed on the square in Figure 6.8.

Each corner of the square contains two reaction times. The first reaction time is the result from the first replication (the first blend of ingredient X), and the second is the result from the second replication. For each replication, there are two paired comparisons of the effect of the temperature. The first comparison is performed by comparing the results at the top corners of the square for the first replication. This is done quantitatively by subtracting the reaction time for the low temperature from the reaction time for the high temperature, with X held constant at 48%, that is

$$1.5 - 9.0 = -7.5$$

When the concentration of X is 42%, the effect of temperature is:

$$5.5 - 9.0 = -3.5$$

Similarly, the effect of the percentage of X when temperature is held constant at 175°C comes from the left corners of the square:

$$9.0 - 9.0 = 0.0$$

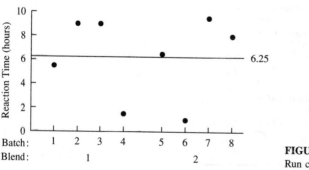

FIGURE 6.7
Run chart for Example 6.1

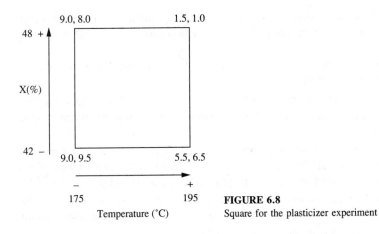

FIGURE 6.8
Square for the plasticizer experiment

The effect of the percentage of X when temperature is held constant at 195°C comes from the right corners:

$$1.5 - 5.5 = -4.0$$

The same comparisons can be made for the second replication. The results for both replications are given in Table 6.3.

TABLE 6.3
Results of paired comparisons

Difference in reaction time for change from 175°C to 195°C	Percentage of X at which comparison is made
Replication 1 (blend 1)	
$1.5 - 9.0 = -7.5$	48
$5.5 - 9.0 = -3.5$	42
Replication 2 (blend 2)	
$1.0 - 8.0 = -7.0$	48
$6.5 - 9.5 = -3.0$	42

Difference in reaction time for change from 42% to 48%	Temperature (°C) at which comparison is made
Replication 1 (blend 1)	
$9.0 - 9.0 = 0.0$	175
$1.5 - 5.5 = -4.0$	195
Replication 2 (blend 2)	
$8.0 - 9.5 = -1.5$	175
$1.0 - 6.5 = -5.5$	195

Based on these comparisons, some important observations can be made about how temperature and the percentage of X affect reaction time:

- The replications produced very similar results, indicating that there were no important effects of blends.
- All four cases resulted in lower temperature associated with longer reaction time. The magnitude of the difference was large enough to be of importance in the process.
- The temperature comparisons when X was at 48% produced larger differences than when X was at 42%. This observation along with the consistency of the replications suggests an interaction between temperature and percentage of X.
- Three of the four comparisons of the effect of high verses low percentage of X on reaction time resulted in longer reaction time associated with 42% of X. In one comparison, no effect was seen.

Since this is an analytic study, the important uncertainty associated with the experiment is whether the results can be repeated in future lab tests and then later in the plant. It is up to those familiar with the process to decide what action should be taken next based on their degree of belief. The next cycle could be further lab tests under different conditions at confirmation of the results in the plant.

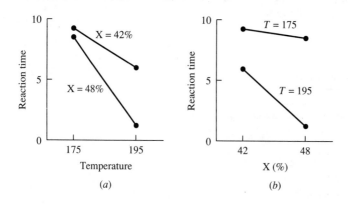

(a)   (b)

Average of replication tests for the four conditions

X (%)	Temperature (°C)	
	175	195
42	9.25	6.00
48	8.50	1.25

(c)

FIGURE 6.9
Response plots

Once the important effects have been determined, the relationship between the response and the important factors should be graphically displayed. This will provide the analyst with further insight into the cause-and-effect relationships (especially when interaction is present) and simplify the presentation of conclusions. Since there is an interaction between temperature and percentage of X, the relationship between reaction time and temperature should be plotted separately for each percentage of X. These plots are shown in Figure 6.9a.

The points on the plot are obtained by averaging the two replications for each of the four conditions in the experimental pattern (i.e., averaging the two results at each corner of the square in Figure 6.8). The relationship between reaction time and percentage of X for the two levels of temperature is plotted in Figure 6.9b. The plots in Figure 6.9a and b contain the same information displayed in two different ways, so only one is actually needed. (It is sometimes helpful, however, to view both plots.) The plots in Figure 6.9 are especially useful in presenting the results of the study to those who are not very knowledgeable in the analysis of factorial experiments. Plots of this type will be called *response plots*.

The next example extends these concepts to an experiment with three factors.

Example 6.2   2³ Design for a Dye Process. The data in Table 6.4 resulted from a 2^3 factorial design on a dye process. The aim of the experiment was to quantify the effects of three factors on the shade of dyed material and use the information to determine settings of the factors. The three factors studied were:

material quality (*M*),

oxidation temperature (*T*),

oven pressure (*P*).

Important background variables were identified and held constant. The response variable was a measure of shade using an optical instrument. It was desired to choose

TABLE 6.4
Data on shade of the dyed material

	Material quality			
	A		B	
	Oxidation temperature		Oxidation temperature	
Oven pressure	Low	High	Low	High
Low	189 (4)	195 (8)	228 (7)	200 (6)
High	218 (3)	238 (2)	259 (5)	241 (1)

Run order indicated in parentheses

settings of the factors to obtain a shade reading of 220. (*Note*: No units are given because the unit of measurement is somewhat arbitrary and depends on calibration of the instrument to reference standards.) In addition, it was desired to choose conditions to make the process as insensitive to variations in material as possible.

To begin the analysis, a run chart was made; it appears in Figure 6.10. For an unreplicated 2^3 design, only eight points are available to plot, so that only gross trends or outlying values can be identified from the run chart. No such values appear to be present in the chart in Figure 6.10.

Figure 6.11 contains the cube labeled with the three factors studied in this experiment. At each corner, the value of the response variable at that set of combinations of the factors is given.

By appropriate analysis of the cube, the effects of the factors can be estimated. To study the effect of material quality, the four values on the left side of the cube are compared to the four values on the right side. The effect of oxidation temperature is obtained by comparing the top and bottom of the cube. Comparison of the back and front of the cube provides information on the effect of oven pressure.

A more detailed analysis of the data must be performed if interactions between factors or special causes of variation in the data are to be found. The 2^3 factorial design can also be thought of as a series of paired comparisons. Each edge of the cube represents one of these comparisons. For example, the bottom front edge connects two corners between which the only difference in the test conditions is that material of quality A is used at the left corner and material of quality B is used at the right corner. At both corners, oxidation temperature is low and oven pressure is low. There are three other comparisons of the effect of material quality during which the other factors were held constant. These comparisons are performed by comparing the corners connected by the top front edge, the corners connected by the top back edge, and the corners connected by the bottom back edge.

Because of the symmetry of the 2^3 design, there are also four pairs of tests that can be used to study the effect of oxidation temperature. These pairs are found

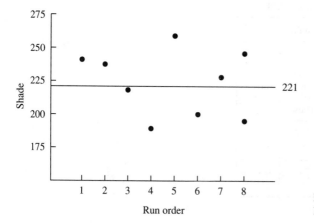

FIGURE 6.10
Run chart for Example 6.2

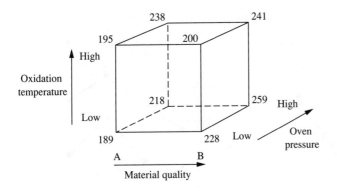

FIGURE 6.11
Cube for the dye process experiment

at the corners connected by edges running from the top to the bottom of the cube. The four pairs used to study the effect of oven pressure are found along the edges connecting the front of the cube to the back. The computation of these effects from the paired comparisons are contained in Table 6.5. The paired comparisons could also be displayed graphically. These graphs are shown in Figure 6.12.

TABLE 6.5
Computation of effects

Effect of oven pressure	Combination for which comparison is made	
	Oxidation temperature	**Material**
218 − 189 = 29	Low	A
238 − 195 = 43	High	A
259 − 228 = 31	Low	B
241 − 200 = 41	High	B

Effect of material quality	**Oxidation temperature**	**Oven pressure**
228 − 189 = 39	Low	Low
200 − 195 = 5	High	Low
259 − 218 = 41	Low	High
241 − 238 = 3	High	High

Effect of oxidation temperature	**Oven pressure**	**Material**
195 − 189 = 6	Low	A
238 − 218 = 20	High	A
200 − 228 = −28	Low	B
241 − 259 = −18	High	B

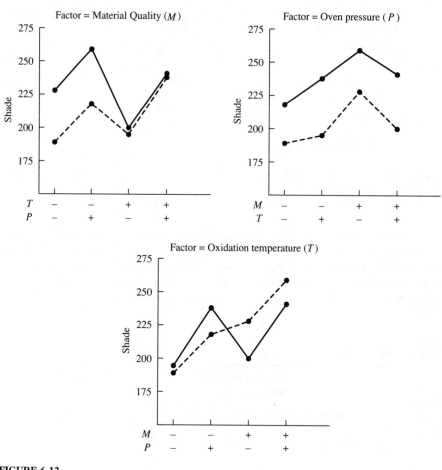

FIGURE 6.12
Plot of paired comparisons

From Table 6.5 and Figure 6.12, it is apparent that as oven pressure increases, the measure of shade increases and that this increase is reasonably consistent (from 29 to 43) over the four sets of conditions. Since the results are consistent, the average effect of oven pressure is a useful summary of the data. The average effect of oven pressure is 36, computed by subtracting the average shade when oven pressure is low from the average shade when oven pressure is high:

$$239 - 203 = 36$$

This average effect can also be obtained by averaging the four paired comparisons.

TABLE 6.6
Computation of interaction effect

Average effect of
material when oxidation = 4 ⎫
temperature is high ⎬ Interaction = 4 − 22 = −18
Average effect of = 22 ⎭
material
Average effect of
material when oxidation = 40
temperature is low

On average, the shade value when material B is used is 22 units higher than when material A is used. That is, the average effect of material is 22. As shown in Table 6.5 and Figure 6.12, the results are not consistent over all the conditions. The effect of material is great when oxidation temperature is low. When oxidation temperature is high, material has little effect. That is, there is an interaction between material quality and oxidation temperature. The effect of material cannot be given without first specifying the oxidation temperature. The average effect of material when oxidation temperature is low is 40. When oxidation temperature is high, the average effect of material is 4. The magnitude of the interaction is −18 and is computed according to the convention outlined in Table 6.6.

Next the effect of oxidation temperature is studied. In Table 6.5 and Figure 6.12, it can be seen that shade increases as temperature increases in two cases and shade decreases as temperature increases in two cases. The positive effects of temperature were observed when material quality A was used, and the negative effects of temperature were seen when material of quality B was used. This is another way of viewing the material-temperature interaction.

Estimating Effects Using the Design Matrix

The computation of the effects of the factors and the interactions can be performed using an algorithm based on the extended design matrix. Table 6.7 illustrates the standard form of a design matrix for a 2^3 design and the estimated effects. The columns corresponding to the various interactions are obtained by multiplying the signs for the factors contained in the interactions.

Each of the effects is estimated by adding or subtracting the value of the response variable, depending on whether the sign of the appropriate column is plus or minus. For example, the average effect of oven pressure is

$$\frac{-189 - 228 + 218 + 259 - 195 - 200 + 238 + 241}{4} = 36$$

It is easy to see that this computation is the same as subtracting the average shade

TABLE 6.7
Design matrix for 2^3 factorial pattern

Test	Run order	M	P	T	MP	MT	PT	MPT	Response
1	4	−	−	−	+	+	+	−	189
2	7	+	−	−	−	−	+	+	228
3	3	−	+	−	−	+	−	+	218
4	5	+	+	−	+	−	−	−	259
5	8	−	−	+	+	−	−	+	195
6	6	+	−	+	−	+	−	−	200
7	2	−	+	+	−	−	+	−	238
8	1	+	+	+	+	+	+	+	241
Divisor = 4									
Effect		22	36	−5	0	−18	6	−1	

value when oven pressure is low from the average shade value when oven pressure is high.

The estimate of the interaction between material and oxidation temperature is

$$\frac{+189 - 228 + 218 - 259 - 195 + 200 - 238 + 241}{4} = -18$$

This is the same as the estimate obtained in Table 6.6.

Once the estimates are obtained, they can be plotted to help determine which are the most important effects. The effects are plotted in Figure 6.13. This plot is called a *dot diagram* by Box, Hunter, and Hunter (1978, p. 25). Effects clustered near zero on the diagram cannot be distinguished from variation due to nuisance variables. It is clear from the plot and the analysis of the cube that the most important effects are the average effect of oven pressure, the average effect of material, and the interaction between material quality and oxidation temperature.

During the analysis of data from any experiment, a close watch must be kept for evidence of special causes of variation due to background or nuisance variables. Since the effects computed from the design matrix are averages, they can be distorted by special causes. These estimates should be verified by a more detailed analysis of the data. The estimates of average effects from the design matrix should initially be used to provide some preliminary sense of which factors or interactions might be important. Usually in studying two or three factors, this method of estimating

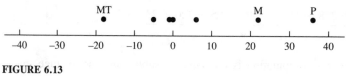

FIGURE 6.13
Dot diagram

effects is not necessary since the analysis can be done using a square or cube. However, the approach is very useful for four or more factors.

Once the preliminary estimates are obtained, the data in the cube should be studied for consistency between the individual comparisons contained in the average effects. Plots of the individual comparisons such as those in Figure 6.12 are also important for verifying the validity of the estimates of average effects.

Response Plots

Once the important effects have been identified and estimated, the relationships can be graphically summarized using simple response plots such as those in Figure 6.14. This figure shows three plots, one for each of the three combinations of two factors.

Construction of the plots involves only some simple arithmetic. For example, Figure 6.14a shows the effect of material quality and oxidation temperature on shade. To construct this plot, the 2×2 table shown under the plot is needed. Each entry in the table is the average of the two values of shade corresponding to the same level of material quality and oxidation temperature. For example, 203.5 is the average of the two values (189 and 218) obtained when temperature was low and material quality was A. Referring to the cube for the experiment should assist in determining the values that correspond to the same levels of the factors.

The response plot is then constructed by plotting shade versus oxidation temperature separately for each material quality. The numbers 203.5 and 216.5 are connected by a straight line to display the relationship between shade and oxidation temperature when material quality A is used. A straight line between 243.5 and 220.5 displays the same relationship when material B is used. The usefulness of the plot depends on the assumption that shade increases approximately linearly between the two extremes of oxidation temperature for both levels of material quality.

The remaining two plots shown in Figure 6.14 are constructed in the same manner. Each of the three plots could be shown with the factors reversed, as in Figure 6.9.

The lines in the response plot shown in Figure 6.14a are not parallel. This indicates that the factors—material quality and oxidation temperature—interact. Since the lines in the response plots shown in Figures 6.14b and c are approximately parallel, this indicates that oven pressure does not interact with either material quality or oxidation temperature. Oven pressure, however, does have an important effect by itself on shade. These results are consistent with those determined from the design matrix in Table 6.7.

Since oven pressure does not interact with either material quality or oxidation temperature, its effect can be shown alone on a response plot. This response plot is obtained by plotting the average shade versus oven pressure when oven pressure is low and high and connecting the points with a straight line. The average shade when oven pressure is low, 203, is computed by averaging the four shade figures on the front face of the cube. The average shade when oven pressure is high is 239, obtained by averaging the four shade figures on the back side of the cube. This plot is shown in Figure 6.15a.

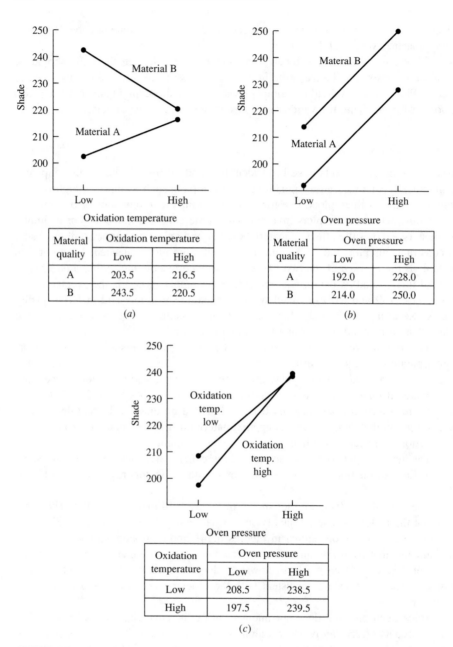

| Material | Oxidation temperature | |
quality	Low	High
A	203.5	216.5
B	243.5	220.5

(a)

| Material | Oven pressure | |
quality	Low	High
A	192.0	228.0
B	214.0	250.0

(b)

| Oxidation | Oven pressure | |
temperature	Low	High
Low	208.5	238.5
High	197.5	239.5

(c)

FIGURE 6.14
Response plots

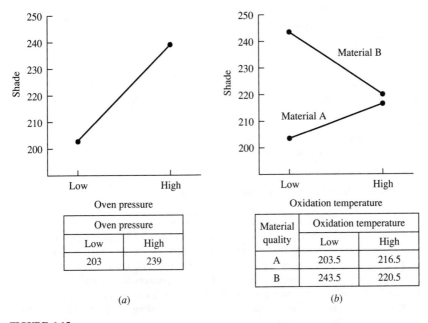

Oven pressure	
Low	High
203	239

(a)

Material quality	Oxidation temperature	
	Low	High
A	203.5	216.5
B	243.5	220.5

(b)

FIGURE 6.15
Response plots

As the number of factors included in the experiment increases, the number of response plots needed to plot all combinations of two factors becomes rather large. In order to reduce the number of plots necessary, only those effects determined to be important from the design matrix are plotted. Therefore, in this example, one plot of average shade versus oven pressure is needed since oven pressure is an important effect but does not interact with either of the other two factors. Another plot is needed to display the interaction between material quality and oxidation temperature. These plots are both shown in Figure 6.15.

Conclusions for Example 6.2

1. Increases in oven pressure increase shade at a rate independent of material quality and oxidation temperature (within the bounds of the experiment).
2. There is an interaction between oxidation temperature and material quality. Increases in oxidation temperature increase shade with material quality A but decrease shade with material B.
3. Running the process at high oxidation temperature would make the process less sensitive to variation in material quality and result in a more uniform shade.
4. Setting oven pressure approximately halfway between the levels used in the experiment would center the process at the desired level of 220.
5. Follow-up to this study should be a verification of the average shade level and of the reduced variation in shade as a result of using a high oxidation temperature to dampen the effect of variation in material.

TABLE 6.8
Factors for the solenoid experiment

	Level	
Factor	−	+
A = Length of the armature (in)	0.595	0.605
S = Spring load (g)	70	100
B = Bobbin depth (in)	1.095	1.105
T = Length of the tube (in)	0.500	0.510

The next example illustrates a study with four factors.

> **Example 6.3   A 2^4 experiment for the design of a solenoid valve.** A manufacturer of solenoid valves designed a solenoid to be used with a pollution control device on an automotive engine. The solenoid was used to turn the pollution control device on and off. Engineers responsible for the design of the solenoid ran a 2^4 factorial design to determine the effect of some of the important components in the solenoid valve on the flow (measured in cubic feet per minute, or cfm) of air from the valve. Flow is an important quality characteristic of the valve. The results of the study would be used to set specifications for components of the solenoid. The four factors that were studied and their levels are shown in Table 6.8. Figure 6.16 contains the planning form for the study.

Run Charts

The run charts for the averages and standard deviations of flow from the four tests of each of the 16 combinations appear in Figure 6.17. Both of the charts have some patterns worth noting. The chart for averages shows eight points all near 0.72. All of the eight are associated with bobbin depth of 1.105, indicating its important effect on flow. The relatively small variation of the eight points led the engineers to hypothesize

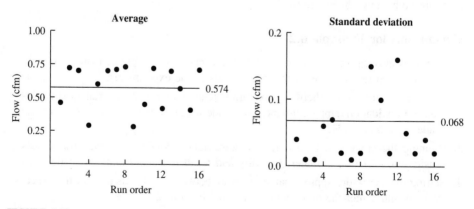

FIGURE 6.17
Run charts for the solenoid study

FIGURE 6.16
Documentation of solenoid experiment

1. **Objective:**
   Study the effects that four important components of the solenoid valve have on flow. This information will be used to determine manufacturing specifications for the components.

2. **Background information:**
   Previous experiments using fractional factorial designs (see Chapter 7) indicated that the four components chosen for this experiment had the largest effect on flow. Experience in manufacturing has shown that bobbin depth is one of the most difficult dimensions to control.

3. **Experimental variables:**

   A. | Response variables | Measurement technique |
   |---|---|
   | 1. Average and standard deviation of flow (cfm) | Flow meter 65c |

   B. | Factors under study | Levels | |
   |---|---|---|
   | 1. Armature length (in) | 0.595 | 0.605 |
   | 2. Spring load (g) | 70 | 100 |
   | 3. Bobbin depth (in) | 1.095 | 1.105 |
   | 4. Tube length (in) | 0.500 | 0.510 |

   C. | Background variables | Method of control |
   |---|---|
   | 1. Environment | Pressure, humidity, and temperature recorded at time of test |
   | 2. Resistance of wire | Control chart used at the winders to monitor resistance |
   | 3. Flow tester | Calibration checked at the beginning, middle, and end of the test |

4. **Replication:**
   Four solenoids were assembled for each of the 16 conditions in the study.

5. **Methods of randomization:**
   The order in which the 16 combinations were assembled was randomized using a random permutation table. The four solenoids for each combination were all assembled at the same time. The order of test was the same as the order of assembly to save time.

6. **Design matrix:** (attach copy)
   (See Table 6.9.)

7. **Data collection forms:**
   (Not shown here)

8. **Planned methods of statistical analysis:**
   Compute the average and standard deviation of the four flow readings for each of the 16 combinations. Analyze the effects of the factors on both these statistics.

8. **Estimated cost, schedule, and other resources:**
   Study can be run in one day along with normal production.

TABLE 6.9
Design matrix for the solenoid study

Test	Run order	A	S	B	T	A S	A B	A T	S B	S T	B T	A S B	A S T	A B T	S B T	A S B T	Flow (cfm) $\bar{X}$	s
1	1	−	−	−	−	+	+	+	+	+	+	−	−	−	−	+	0.46	0.04
2	12	+	−	−	−	−	−	−	+	+	+	+	+	+	−	−	0.42	0.16
3	14	−	+	−	−	−	+	+	−	−	+	+	+	−	+	−	0.57	0.02
4	10	+	+	−	−	+	−	−	−	−	+	−	−	+	+	+	0.45	0.10
5	8	−	−	+	−	+	−	+	−	+	−	+	−	+	+	−	0.73	0.02
6	7	+	−	+	−	−	+	−	−	+	−	−	+	−	+	+	0.71	0.01
7	13	−	+	+	−	−	−	+	+	−	−	−	+	+	−	+	0.70	0.05
8	3	+	+	+	−	+	+	−	+	−	−	+	−	−	−	−	0.70	0.01
9	15	−	−	−	+	+	+	−	+	−	−	−	+	+	+	−	0.42	0.04
10	9	+	−	−	+	−	−	+	+	−	−	+	−	−	+	+	0.28	0.15
11	5	−	+	−	+	−	+	−	−	+	−	+	−	+	−	+	0.60	0.07
12	4	+	+	−	+	+	−	+	−	+	−	−	+	−	−	−	0.29	0.06
13	6	−	−	+	+	+	−	−	−	−	+	+	+	−	−	+	0.70	0.02
14	16	+	−	+	+	−	+	+	−	−	+	−	−	+	−	−	0.71	0.02
15	11	−	+	+	+	−	−	−	+	+	+	−	−	−	+	−	0.72	0.02
16	2	+	+	+	+	+	+	+	+	+	+	+	+	+	+	+	0.72	0.01

Divisor = 8

$\bar{X} = 0.574$ $\bar{s} = 0.068$

$\bar{X}$ = Overall average
$\bar{s}$ = Pooled standard deviation

136

that some other component in the solenoid was preventing the flow from exceeding 0.73. Further tests were planned to investigate this possibility.

Three large standard deviations appear in the run chart. A check of data on the background variables did not identify any special causes of variation. The three were all associated with long armature length and short bobbin depth, so the engineers attributed them to an interaction between the factors. This would be substantiated during subsequent analysis.

Design Matrix and Dot Diagrams

Table 6.9 contains the design matrix for the experiment. Figure 6.18 shows the effects of the factors computed from the design matrix as well as the dot diagrams. Effects of the factors on both the average flow and the standard deviation of flow were estimated.

From Figure 6.18, it is seen that bobbin depth had a large effect on average flow. Bobbin depth, armature length, and their interaction were the most important factors affecting the standard deviation of flow. This interaction is what was observed in the run chart.

For a 2^4 design, the dot diagrams can be modified to provide a display of the variation due to nuisance variables. The magnitudes of the four three-factor interactions and the single four-factor interaction can usually be assumed to be primarily the result of nuisance variables. The smaller the effect of nuisance variables, the closer these higher-order interactions will be to zero. Figure 6.19 contains a modified dot diagram of the effects on the standard deviation of flow. In the figure, the effects are spread out horizontally, beginning with the effects of the individual factors and ending with the four-factor interaction. The vertical dashed line alerts the analyst to the range of variation in the effects that could be expected due to nuisance variables.

Analysis of the Cubes

Figure 6.20 shows the two cubes representing the 16 combinations in the experiment. The paired comparisons are plotted in Figure 6.21.

The paired comparisons do not indicate the presence of any special causes of variation and therefore confirm the effects that were found using the design matrix. The eight paired comparisons for each factor (four for each cube) are obtained in the same manner as for the 2^3 design except for the factor, tube length. The paired comparisons for tube length are found by comparing the corresponding corners of the two cubes: for example, compare the result at the bottom front left corner of the top cube, 0.46, to the result at the corresponding corner of the bottom cube, 0.42. The consistency of the effects can be evaluated using these plots.

The following are examples of observations based on evaluating the consistency of the effects from the paired comparisons plotted in Figure 6.21.

1. As bobbin depth increases, the average flow increases for each of the eight combinations of the factors. This indicates an important positive effect of bobbin depth on the average flow.

Factor or interaction	Estimate (cfm)		Interaction	Estimate (cfm)	
	$\overline{X}$	s		$\overline{X}$	s
A	− .08	.03	ST	.01	.00
S	.04	− .02	BT	.04	.00
B	.28	− .06	ASB	.03	.02
T	− .04	.00	AST	− .01	− .01
AS	− .03	− .02	ABT	.04	.02
AB	.07	− .05	SBT	.00	− .01
AT	− .03	− .01	ASBT	.01	.01
SB	− .04	.02			

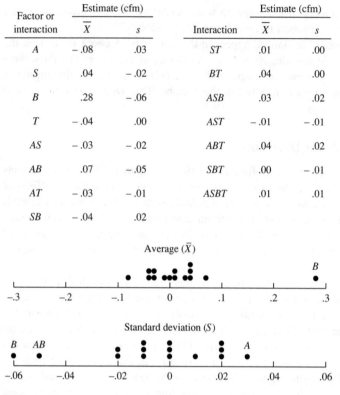

FIGURE 6.18
Effects of factors on flow

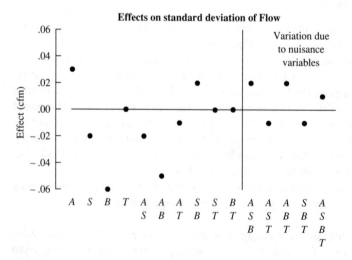

FIGURE 6.19
Modified dot diagram

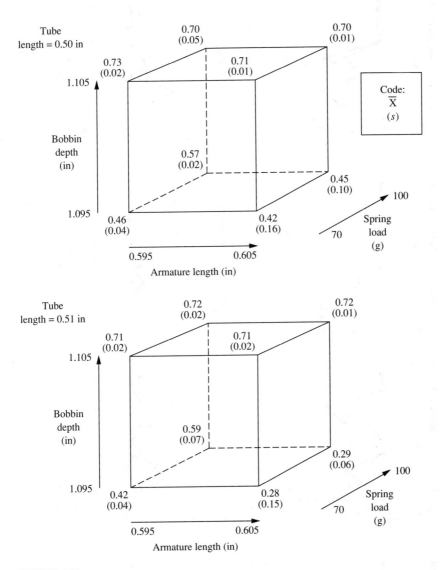

FIGURE 6.20
Cubes for the solenoid experiment

2. As bobbin depth increases, there is no effect on the standard deviation of flow for low levels of armature length, and a consistent positive effect for high levels of armature length. This indicates an important armature length–bobbin depth interaction on the standard deviation of flow.

3. As spring load increases, the standard deviation of flow increases for two conditions, decreases for five conditions, and is approximately zero for the remaining condition. This indicates that there is probably not an important effect of spring load on the standard deviation of flow.

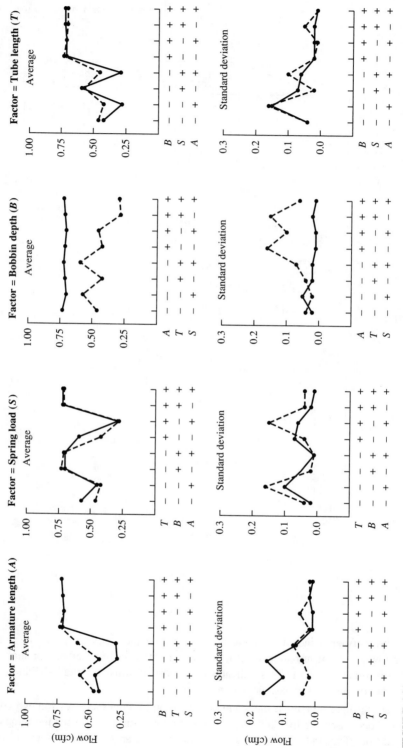

FIGURE 6.21
Plot of paired comparisons

Response Plots

Figure 6.22 shows the response plots summarizing the results of the experiment. The response plots are constructed just as they were for the 2^3 design. Since bobbin depth was the only factor with a substantial effect on average flow, there is a single response plot for this factor. When the standard deviation of flow is the response variable, there is an interaction between bobbin depth and armature length. Therefore, two plots of the standard deviation of flow versus bobbin depth are needed to summarize the data (one for each level of armature length).

We do not compute the the arithmetic average of the four standard deviations corresponding to a given level of bobbin depth and armature length; rather, the standard deviations are pooled together. (For more detail on pooling standard deviations, see Appendix to Chapter 8.)

Conclusions for Example 6.3

1. There were no special causes of variation detected in the run chart or the analysis of the paired comparisons in the cube.
2. Bobbin depth is the only factor in the experiment that had a substantial effect on the average flow of the solenoid. The longer the bobbin depth, the greater the flow of the valve.
3. The standard deviation of flow was affected by bobbin depth and armature length. These two factors interacted with each other. The combination of a long armature and a short bobbin produced large variations in flow.
4. The control of bobbin depth will be critical in the manufacture of the solenoid. The armature length should be specified at 0.595 inches to minimize variation.

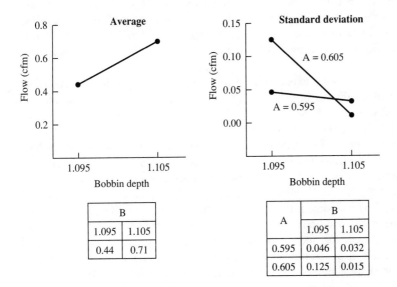

FIGURE 6.22
Response plots for the solenoid experiment

6.2   DESIGN OF FACTORIAL EXPERIMENTS

In Chapter 4, the principles of experimental design were discussed. These include:

defining the objective of the experiment,
providing background information,
choosing the number of factors,
choosing the levels for each factor,
considering background variables in the design,
selecting the amount of replication,
randomizing the order of the tests.

In this section, the discussion will center on how the last five items relate to factorial designs.

Choosing the Number of Factors

The number of factors to be included in the experiment will obviously depend on the objective and the resources available. In the early stages of experimentation, when little is known about the process or the product, the objective of the experiment may be to screen out the unimportant variables. If it is reasonable to assume that only a few of the variables will be important, a fractional factorial design should be considered. These designs will be discussed in Chapter 7. There are some particularly useful fractional factorial designs for screening up to 16 factors.

As discussed in Chapter 4, if it is desired to study in depth the relationship between the factors and the response variables, including interactions between the factors, two to five factors should be chosen for the study. A full factorial design is most useful for two to four factors. A fractional factorial design should be considered for five factors.

If the objective of the experiment includes the study of the important factors under a range of conditions in order to increase degree of belief, a chunk variable could be included as one of the factors. A chunk factor was defined in Chapter 4 as a combination of background variables.

Another important consideration for choosing the number of factors is how difficult each of the factors is to change. To change some factors may require little more than a turn of a dial, and their effect on the response is seen very quickly. Other factors are hard to change. A physical change to the equipment may take hours or days to accomplish. For some processes, it may take hours to reach equilibrium after changes in operating conditions have been made. In choosing the number of factors to be included in the experiment, consideration should be given to the difficulty of changing the levels of the factors. It is usually unwise to run a study in which more than half of the factors are hard to change. In that case, it would be advisable to run a series of smaller experiments.

Table 6.10 contains a summary of these ideas.

TABLE 6.10
Choosing the number of factors

Objective	Number of factors	Design
Screen out unimportant factors	5 or more	Fractional factorial
Study important factors in depth	2–4 5	Factorial Fractional factorial
Increase degree of belief	2–4 5	Factorial with a a chunk variable Fractional factorial with a chunk variable

Choosing the Levels for Each Factor

Factorial designs at two levels have been emphasized. The reasons for this were given at the beginning of this chapter. Choice of the specific two levels used for each factor will be based on knowledge of the process or product and the conditions of the study. As discussed in Chapter 4, it is desirable to set the levels of the factors far enough apart so that the effects of the factors will be large relative to the variation caused by the nuisance variables. However, the levels should not be so far apart that there is a good chance for trouble to develop. Examples of such trouble are the following:

- conditions that make the experimental run unsafe
- conditions that cause substantial disruption of a manufacturing facility
- important nonlinearities or discontinuities hidden between the levels
- substantially different cause-and-effect mechanisms for the different conditions in the experiment

In some experiments, it will not be possible to hold a factor constant at the desired level because the factor can not be controlled that precisely. Although it is desirable, it is not necessary that a factor be held constant at the planned level. However, the variation in the factor should be small relative to the distance between the planned levels. If this cannot be achieved, the factor should be made a background variable and measured during the experiment.

When continuous variables such as temperature, pressure, and line speed are used as factors, then a run at the center of the design can be used to check for discontinuities or nonlinear effects. For example, suppose the following factors and levels were used in a factorial design: (See table below and on following page.)

Factor	Level	
	−	+
Temperature (°C)	220	240
Pressure (psi)	50	80
Concentration (%)	10	12

The center point of this design is:

	Center point
Temperature	230
Pressure	65
Concentration	11

This point would be in the middle of the cube. (See Chapter 9 for more on the use of center points.)

The experimenter will also need to take into account the amount of unusable material generated during the study and the safety of those conducting the experiment. Box and Draper (1969) discuss ways of designing experiments for processes already in production. In their approach, levels are set close together to minimize unacceptable product. The small effect of a factor when the levels are close together is overcome by running many experimental units per factor combination. This usually can be done when the process is in full production. This approach is especially useful when most of the special causes have been removed from the process. Chapter 9 contains further discussion of this situation.

Considering Background Variables in the Design

As indicated in Chapter 4, there are two important decisions to be made concerning background variables:

1. How to control the background variables so that the effects of the factors are not distorted by them
2. How to use the background variables to establish a wide range of conditions for the study to increase degree of belief.

When a sequential approach to a study is taken, it is often useful to hold background variables constant and study the effects of the factors. To design the next study, the background variables should be used to establish conditions for the study that differ substantially from those of the previous study. This will allow the experimenters to determine if the effects of the factors are consistent over a wide range of conditions. The following example illustrates the use of background variables to widen the range of conditions.

> **Example 6.4   A welding process.** Consider an experiment in a welding operation for assembling an automotive part. Previous improvement cycles using fractional factorial designs indicated that the two most important factors were pressure and vacuum. Improvement cycles using control charts indicated that day-to-day variation in the process and the environmental conditions as well as operator technique were important background factors. The response variable of interest was a particular dimension of the assembled part.
>
> The primary aim of the experiment was to study the effects of pressure and vacuum over a wide range of conditions to increase degree of belief in their effects. Table 6.11

TABLE 6.11
Design for welding experiment

Response variable: Dimension

Factors:	Levels	
Pressure (psi)	30	45
Vacuum (in/hg)	8	10

Background variables (combined into one chunk factor):
 Day-to day-variation
 Operator technique

Chunk factor:
 Level 1: Monday, operator 1
 Level 2: Wednesday, operator 2

Design: 2^3 factorial

Factors:	Levels	
Pressure	30	45
Vacuum	8	10
Chunk	1	2

Note: Students of planned experimentation will recognize this as a split-plot design. The primary reason for considering the design as a split plot is to facilitate selection of the proper error term to use in the analysis of variance. The use of graphical methods of analysis reduces the importance of distinguishing the design as a split-plot arrangement.

identifies the design used to accomplish this aim.

Day-to-day variation and operator technique were incorporated into a chunk factor. This chunk factor was used along with pressure and vacuum to form a 2^3 factorial design. Because of the nature of the chunk factor, randomization would be restricted to each chunk. The four tests in chunk 1 would be done in random order, followed by the four tests in chunk 2 in random order.

If neither pressure nor vacuum are found to interact with the chunk factor, then an increased degree of belief will result that the estimated effects of pressure and vacuum can be used in the future. That is, the effects can be used to set specifications for pressure and vacuum and also be used as a guide to adjust the process (e.g., to counteract a special cause).

If either pressure or vacuum is found to interact with the chunk factor, then the appropriate setting of either pressure or vacuum can be used to mitigate the effect of the chunk factor, that is, day-to-day variation and operator technique. This will lead to improved consistency of the welding process.

The analysis of this experiment is discussed in Section 6.3.

Selecting the Amount of Replication

As stated in Chapter 4, replication is the primary means of studying the stability of the effects of the factors and increasing the degree of belief in the effects. The types of replication that can be used in an experiment were indicated in that chapter.

In factorial experiments, replication is built into the experimental pattern. This is sometimes forgotten by those not familiar with the design. For example, in a 2^4 design there are eight comparisons of the high and low levels of each factor. These comparisons are made under the eight different combinations of the other three factors.

The amount of replication that is feasible depends partly on how difficult it is to change the levels of the factors.

When multiple factors are included in an experiment, between 8 and 16 runs are desirable. If the aim of the study is to improve a process, and the effects of the factors cannot be distinguished from the nuisance variables in 8 to 16 runs, then the wrong factors or levels have probably been chosen.

More important than the number of runs is the range of conditions included in the study. Replication over similar conditions provides little increase in the degree of belief.

Randomizing the Order of the Tests

Randomization is the primary means of controlling the effect of nuisance variables. If blocks are constructed to control background variables, randomization should be performed within each block. The benefits of randomizing the order of the tests in an experiment have been outlined in Chapter 4. Also in Chapter 4, it was indicated that randomization should be considered for choosing the order of the tests, assigning the combinations of the factors to experimental units, and choosing the order in which the measurements are made. (Because of practical considerations, the order of measurements is sometimes constrained to be the same as the order of the tests.)

In factorial experiments, sometimes randomization is not practical. This results from the difficulty of changing the levels of some of the factors. In such a case, it is desirable to repeat one or more of the earlier runs at the end of the experiment. These replications can be used to check for special causes of variation that may have occurred during the experiment.

There is less need to randomize when experiments are conducted on stable processes. For processes dominated by special causes, randomization is an important tool. See Daniel (1976, pp. 22–26) for an excellent discussion of the use of randomization in factorial experiments.

6.3 ANALYSIS OF FACTORIAL EXPERIMENTS

In Section 6.1 analyses of data from factorial designs with two, three, and four factors were illustrated. The aim of this section is to provide more information on the analysis of factorial designs. The emphasis is the analysis of factorial designs when things do not go as planned. Also, some additional detail about the various graphical displays used in the earlier examples will be given.

The approach to the analysis of factorial designs in the previous examples followed four steps:

1. Plot the data in run order to look for trends and obvious special causes. If an uncontrolled background variable has been measured, then the data should also be plotted in increasing order of measurement of the background variable.

The purpose of the remaining three steps is to partition the variation seen in the run chart between the factors, background factors, and nuisance variables.

2. Estimate the effects of the factors using the design matrix, and plot the estimates on a dot diagram. This step provides a preliminary assessment of the factors that contribute most to the variation in the run chart.

3. Analyze the data using a square, a cube, or sets of cubes to study the variation in the paired comparisons that make up the effects estimated in step 2. A study of the paired comparisons will help identify the impact of nuisance variables on the variation and indicate the presence of special causes.

4. Summarize the results of the analysis by preparing response plots for the important factors and interactions.

The aim of these four steps is to partition the variation in the measurements of the response variable among the factors, the background variables, the nuisance variables, and any interactions between them. This partitioning, along with other knowledge of the product or process, allows the experimenter to determine the most advantageous approach to improvement.

This analysis can be carried out on the individual measurements of the response variable or on a statistic chosen based on knowledge of the process or product. If multiple experimental units are measured for each combination of the factors, statistics such as the average or range may be computed. The effects of the factors on these statistics can then be estimated.

For example, in a machining operation it may be simple to obtain the desired average dimension but difficult to find ways to reduce the variation. In this case, several experimental units would be measured for each condition, the range of the measurements computed, and the four-step analysis carried out using the range as the response of interest.

Taguchi (1987) suggests combining various statistics into what he calls a signal-to-noise ratio and using the signal-to-noise ratio as the response of interest. Kackar (1985) provides an excellent summary of these ideas. Box (1988) discusses the use of signal-to-noise ratios and transformation.

Graphical Displays for the Analysis of Factorial Designs

The following graphical displays have proven useful in analyzing data from factorial designs.

- run charts
- dot diagrams

- geometric figures or other displays of paired comparisons
- response plots

Each of these displays has been illustrated in the examples. In this section, some additional insight in the use of these displays will be given.

RUN CHART. The run chart is usually a display of the measurements of the response variable in the order that the tests were made. Other orderings, such as order of measurement or increasing order of a measured background variable, are also useful. The run chart displays the total variation in the response variable that is to be partitioned among the factors, background variables, and nuisance variables.

Analysis of the run chart begins this partitioning. Special causes of variation due to nuisance variables can be spotted. If some type of replication has been included in the design, the individual measurements can be plotted in the run chart. The run chart then can be used to assess the magnitude of either common or special causes of variation resulting from nuisance variables. If the experiment has been run in blocks, the blocks should be designated on the run chart. The variation in the measurements of the response variable that is due to the background variables comprising the blocks can be assessed. (See, for example, Figure 6.7.) Sometimes the run chart will point out that the variation is primarily due to one dominant factor because the data on the run chart stratify into two groups. (See run chart for averages, Figure 6.17.)

DOT DIAGRAMS. The second step in the analysis of data from a factorial experiment is estimating the effects of the factors from the design matrix and plotting the estimates on a dot diagram. The dot diagram is used to obtain a finer partitioning of the variation than is possible in the run chart. The dot diagram identifies factors whose effects are clearly separated from the variation due to nuisance variables.

It is important to remember that data from a factorial design will vary because of the factors, background variables, and nuisance variables. Effects clustered near zero cannot be distinguished from variation due to nuisance variables. Nuisance variables and background variables can impact a study as either common or special causes of variation. If a three- or four-factor interaction is one of the largest (in absolute value) effects in the dot diagram, then:

- it may not be possible to separate any of the effects of the factors from variation due to nuisance variables,
- a special cause may be dominating the estimates of the effects (the run chart and the cubes should be checked), or
- the interaction may be important (the least likely alternative).

For a 2^4 design, a modified dot diagram can be used to better distinguish the effects of nuisance variables. (See Figure 6.19.)

Box, Hunter, and Hunter (1978, p. 328) use reference distributions in combination with dot diagrams to separate variation due to nuisance variables from variation due to factors. Daniel (1976, Chapter 6) uses normal probability plots to graphically accomplish this separation.

Regardless of the method of analysis, caution should be exercised in placing a high degree of belief in effects whose magnitude is small although discernible.

If an effect is small compared to the variation caused by other factors, background variables, or nuisance variables, it is liable to change substantially in the future because of interactions with the conditions of the study. A high degree of belief in small effects can be established only through a synthesis of knowledge of the subject matter and replication of the effect over a wide variety of conditions.

By following these simple guidelines one can make the dot diagram more useful for separating the important effects from variation due to nuisance variables:

1. Construct the scale on the diagram such that zero is at the center and the endpoints of the scale are symmetric (e.g. −50, 50). The endpoints should be chosen such that all effects can be plotted on a linear scale.
2. Effects too close together to be plotted side by side should be plotted one above the other, as in a histogram.
3. All effects that are separated from the cluster around zero should be labeled.
4. If a complete factorial design is replicated in two or more blocks, the effects should be computed separately for each block. The effects from each of the blocks are all plotted on the same dot diagram.

GEOMETRIC FIGURES AND PLOTS OF PAIRED COMPARISONS. Factorial designs can be depicted using a square or one or more cubes. From these geometric figures, the paired comparisons that make up the factorial designs can be identified and studied. Analysis of the paired comparisons determines the factors with the greatest effect on the response variable. This is the same information obtained from the dot diagram. However, study of the paired comparisons provides some important additional information.

By studying the individual paired comparisons for consistency, the experimenter can ascertain the presence of special causes in the data. The existence of special causes in the data may not be evident in the run chart because of the variation attributable to changes in the factors. If special causes of variation exist, they may also not be evident in the dot diagram, but they will influence the estimates of the effects. Example 6.7 illustrates the existence of a special cause that was identified by analysis of the paired comparisons.

In Example 6.2, the paired comparisons were analyzed by identifying them on the cube, listing them (Table 6.5), and plotting them (Figure 6.12). Whether the effects are listed or plotted, the experimenter can analyze them for consistency; it is not necessary to do both.

Another means of graphically displaying the paired comparisons is difference diagrams. Figure 6.23 shows the difference diagrams for the data in Example 6.3. Each point plotted corresponds to a paired comparison. For example, one of the paired comparisons to study the effect of armature length on flow in Example 6.3 is

Test	Run order	A	S	B	T	Flow (cfm)
1	1	−	−	−	−	0.46
2	12	+	−	−	−	0.42

The difference, $0.42 - 0.46 = -0.04$, is the effect on flow of changing armature length, A, from $-$ to $+$. This difference is then plotted on the difference diagram in the position corresponding to B -, T -, S -. The difference diagrams in Figure 6.23 are seen to be simply the difference between the solid and dotted lines in Figure 6.21.

The difference diagrams should be interpreted as follows:

1. Differences centered around zero indicate that any effect of the factor cannot be separated from the variation due to nuisance variables. For example, see the difference diagrams for spring load in Figure 6.23.

2. Differences centered away from zero indicate an effect of a factor that can be separated from variation due to nuisance variables. See, for example, the difference diagram for the effect of bobbin depth on average flow.

3. Differences that stratify into two groups indicate an interaction between the factor under study and some other factor. The analyst should attempt to identify the other factor that is related to the two groups. For example, consider the difference diagram for the effect of bobbin depth on standard deviation. The differences stratify into two groups. When armature length is $-$, the effect of bobbin depth on standard deviation is negligible, as indicated by the differences close to zero. When armature length is $+$, the differences average approximately -0.1. This indicates that increasing bobbin depth decreases the standard deviation when armature length is $+$. This is another way of viewing the interaction between armature and bobbin that was discussed in Example 6.3.

4. If no interaction is present, the magnitude of the variation in the differences is the amount of variation attributable to nuisance variables. This variation can result from either common or special causes. See, for example, the diagrams for tube length or the diagram for the effect of bobbin depth on average flow.

Three methods of studying the consistency of the paired comparisons that make up a factorial design have been given: a listing of the differences, a plot of the differences on a difference diagram, and a plot of the paired comparisons. Which of the methods to use for a particular application is left to the preference of the experimenter. The aim of the analysis is the same in each case: to confirm that special causes of variation are not distorting the effects estimated from the design matrix.

RESPONSE PLOTS. Response plots are used to help experts in the subject matter visualize the impact of the effects of the factors found to be important in steps 2 and 3. The number of response plots that are constructed depends on the factors and interactions found to be important in steps 1, 2, and 3. The following items should be considered when constructing response plots.

- If a factor is found to be important but does not interact with any other factor, the response plot consists of one line approximating the relationship between the factor and the response. (See, for example, Figure 6.15a.)

- If two factors, A and B, interact, their relationship to the response variable is displayed on one response plot by using two lines. The relationship between

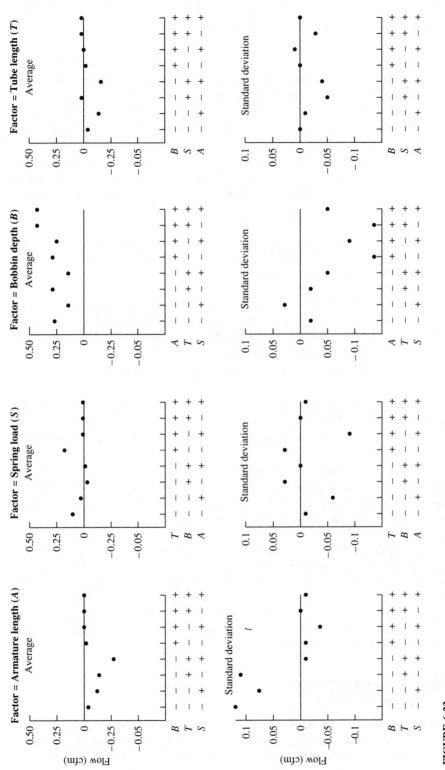

FIGURE 6.23
Difference diagrams for the solenoid experiment

151

the response variable and factor A when B is at the $-$ level is displayed by one line. The relationship between the response variable and factor A when B is at the $+$ level is displayed by another line. (See, for example, Figure 6.15b.) Once this response plot has been constructed, there is no need to construct individual response plots of the type in Figure 6.15a for each of the factors A and B.

- If two factors, A and B, interact and, in addition, one or both of them interact with a third factor, C, the response plot displaying the interaction of A and B must be done separately for the two levels of C. (See, for example, Figure 7.11 in Chapter 7.)

The response plots constructed for a particular experiment provide a visual model of the relationships between the important factors and the response variable. The response plots are interpreted as follows:

1. The response plots are only a linear approximation of the relationship; therefore, their usefulness is limited based on the linear relationship of the data. The use of center points to check for lack of linearity has been mentioned in this chapter and will be discussed in more detail in Chapter 9.
2. The slope of the response plot is proportional to the effect estimated from the design matrix.
3. If two factors interact and the interaction is displayed using two lines on the response plot (as in Figure 6.15b), then the difference in the slopes is proportional to the interaction between A and B. If A and B do not interact, then the two lines will be parallel.
4. The slopes of the lines in a response plot are their most important aspect. The intercepts of the lines depend on the levels of other factors in the study that are not in the response plot. As constructed in this chapter, the intercepts correspond to the other factors being set midway between the $-$ and $+$ levels.

In some cases, the analysis of data from a factorial design can proceed in a straightforward manner through the four steps listed at the beginning of this section. Often, the analysis of the data is not so straightforward because of problems such as the presence of special causes of variation, a large amount of variation caused by nuisance variables, or one or more data points missing. The primary purpose of the remainder of this section is to provide some guidance for the analysis of data from factorial experiments when things "go wrong."

Analysis of Factorial Designs When Things Go Wrong

SPECIAL CAUSES IN THE RUN CHART. There are times when trends or other obvious special causes are seen in the run charts. It is not advisable to proceed further with the analysis in such cases until the special cause is determined. The potential interactions of the special cause with the factors should also be assessed. If the cause

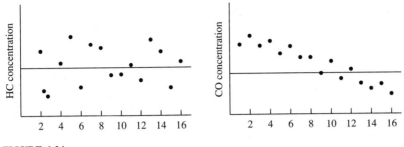

FIGURE 6.24
Run charts for Example 6.5

is identified, the data can possibly be adjusted to remove its effect. Regardless of the method of analysis, extreme care should be taken in extrapolating the results of the experiment. Verification of a conclusion by future experiments is almost always necessary.

Example 6.5 Study of automotive emissions. A 2^4 factorial design was run to study the effects of four factors on emissions from an automotive engine. The run charts for hydrocarbon (HC) and carbon dioxide (CO) concentrations are shown in Figure 6.24. The plot of HC concentration shows no trends or special causes. The plot for CO concentration shows an obvious downward trend. This trend accounts for a large portion of the variation in the run chart.

　　Analysis of the data for HC concentration should proceed only on the advice of experts in the process. It is their responsibility to determine if the nuisance variable affecting the CO concentration could also be affecting the measurements of HC concentration.

　　If the nuisance variable affecting the CO readings can be identified, its effect could be quantified and the CO data adjusted to remove the trend.

Example 6.6 Design of a throttle return mechanism. In a study to determine the durability of a throttle return mechanism, the following three factors, which related to the pin in the mechanism, were included:

Factor	Level	
Pin coated	No	Yes
Pin length	−	+
Pin diameter	−	+

　　A 2^3 design was used, in which the response variable was the number of cycles to failure on a stress test. Figure 6.25 contains the run chart. Figure 6.26 shows the data displayed on a cube.

From the run chart it is seen that the number of cycles obtained on the first run (corresponding to the conditions of pin coated, + length, and + diameter) differ

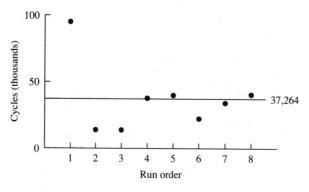

FIGURE 6.25
Run chart for Example 6.6

substantially from the rest of the data. These conditions represent an extreme case because each of the factors is set at the level that the engineer believed would provide increased durability. It is not unusual that extreme points produce results of greater magnitude than the sum of the individual effects.

In this experiment a special fixture had to be set up to run the tests. Since the extreme result was the first run, questions were raised about the setup and break-in of the fixture. It was decided to verify the results by running a paired comparison of the two extreme conditions, $+ + +$ and $- - -$.

SPECIAL CAUSES BURIED IN THE DATA. A special cause of variation in the data from a factorial experiment may not be evident in a run chart. Since the experimental conditions differ for each run, a point that appears in agreement with the rest of the

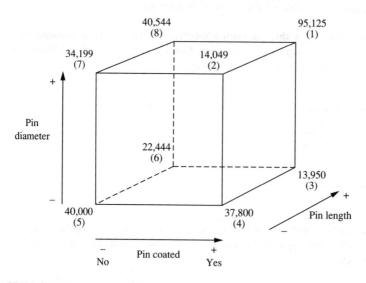

FIGURE 6.26
2^3 Factorial design for Example 6.6

data from the experiment may be quite abnormal. Analysis of the data using a square or a cube is the primary means of determining this type of special cause. An example will help clarify this idea.

Example 6.7 Pilot line for a tile process. On a pilot line built to study a new process of making flooring tile, a 2^3 design was used to determine the effect of three factors on an important quality characteristic of the tile. The factors are related to the setup of the production line.

The pilot line was run for an hour at each of the eight conditions. During each hour, 10 tiles were selected and the quality characteristic measured. Based on an analysis of the run chart of the readings for each condition, it was deemed appropriate to summarize the data for each condition by an average and a standard deviation.

Figure 6.27 is the run chart of the averages. No obvious special causes are seen in the run chart. The data are displayed on a cube in Figure 6.28. The paired comparisons, plotted in Figure 6.29, show that of the four comparisons for each factor, three are consistent and one is quite different. As they are plotted in Figure 6.29, the first, third, and third comparisons for factors A, B, and C, respectively, are quite different from the other three in each plot. These comparisons have one point in common; it is the fifth run. This run was performed at the conditions $A = +$, $B = -$, $C = -$ and resulted in an average response of 16.1. Based on an examination of the other comparisons, an average response of 23 to 24 would have been consistent with the rest of the data. After a check of the notes kept during the experiment, it was found that some trouble was encountered running the line at these conditions.

The analysis of the cube uncovered the possible existence of a special cause, although none was evident from the run chart. This provides an increased opportunity to learn about the process. This example illustrates the importance of detailed analysis of the cube prior to estimating the factor effects. If only the average effects estimated from the design matrix are used, some important information concerning special causes may be lost.

MISSING DATA. It occasionally happens that the data from one of the experimental conditions are missing. When this happens, it is important to determine whether or not the fact that the data are missing is related to the experimental conditions. For

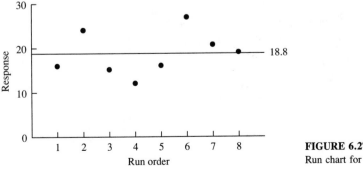

FIGURE 6.27
Run chart for Example 6.7

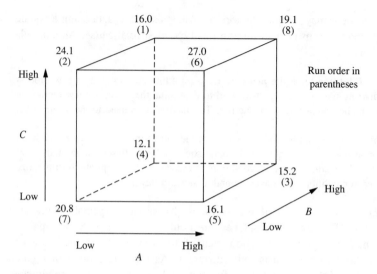

FIGURE 6.28
Cube for the tile process

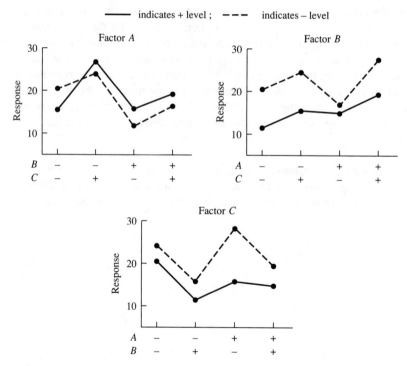

FIGURE 6.29
Plot of paired comparisons

example, the data may be missing because the product could not be made under those conditions.

If the data are missing because of the experimental conditions, the analysis should proceed using the cube to determine better conditions under which to run the process.

Sometimes data are missing for reasons that have nothing to do with the experimental conditions, (e.g., lost or damaged samples). In this case, the missing data can be estimated using the highest-level interaction in the design matrix (e.g., a three-factor interaction for a 2^3 design. Table 6.12a contains a portion of the design matrix from Example 6.2, the 2^3 design for a dye process, which is taken from Table 6.7.

The three-factor interaction from Table 6.12a is -1. Table 6.12b shows a procedure for estimating a missing value. Assume, for example, that the data for test 5 were missing, for reasons that had nothing to do with the experimental conditions of that test. Since the data are missing, the design matrix cannot be used to estimate effects. Since the value of the 123 interaction is expected to be small, the missing data value x is estimated by setting x equal to a value that will make the 123 interaction equal to 0. In this case, that value is 199. This is close to 195, the actual value measured. After the missing value is estimated, the analysis can proceed as usual.

NUISANCE VARIABLES DOMINATE.   There are occasions when the variation in the data from a factorial experiment is caused primarily by nuisance variables rather than changes in the factors under study. The nuisance variables may be either common or special causes of variation.

The experimenter can be alerted to the presence of nuisance variables dominating the variation in the data by the presence of one or more of the following conditions:

- Variation in the run chart that is of similar magnitude to the variation seen in an existing control chart for the process.
- A measurement system with variation that is large relative to the variation in the data

TABLE 6.12
Estimation of missing value

Test	(a) 123	(a) Response	(b) 123	(b) Response
1	−	189	−	189
2	+	228	+	228
3	+	218	+	218
4	−	259	−	259
5	+	195	+	x
6	−	200	−	200
7	−	238	−	238
8	+	241	+	241
Divisor = 4	−1		Set = 0 and solve for x (x = 199)	

- Three- or four-factor interactions that are large relative to average effects of the factors
- Unexplained inconsistencies in the paired comparisons embedded in the cube
- Estimates of effects that run counter to the theory that is held with a high degree of belief.

When it is suspected that nuisance variables dominate the variation, the following steps can be taken:

- Examine the measurement system. If it is stable but too imprecise, average multiple measurements per experimental unit. If it is unstable, remove the special causes and repeat the experiment.
- If the variation caused by the nuisance variables is from common causes (use the run chart and the cube to determine this), average multiple experimental units per combination of the factors.
- If there are special causes present, identify them and control them in subsequent experiments.
- Widen the range between the levels of the factors.
- Use statistical process control to reduce variation.

The following example illustrates a situation where the nuisance variables in an experiment are suspected of dominating the variation.

Example 6.8 Study of a process for batch mixing.   An experiment was conducted in a process of batch mixing to determine the effects of three factors on the increase in viscosity of the material during mixing. The three factors were mixing speed (S), mixing time (T), and the viscosity of the material before mixing (V). Figure 6.30 contains the run chart. No obvious time trends or outlying values are observed on the run chart. Table 6.13 contains the design matrix and Figure 6.31 shows the dot diagram. The data are displayed on a cube in Figure 6.32, and the paired comparisons are plotted in Figure 6.33.

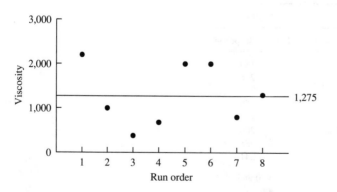

FIGURE 6.30
Run chart for Example 6.8

TABLE 6.13
Design matrix for the experiment on batch mixing

Test	Run order	S	T	V	ST	SV	TV	STV	Viscosity
1	3	−	−	−	+	+	+	−	380
2	1	+	−	−	−	−	+	+	2,200
3	5	−	+	−	−	+	−	+	2,000
4	2	+	+	−	+	−	−	−	1,000
5	8	−	−	+	+	−	−	+	1,300
6	6	+	−	+	−	+	−	−	2,000
7	4	−	+	+	−	−	+	−	680
8	7	+	+	+	+	+	+	+	800
Divisor = 4									
Effect		410	−350	−200	−850	0	−560	560	

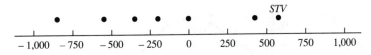

FIGURE 6.31
Dot diagram for Example 6.8

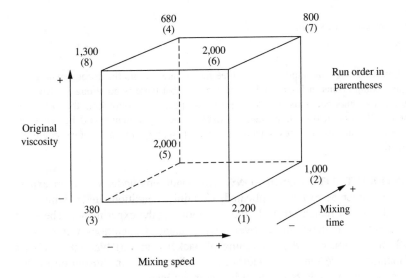

FIGURE 6.32
Cube for the experiment on batch mixing

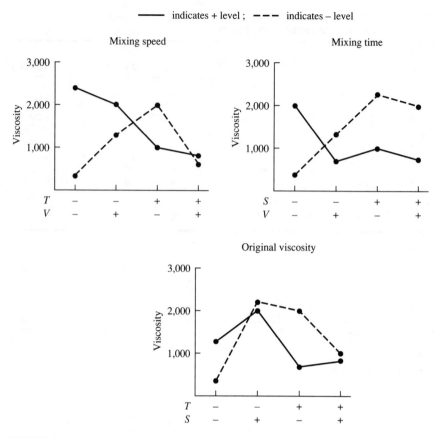

—— indicates + level ;   – – – indicates – level

FIGURE 6.33
Plot of paired comparisons for Example 6.8

Based on the magnitude of the three-factor interaction, the inconsistencies of the paired comparisons in Figure 6.33, and the fact that there is no strong evidence of one run being abnormal, nuisance variables are suspected of dominating the variation. To confirm the domination of nuisance variables, the experiment could be repeated, or a control chart for batches mixed at constant settings of the factors could be developed.

FACTORS INTERACT WITH CONDITIONS. The conditions under which an experiment is run can never be exactly duplicated. Certainly, conditions under which the results are used will not match precisely the conditions of the experiments. The need to run experiments in analytic studies over a wide range of conditions was stressed in Chapter 3. In Section 6.2, the combining of background variables into a chunk variable to include a wide range of conditions was discussed. An illustration of this method was given in Example 6.4, which is continued below.

Example 6.4 (continued) A Welding Process. This example examines two factors, vacuum and pressure, to determine their effects on a dimension (in thousandths of an inch) of a welded part. In addition, a chunk variable was formed from the background variables, operator technique and day-to-day variation. The chunk variable was used as the third factor in a 2^3 design. For each of the eight combinations of the factors, parts were welded.

Figure 6.34 contains the run chart. No evidence of trends or special causes unrelated to the combinations of the factors is seen. However, significant variation between the eight parts within each combination of factors is present.

Since the variation within each combination of factors is rather large, the analysis was carried out on both the average and standard deviation of the eight parts. The run charts for the averages and standard deviations appear in Figure 6.35. No special causes seem to be present.

Table 6.14 shows the design matrix and the estimated effects, and Figure 6.36 contains the dot diagrams. The averages and standard deviations for each run are displayed on a cube in Figure 6.37. The paired comparisons are plotted in Figure 6.38.

For the averages, it is seen in Table 6.14 and Figure 6.36 that pressure, the chunk factor, and the interaction between the chunk and pressure are the largest effects. In all cases, increased pressure resulted in a larger average dimension. However, the magnitude of the effect depends on the setting of the chunk factor. The chunk factor has less of an effect at the − level of pressure (30 psi) than at the + level (45 psi). Therefore, the effect of the chunk factor could be mitigated by setting the pressure at 30 psi.

In Table 6.14 and Figure 6.36, it can also be seen that pressure is the factor that has the largest effect on the standard deviation of the dimension. Since pressure does not interact with the chunk, a high degree of belief is established that increasing pressure will reduce short-term variation. Unfortunately, this is the opposite of the pressure setting needed to mitigate the effect of the chunk factor.

The most desirable result would have been that the same setting of pressure would reduce both the effect of the chunk factor and the short-term variation. A better consequence than the present result would have been if pressure were an important factor in only one of these situations; then, at least, a setting for pressure could be determined to take advantage of that fact.

However, based on the results of this experiment, an easy decision on the setting for pressure is not possible. It might be advisable to begin by setting the pressure at the + level (30 psi) to reduce the short-term variation, and then monitor the variation caused by the chunk factor (day-to-day variation and operator technique) using a control chart. The variation due to the chunk factor would then be reduced if special causes of variation could be identified and removed.

Figure 6.39 contains the response plots for the averages and standard deviations, which summarize the results of the experiment.

When a factor interacts with a chunk variable, as was the case in the preceding example, there are several possible courses of action. It may be the case that the effects of background and nuisance variables on the process must be reduced before a reason-

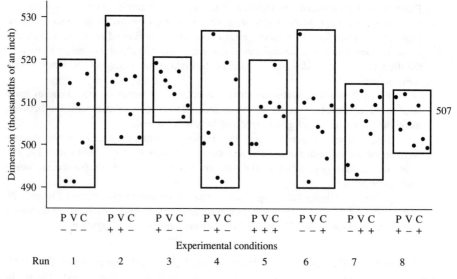

FIGURE 6.34
Run chart for Example 6.4

TABLE 6.14
Design matrix for welding experiment

Test	Run order	P	V	C	PV	PC	VC	PCV	$\overline{X}$	s
1	1	−	−	−	+	+	+	−	505	8.8
2	3	+	−	−	−	−	+	+	514	5.5
3	4	−	+	−	−	+	−	+	504	9.8
4	2	+	+	−	+	−	−	−	511	6.8
5	6	−	−	+	+	−	−	+	505	9.4
6	8	+	−	+	−	+	−	−	508	4.5
7	7	−	+	+	−	−	+	−	503	6.4
8	5	+	+	+	+	+	+	+	505	4.7
Divisor = 4										
Effect: $\overline{X}$		5.3	−2.3	−3.3	−0.8	−2.8	−0.3	0.3		
s		−3.2	−0.1	−1.5	0.9	−0.1	−1.3	0.7		

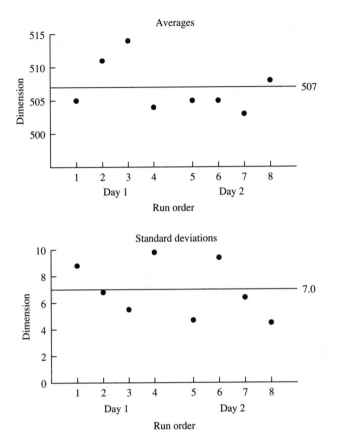

FIGURE 6.35
Run charts for averages and standard deviations, Example 6.4

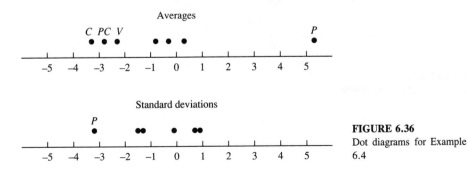

FIGURE 6.36
Dot diagrams for Example 6.4

Averages

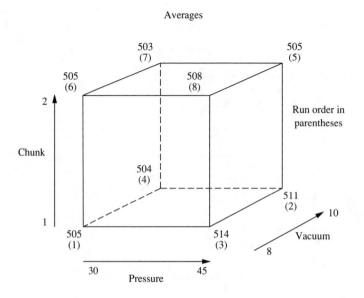

Standard deviations

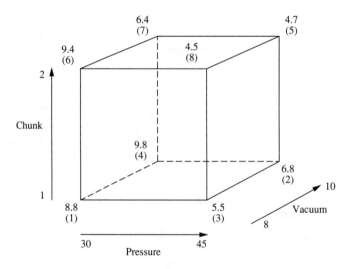

FIGURE 6.37
Cubes for welding experiment

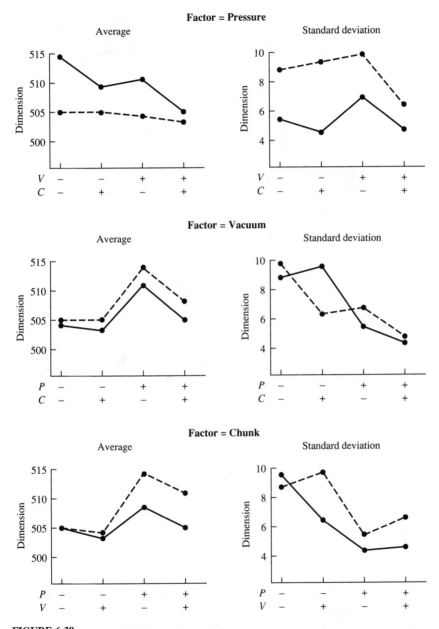

FIGURE 6.38
Plots of paired comparisons for Example 6.4

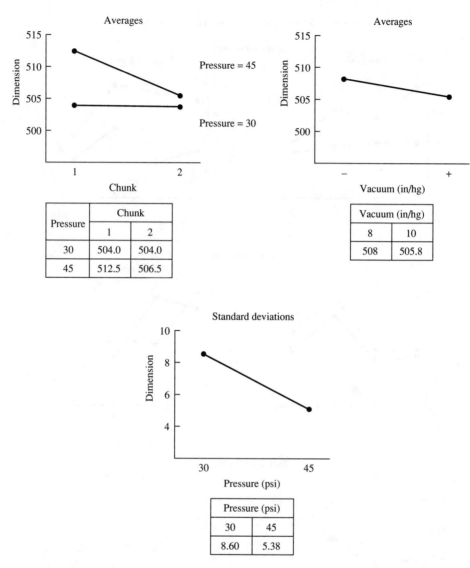

Chunk

Pressure	Chunk	
	1	2
30	504.0	504.0
45	512.5	506.5

Vacuum (in/hg)	
8	10
508	505.8

Pressure (psi)	
30	45
8.60	5.38

FIGURE 6.39
Response plots for Example 6.4

able degree of belief in the estimates of the effects of the factors can be established. This usually can be done through the use of control charts.

It may also be the case that certain settings of the factors can mitigate the effect of background and nuisance variables. This idea was pioneered by Taguchi (1979). In Figure 6.39, the response plots show the result that the chunk factor has less of an effect when pressure is low. If it is feasible in this case to set the pressure at the low level and confirm this result by future experiments, the effect of variation in the chunk factor could then be mitigated by low settings of pressure.

6.4 SUMMARY

In this chapter, factorial designs of experiments were presented. Some of the important features of factorial designs that were discussed are:

1. They are most useful to determine the effects of multiple factors (2, 3, or 4) on a response variable.
2. Interactions between the factors can be studied.
3. There are a number of advantages to studying the factors at two levels.
4. The use of graphical displays (run charts, cubes, paired comparisons, and response plots) can be used to simplify the analysis.

The references that follow contain additional discussion of factorial designs.

REFERENCES

Box, G. (1988): "Signal-to-Noise Ratios, Performance Criteria, and Transformations," *Technometrics*, vol. 30, no. 1, February.

Box, G., and N. Draper, (1969): *Evolutionary Operation*, John Wiley & Sons, New York.

Box, G., W. Hunter, and J.S. Hunter, (1978): *Statistics for Experimenters*, John Wiley & Sons, New York.

Daniel, C. (1976): *Applications of Statistics to Industrial Experimentation*, John Wiley & Sons, New York.

Kackar, R. N. (1985): "Offline Quality Control, Parameter Design, and the Taguchi Method," *Journal of Quality Technology*, vol. 17, pp. 176–188.

Taguchi, G. (1987): *System of Experimental Design*, Unipub/Krause International Publications, White Plains, N. Y.

EXERCISES

6.1. Choose a product or process with which you are familiar. For an important quality characteristic, list potential causes that affect either the average level or the variation of the quality characteristic. (Consider use of a cause-and-effect diagram to do this.) Plan a factorial experiment to determine the effects of some of the potential causes. Use the planning form.

6.2. The data in the following table were the result of an experiment to determine the effect that three factors—-holding pressure, booster pressure, and screw speed—had on part shrinkage in an injection molding process. The data in the table are coded values of part shrinkage.

	Holding pressure			
	−		+	
Screw speed	Booster pressure		Booster pressure	
	−	+	−	+
−	21.9 20.3 (2)	15.9 16.7 (8)	22.3 21.5 (1)	17.1 17.5 (6)
+	16.8 15.4 (7)	14.0 15.0 (3)	27.6 27.4 (5)	24.0 22.6 (4)

(a) Analyze the data to determine the important effects.

(b) Within the bounds of the study, which settings of the three factors minimize shrinkage?

(c) Based on the analysis in (b), suggest another design to see if better settings can be found.

(d) Assume that uniformity of the parts is the goal, with minimum shrinkage a secondary objective. Assume also that the inability to keep holding pressure constant during operation is a major cause of variation in the dimension of the parts. Choose settings to increase uniformity of the parts by desensitizing the process to variations in holding pressure.

6.3. In the solenoid experiment (Example 6.3) an additional response variable was measured. The response variable was pull-in voltage, which is the voltage needed to activate the solenoid. The average and standard deviation of the pull-in voltage for the four solenoids at each combination of the factors are indicated in the design matrix below. Analyze the data to determine the effects of the four factors.

Test	Run order	A	S	B	T	Pull-in voltage Average	Pull-in voltage Standard deviation
1	1	−	−	−	−	5.66	0.26
2	12	+	−	−	−	5.77	0.15
3	14	−	+	−	−	7.96	0.21
4	10	+	+	−	−	8.18	0.52
5	8	−	−	+	−	7.33	0.18
6	7	+	−	+	−	6.92	0.15
7	13	−	+	+	−	10.79	0.39
8	3	+	+	+	−	10.30	0.54
9	15	−	−	−	+	5.50	0.22
10	9	+	−	−	+	6.19	0.56
11	5	−	+	−	+	8.27	0.44
12	4	+	+	−	+	8.42	0.11
13	6	−	−	+	+	8.83	0.22
14	16	+	−	+	+	6.93	0.11
15	11	−	+	+	+	10.57	0.42
16	2	+	+	+	+	10.42	0.20

6.4. List some factors related to your customers (internal or external) that could potentially affect their needs. These factors could be demographic, such as age, sex, work experience, and place of residence. The factors might pertain to how the customer uses your product or service. Choose two to four of these factors and define two levels for each. Construct a factorial pattern using the factors and add a centerpoint. Use the pattern as a basis for selecting customers to interview to determine their present and future needs.

6.5. The following artificial data are given in Daniel (1976, p. 65) to illustrate the effect of one "bad value" in a factorial design. Find the test that seems to be influenced by a special cause. What value would be more consistent with the rest of the data?

Test	A	B	C	Response
1	−	−	−	158
2	+	−	−	132
3	−	+	−	212
4	+	+	−	136
5	−	−	+	264
6	+	−	+	188
7	−	+	+	318
8	+	+	+	242

CHAPTER
7

REDUCING
THE SIZE OF
EXPERIMENTS

The number of runs required by a full factorial design increases geometrically as the number of factors increases. Factorial designs with two or three factors make efficient use of the experimental resources by using all of the data to estimate the average effects and interactions. This is one of the strengths of factorial designs. However, as the number of factors increases, an increasing proportion of the data is used to estimate higher-order interactions. These interactions are usually negligible and therefore are of little interest to the experimenter.

For example, for a 2^4 factorial design, estimates of four main effects, six 2-factor interactions, four 3-factor interactions, and one 4-factor interaction can be made from the design. Only 10 of the 15 estimates available are likely to be of interest. Fractional factorial designs are an important class of experimental designs that allow the size of factorial experiments to be kept practical while still enabling the estimation of important effects.

The notation used for the experimental patterns in fractional factorial designs differs slightly from that used for full factorial designs. Whereas 2^7 indicates a full factorial pattern using seven factors in 128 runs, 2^{7-4} indicates a 1/16 fraction of a 2^7 pattern, using seven factors in 8 runs. The notation for describing fractional factorial patterns comes from the following:

$$1/16 \times 2^7 = 2^7/2^4 = 2^{7-4}$$

Figure 7.1 shows two tabular displays for fractional factorial patterns involving seven factors. These patterns are (1) a 1/8 fraction of a 2^7 pattern (2^{7-3}) in 16 runs, and (2) a 1/16 fraction of a 2^7 pattern (2^{7-4}) in 8 runs. The letters A through G indicate the seven factors, and the symbols $-$ and $+$ indicate the two levels of each factor. The darkened squares indicate the appropriate tests.

170

A 1/8 fraction of a 2^7 factorial design (2^{7-3})

E	F	G	A− −B− −C− −D+								A+ +B+ +C+ +D+							
			−D+	−D+	−D+	−D+	−D+	−D+	−D+	−D+	−D+	−D+	−D+	−D+	−D+	−D+	−D+	−D+
−	−	−	■															■
−	−	+				■									■			
−	+	−						■						■				
−	+	+							■			■						
+	−	−							■			■						
+	−	+						■						■				
+	+	−				■									■			
+	+	+	■															■

A 1/16 fraction of a 2^7 factorial design (2^{7-4})

E	F	G	A− −B− −C− −D+								A+ +B+ +C+ +D+							
			−D+	−D+	−D+	−D+	−D+	−D+	−D+	−D+	−D+	−D+	−D+	−D+	−D+	−D+	−D+	−D+
−	−	−													■			
−	−	+			■													
−	+	−						■										
−	+	+										■						
+	−	−							■									
+	−	+				■												
+	+	−		■														
+	+	+																■

FIGURE 7.1
Tabular displays for fractional factorial patterns

The choice to use a full factorial design or a specific fractional factorial design depends on the level of the current knowledge of the process or product. A high level of knowledge is indicated when all of the factors in the experiment are known to have a substantial effect on the response. When there is a high level of knowledge, the aim of the study is usually to obtain detailed information on the effects of the factors and their interactions. This is best done with a full factorial design.

A moderate level of knowledge is indicated by a strong belief that most but probably not all of the factors have a substantial effect on the response. Fractional factorial designs for these cases will be discussed in this chapter. A low level of knowledge is indicated by a belief that only a small proportion of the factors have a substantial effect on the response, and it is not known which ones they are. Fractional factorial designs that are useful for screening out the unimportant factors will also be discussed in this chapter.

The analysis of data from a fractional factorial design follows the methods used for a full factorial design given in Chapter 6. The run chart is used to screen for special causes of variation from nuisance variables. Next effects are estimated from the design matrix, and then geometric displays (e.g., cubes) are used to identify special causes and response plots are prepared for the important factors.

Section 7.1 presents the conceptual development of the experimental pattern for fractional factorial designs. An example of a study discussed in Chapter 6 is used to illustrate the analysis of fractional factorial designs. Section 7.2 presents fractional factorial designs to be used when the current knowledge of the significance and interrelationship of the factors under study is at a moderate level. Section 7.3 presents screening designs for use when the level of current knowledge is low. Section 7.4 discusses the use of incomplete blocks with factorial and fractional factorial designs.

7.1 INTRODUCTION TO FRACTIONAL FACTORIAL DESIGNS

Suppose the size of the 2^4 pattern in Table 7.1 is to be reduced to eight runs. It is important to choose the eight runs wisely so that as much information as possible about the effects of the factors is preserved. A simplistic approach is to choose the first eight runs. Inspection of the design matrix reveals that this results in a 2^3 factorial pattern for factors 1, 2, and 3, with factor 4 held constant at its minus level. Choosing the even-numbered runs results in a 2^3 factorial pattern in factors 2, 3, and 4 with factor 1 set at the plus level.

Both of these approaches lead to unsatisfactory loss of information about one of the four factors. How should the eight runs be chosen? Intuitively, it seems desirable that the eight runs should be chosen so that each of the columns in the design matrix has four minuses and four pluses. This selection would provide a balance to the design. It turns out that this balance can be achieved for 14 out of the 15 columns

TABLE 7.1
Design matrix for a 2^4 factorial pattern

Test	1	2	3	4	12	13	14	23	24	34	123	124	134	234	1234
1	−	−	−	−	+	+	+	+	+	+	−	−	−	−	+
2	+	−	−	−	−	−	−	+	+	+	+	+	+	−	−
3	−	+	−	−	−	+	+	−	−	+	+	+	−	+	−
4	+	+	−	−	+	−	−	−	−	+	−	−	+	+	+
5	−	−	+	−	+	−	+	−	+	−	+	−	+	+	−
6	+	−	+	−	−	+	−	−	+	−	−	+	−	+	+
7	−	+	+	−	−	−	+	+	−	−	−	+	+	−	+
8	+	+	+	−	+	+	−	+	−	−	+	−	−	−	−
9	−	−	−	+	+	+	−	+	−	−	−	+	+	+	−
10	+	−	−	+	−	−	+	+	−	−	+	−	−	+	+
11	−	+	−	+	−	+	−	−	+	−	+	−	+	−	+
12	+	+	−	+	+	−	+	−	+	−	−	+	−	−	−
13	−	−	+	+	+	−	−	−	−	+	+	+	−	−	+
14	+	−	+	+	−	+	+	−	−	+	−	−	+	−	−
15	−	+	+	+	−	−	−	+	+	+	−	−	−	+	−
16	+	+	+	+	+	+	+	+	+	+	+	+	+	+	+

in the design matrix in Table 7.1. To obtain this balance we must give up the ability to estimate one of the 15 effects represented by the column headings of the design matrix.

The least important estimate obtained from a full 2^4 design is the four-factor interaction. This estimate would usually be readily given up to reduce the size of the experiment. Fortunately, there is a simple way to choose the eight runs to provide the desired balance. The eight runs are chosen by selecting only the rows in the design matrix in which the sign in the column headed 1234 is + (or, alternatively, −). This results in the design matrix in Table 7.2. From this design matrix, it is seen that the desired balance, four minuses and four pluses in each column, has been obtained.

It would be fortunate if the only information lost by reducing the pattern in this manner was the estimate of the four-factor interaction and some precision of the remaining estimates. This is not the case. From Table 7.2 it is seen that columns 14 and 23 are identical, which means that the estimates of the two effects will be identical. The estimate provided by either column is actually an estimate of the sum of the 14 and the 23 interactions. These two interactions are said to be confounded or confused with each other.

Further inspection of Table 7.2 reveals other pairs of columns that are identical and therefore produce estimates that are actually sums of two effects. The confounding pattern for a one-half fraction of a 2^4 design is:

$$
\begin{array}{ll}
1 + 234 & 12 + 34 \\
2 + 134 & 13 + 24 \\
3 + 124 & 14 + 23 \\
4 + 123 &
\end{array}
$$

Estimates of the single factor effects can be obtained that are confounded only by three-factor interactions. A more serious confounding pattern exists for the two-factor interactions.

There is another justification for using this pattern besides the fact that the effect of the individual factors can be estimated without serious confounding. The pattern is such that any three of the four factors form a full factorial design. If any one of the factors has a negligible effect on the response, the other three can then be analyzed

TABLE 7.2
Design matrix for a 1/2 fraction of a 2^4 pattern

Test	1	2	3	4	12	13	14	23	24	34	123	124	134	234	1234
1	−	−	−	−	+	+	+	+	+	+	−	−	−	−	+
4	+	+	−	−	+	−	−	−	−	+	−	−	+	+	+
6	+	−	+	−	−	+	−	−	+	−	−	+	−	+	+
7	−	+	+	−	−	−	+	+	−	−	−	+	+	−	+
10	+	−	−	+	−	−	+	−	−	+	−	+	−	+	+
11	−	+	−	+	−	+	−	−	+	−	+	−	+	−	+
13	−	−	+	+	+	−	−	−	−	+	+	+	−	−	+
16	+	+	+	+	+	+	+	+	+	+	+	+	+	+	+

as a full 2^3 factorial design in the manner laid out in Chapter 6. This analysis is performed under the assumption that the factor having negligible effect also does not interact with any of the other factors. The lack of interaction cannot be determined by analysis of the data but is an assumption that must be made by experts in the subject matter.

Example 7.1 illustrates the analysis of a fractional factorial design.

Example 7.1 The solenoid experiment as a 2^{4-1} design. Suppose that the solenoid experiment in Chapter 6 was run as a 2^{4-1} rather than as a full 2^4. The experiment would then have included eight runs rather than sixteen runs. Table 7.3 contains the design matrix assuming a 2^{4-1} was run. The headings of the design matrix give the confounding pattern for the design. The property that any three factors form a full factorial pattern is given in parentheses below the title. The standard deviations of flow of the four solenoids at each of the eight factor combinations that would have been run if the pattern had only been a half fraction are listed. The run order from the full factorial experiment is also provided.

Figure 7.2 contains a list of effects estimated from the design matrix in Table 7.3 and a plot of these effects on a dot diagram. The substantial effect of bobbin depth on the standard deviation of flow is seen, as is the interaction of bobbin depth and armature length. This interaction is now confounded with the ST interaction.

Figure 7.2 also contains the dot diagram from the full factorial design of Chapter 6. Comparison of the two diagrams illustrates what is lost when the experimental design uses a 2^{4-1} pattern rather than a full 2^4 pattern.

1. The dots corresponding to the third- and fourth-order interactions (clustered around zero in the 2^4 dot diagram) are not available. These dots provided information on the effects of nuisance variables and are an aid to determine which are the important factors.

TABLE 7.3
Design matrix for the 2^{4-1} solenoid experiment (any three factors form a full factorial pattern)

Test	Run order	A	S	B	T	AT SB	ST AB	BT AS	Standard deviation of flow (cfm)
1	1	−	−	−	−	+	+	+	.04
2	9	+	−	−	+	+	−	−	.15
3	5	−	+	−	+	−	+	−	.07
4	10	+	+	−	−	−	−	+	.10
5	6	−	−	+	+	−	−	+	.02
6	7	+	−	+	−	−	+	−	.01
7	13	−	+	+	−	+	−	−	.05
8	2	+	+	+	+	+	+	+	.01

Divisor = 4

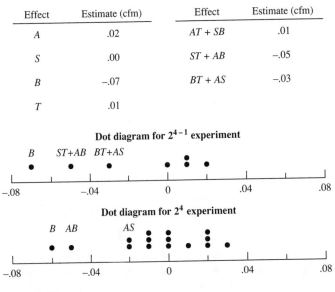

Effect	Estimate (cfm)
A	.02
S	.00
B	−.07
T	.01

Effect	Estimate (cfm)
AT + SB	.01
ST + AB	−.05
BT + AS	−.03

Dot diagram for 2^{4-1} experiment

Dot diagram for 2^4 experiment

FIGURE 7.2
Effects of factors on standard deviation of flows.

2. Two-factor interactions are confounded with each other. To allow further analysis of interactions, it will take an assumption by experts in the subject matter that one of the two interactions added together is small. Methods to separate the confounded interactions using additional tests are given in Box, Hunter, and Hunter (1978, p. 413).

3. Since there are only 8 tests in the fractional factorial pattern rather than the 16 tests in the full factorial pattern, there is less of an opportunity to average out the common causes of variation. This will make it more difficult to separate the effects of the nuisance variables from the factors. (*Note*: Comparison of the two dot diagrams in Figure 7.2 does not give a complete picture of the impact of this loss because the data for the fractional factorial pattern are a subset of the full factorial pattern, and therefore the two diagrams tend to look more alike than if the tests using the two patterns were independent.)

There are several ways for the experimenter to proceed from this point, and these alternatives will be discussed later in this chapter. Knowledge of the product or the process and analysis of the data are needed to decide among the alternatives. At this point it seems reasonable that the experimenter should be pleased to have detected that the combination of a small bobbin depth along with a long armature is to be avoided. Figure 7.3 contains the square obtained by only considering bobbin depth and armature length. (Either tube length or spring load could also have been included and the data displayed on a cube.) Figure 7.4 contains the response plot for the *AB* interaction.

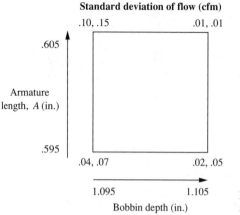

FIGURE 7.3
Square for the solenoid experiment

Future experiments could be used to confirm that the large interaction is in fact *AB* rather than *ST*. Also the experimenter may use future experiments to determine more about the moderate interaction $(-.03)$ associated with $BT + AS$.

7.2 FRACTIONAL FACTORIAL DESIGNS — MODERATE CURRENT KNOWLEDGE

A team of experimenters is said to have a moderate level of current knowledge if they possess a strong belief that most but probably not all of the factors have a substantial effect on the response. This belief may turn out to be wrong. Then the experiment that is designed assuming a moderate level of knowledge may not be the best choice. Nevertheless, when an experimenter is choosing a particular fractional factorial design, he or she should try to assess the current level of knowledge and base the choice in part on this assessment.

Given a moderate level of knowledge about the process or product, there are four fractional factorial patterns that are very useful. These patterns are recommended for three reasons:

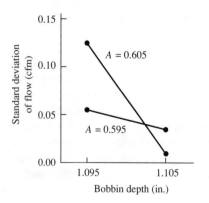

FIGURE 7.4
Response plots for the solenoid experiment

TABLE 7.4
Fractional factorial designs—moderate current knowledge

Experimental pattern	Number of factors	Number of runs	Confounding[a]	Number of factors forming a full factorial design
2^{4-1}	4	8	2fi with 2fi	Any 3
2^{5-1}	5	16	None	Any 4
2^{8-4}	6, 7, 8	16	2fi with 2fi	Any 3 and some groups of 4
2^{16-11}	9–16	32	2fi with 2fi	Any 3 and some groups of 4 or 5

[a] Only confounding of individual factors or their two-factor interactions (2fi) has been considered.

1. They can accommodate 4 to 16 factors.
2. They can be performed in 16 runs or less for up to eight factors and 32 runs for nine to sixteen factors.
3. The effects of the individual factors are only confounded with third-order or higher interactions.

The four patterns along with some of their properties are listed in Table 7.4.

The 2^{4-1} pattern has been discussed previously and its design matrix has been shown in Table 7.2. There are numerous alternatives after the data from this design have been analyzed. Some of these alternatives are contained in Table 7.5.

The design matrix for the 2^{5-1} pattern is given in Table 7.6. This design is a particularly useful one. Any four factors form a full factorial design. Also, individual factors and two-factor interactions are only confounded with three factor or higher interactions. Because this design is so powerful, it is rare that a full 2^5 design would be a better choice. The 2^{5-1} design can be used for situations in which the experi-

TABLE 7.5
Follow-up to a 2^{4-1} experiment

Number of Important Factors[a]	Alternatives
Less than 4	1. Analyze as a full factorial and run other full factorial designs to study interactions and find better operating conditions. 2. Run another 2^{4-1} at different levels of the factors to increase degree of belief in the results and find better operating conditions.
All four	1. Run a 2^4 design at different levels of the factors to find better operating conditions and estimate interactions. 2. Run the other half of the original 2^{4-1} to estimate interactions. This results in a full 2^4 design in two blocks of eight. (See Section 7.4 for details on blocking.)

[a] The number of factors found to have a substantial effect on the response variable based on estimating these effects from the design matrix.

TABLE 7.6
Design matrix for a 2^{5-1} pattern (any four factors form a full factorial pattern)

Test	1	2	3	4	5	12	13	14	15	23	24	25	34	35	45
1	−	−	−	−	+	+	+	+	−	+	+	−	+	−	−
2	+	−	−	−	−	−	−	−	−	+	+	+	+	+	+
3	−	+	−	−	−	−	+	+	+	−	−	−	+	+	+
4	+	+	−	−	+	+	−	−	+	−	+	+	+	−	−
5	−	−	+	−	−	+	−	+	+	−	+	+	−	−	+
6	+	−	+	−	+	−	+	−	+	−	+	−	−	+	−
7	−	+	+	−	+	−	−	+	−	+	−	+	−	+	−
8	+	+	+	−	−	+	+	−	−	+	−	−	−	−	+
9	−	−	−	+	−	+	+	−	+	+	−	+	−	+	−
10	+	−	−	+	+	−	−	+	+	+	−	−	−	−	+
11	−	+	−	+	+	−	+	−	−	−	+	+	−	−	+
12	+	+	−	+	−	+	−	+	−	−	+	−	−	+	−
13	−	−	+	+	+	+	−	−	−	−	−	−	+	+	+
14	+	−	+	+	−	−	+	+	−	−	−	+	+	−	−
15	−	+	+	+	−	−	−	−	+	+	+	−	+	−	−
16	+	+	+	+	+	+	+	+	+	+	+	+	+	+	+

Divisor = 8

menters have either moderate or high current knowledge of the process. Subsequent experiments are usually needed only to establish optimal settings of the factors or to increase the degree of belief by including different background variables.

The design matrix for a 2^{8-4} pattern is contained in Table 7.7. Any three factors form a full factorial pattern. Although not all subsets of four factors form a full factorial pattern, there are some subsets of four that do. One of these subsets contains the four factors associated with the first four columns of the design matrix. Therefore, the four factors thought to be most likely to have a substantial effect on the response should be listed in the first four columns. The individual factors are confounded only with three-factor or higher interactions. The two-factor interactions are confounded with other two-factor interactions.

Although the design matrix stipulates eight factors, this experimental pattern can be used for six, seven, or eight factors. For example, if the pattern is used to study seven factors, the 16 tests are given by the first seven columns. The pattern is then a 2^{7-3}. Column 8 becomes a measure of the effect of nuisance variables since the effect estimated from this column consists of three-factor or higher interactions. The confounding pattern is studied by ignoring any two-factor interactions that contain an 8.

If the pattern is used for six factors, the first six columns provide the 16 tests and the pattern is a 2^{6-2}. Columns 7 and 8 provide a measure of the effect of nuisance variables. The confounding pattern is studied by ignoring any interactions containing a 7 or an 8.

There are numerous alternatives after the data from this design have been analyzed. Some of these alternatives are contained in Table 7.8.

TABLE 7.7
Design matrix for a 2^{8-4} pattern (any three factors form a full factorial pattern)

Test	1	2	3	4	5	6	7	8	12 37 56 48	13 27 46 58	14 36 57 28	15 26 47 38	16 25 34 78	17 23 45 68	24 35 67 18
1	−	−	−	+	+	+	−	+	+	+	−	−	−	+	−
2	+	−	−	−	−	+	+	+	−	−	−	−	+	+	+
3	−	+	−	−	+	−	+	+	−	+	+	−	+	−	−
4	+	+	−	+	−	−	−	+	+	−	+	−	−	−	+
5	−	−	+	+	−	−	+	+	+	−	−	+	+	−	−
6	+	−	+	−	+	−	−	+	−	+	−	+	−	−	+
7	−	+	+	−	−	+	−	+	−	−	+	+	−	+	−
8	+	+	+	+	+	+	+	+	+	+	+	+	+	+	+
9	+	+	+	−	−	−	+	−	+	+	−	−	−	+	−
10	−	+	+	+	+	−	−	−	−	−	−	−	+	+	+
11	+	−	+	+	−	+	−	−	−	+	+	−	+	−	−
12	−	−	+	−	+	+	+	−	+	−	+	−	−	−	+
13	+	+	−	−	+	+	−	−	+	−	−	+	+	−	−
14	−	+	−	+	−	+	+	−	−	+	−	+	−	−	+
15	+	−	−	+	+	−	+	−	−	−	+	+	−	+	−
16	−	−	−	−	−	−	−	−	+	+	+	+	+	+	+

Divisor = 8

The design matrix for a 2^{16-11} pattern is contained in Table 7.9. Any three factors form a full factorial pattern. Also, some subsets of four or five factors form a full factorial pattern. In particular, the five factors corresponding to the first five factors in the design matrix form a full factorial pattern. Therefore, the five factors thought to be the most likely to have a substantial effect on the response should be listed in the first five columns. The individual factors are confounded only with three-factor or higher interactions.

TABLE 7.8
Follow-up to a 2^{8-4} experiment

Number of Important Factors[a]	Alternatives
Any 3 or any subsets of 4 that form a full factorial pattern	Analyze as a full factorial and run other full factorial designs to improve levels or increase degree of belief.
Any 4 not forming a full factorial	Run a 2^{4-1} to study better levels or a 2^4 to study better levels and interations.
5	Run a 2^{5-1} or a full factorial to study better levels and interactions.
6, 7, or 8	Run a 2^{8-4} to study better levels or focus on a subset of 5 or less to study better levels or interactions.

[a] The number of factors found to have a substantial effect on the response variable based on estimating these effects from the design matrix.

TABLE 7.9
Design matrix for a 2^{16-11} pattern (any three factors form a full factorial design)

```
                  1 1 1 1 1 1 1
Test 1 2 3 4 5 6 7 8 9 0 1 2 3 4 5 6 ......... Two-factor interactions .........
  1  − − − − + + + + + + − − − − + +   − − − − − + + + + + + + − − − − +
  2  + − − − − − − − + + + + + + + −   − + + + − − − − − − + + + + + − −
  3  − + − − − + + − − + + + − + − +   − + − + − − + + − − + + + − + −
  4  + + − − + − − − − − + − − + + +   + + − − + − − − + − − + + +
  5  − − + − + − + − + − + − + + −     + − − + − + − + − + − + + +
  6  + − + − − + − − + − − + − + + +   + + − + − − + − − + − − + − + +
  7  − + + − − − + + − − − + + − + +   − + + − − − + + − − + + − +
  8  + + + − + + − + − − + − − − − +   + + + − + + − + − − + − − − −
  9  − − + + + − + − − − + + + − + −   − − + + + − + − − − + + + −
 10  + − − + − − + + − − + − − + + +   + − − + − − + + − − + − − + +
 11  − + − + − + − − + − + − + − + +   − + − + − + − − + − + − + − +
 12  + + − + + − + − + − − + − − + +   + − + + − + + − + − + − − + −
 13  − − + + + − − − − + + + − − + +   − − + + + − − − − + + + − − +
 14  + − + + + − + − − + − − + − − +   + − + + − + + − + + − + − − + −
 15  − + + + − − − + + + − − − + − +   − + + + − − − + + + − − − + −
 16  + + + + + + + + + + + + + + + +   + + + + + + + + + + + + + + + +
 17  + + + + − − − − − + + + + − − −   − − + + + + + + + − − − − + +
 18  − + + + + + + − − − − + + − + −   − − − − + + + + + + − − − −
 19  + − + + + − − + + + − − − + − +   − + − − + + − − + + + − + −
 20  − − + + − + + + + − + + − + − −   + + − + − − − + − − + + +
 21  + + − + − + − + − + − + − − + −   + − + − + − + − + − + + + −
 22  − + − + + − + + − + + − + − − −   + − + − − + − + − + − − + + +
 23  + − − + + + − − + + + − − + − −   − + + − − + + − − − + + − +
 24  − − − + − − + − + + − + + + + −   + + + − + + − + − − + − − −
 25  + + + − − − + − + + + − − − + −   − − − + + + − + − − − + + + −
 26  − + + − + + − − + + − + + − − −   + − − + − − + + − − + − − + +
 27  + − + − + − + + − + − + − + − −   − + − + − + − − + − + − + − +
 28  − − + − − + − + − + + − + + + −   + + − + + − + + − + − − + − −
 29  + + − − − + + + + − − − + + − −   − − + + + − − − − + + + − − +
 30  − + − − + − + + − + + − + + − +   − + + − + + − − + − − + − − −
 31  + − − − + + + − − − + + + − + −   − + + + − − + + + − − − + −
 32  − − − − − − − − − − − − + + + +   + + + + + + + + + + + + + + +
Divisor = 16
```

The two-factor interactions are confounded with other two-factor interactions. Each of the last 15 columns of the design matrix correspond to the sum of eight 2-factor interactions. The eight interactions for each column have not been enumerated. For other patterns where two-factor interactions were confounded, the confounding patterns were listed. This was done to allow the study of interactions to proceed if the experts in the subject matter were willing to assume that factors having negligible effect on the response do not interact with the important factors. Because there are eight interactions confounded with each other, it is usually impractical to make that assumption. The effects for each of the columns should still be computed and plotted on the dot diagram.

The design matrix stipulates 16 factors but can also be used for 9 to 15 factors. The design matrix is used for less than 16 factors in the same way that the design matrix for a 2^{8-4} pattern was used for less than 8 factors. The alternatives for follow-up to this design are given in Table 7.10.

The example that follows illustrates the analysis of a 2^{5-1} fractional factorial design.

Example 7.2 2^{5-1} **design for a welding process.** To assemble a part for an automotive engine, two sets of components were welded together. An important quality characteristic of the part was the force it took to pull the two sets of components apart after they had been welded together (pull force). Through the use of a control chart the process had been stabilized with respect to pull force. Steps had been taken to remove special causes of variation associated with the measurement and the cleanliness of the components as they came to the welding station.

Based on the control chart, the process average and standard deviation were determined to be 1100 inch-pounds (in-lbs) and 61 in-lbs respectively. If it remained stable, the process was capable of meeting the lower specification of 900 in-lbs. However the plant personnel desired to raise the average pull force to decrease the chances that a special cause would result in unacceptable product. They attempted to do this by studying the effect that variables in the welding process had on pull force.

The five factors studied were:

Factor	Level	
	$-$	$+$
H = Heat (°C)	80	95
P = Pressure (psi)	50	75
W = Weld Time (sec.)	25	30
T = Hold Time (sec.)	10	20
S = Squeeze Time (sec.)	45	55

The levels of the factors were set approximately equidistant from the current operating conditions.

Figure 7.5 contains the planning form for the experiment.

TABLE 7.10
Follow-up to a 2^{16-11} experiment

Number of important factors	Alternatives
Any 3 or any subset of 4 or 5 that form a full factorial pattern	Analyze as a full factorial design and run other full of fractional factorial designs to improve levels or increase degree of belief.
Any 4 not forming a full factorial pattern	Run a 2^{4-1} to study better levels or a 2^4 to study better levels or interactions.
Any 5 not forming a full factorial pattern	Run a 2^{5-1} to study better levels or interactions.
6 through 16	Run a fractional factorial design to increase degree of belief or study better levels. Run a full factorial design on a subset to study better levels or interactions.

FIGURE 7.5
Documentation of the welding experiment

1. **Objective:**

Study the effects that five factors have on pull force. Use the results to set operating conditions for the process of resistance welding of cans to brackets.

2. **Background information:**

The process is stable with average 1100 in-lbs and standard deviation 61 in-lbs. The measurement process and the lack of cleanliness have been found to be sources of special causes of variation in pull force in the past.

3. **Experimental variables:**

A. Response variables	Measurement technique
1. Pull force	Dillon force gage

B. Factors under study	Levels	
1. Heat (°C)	80	95
2. Pressure (psi)	5.0	7.5
3. Weld Time (sec.)	2.5	3.0
4. Hold Time (sec.)	1.0	2.0
5. Squeeze Time (sec.)	4.5	5.5

C. Background variables	Method of control
1. Conditions of can and bracket	Clean before welding.
2. Condition of electrodes	Start with new ones.
3. Operator pace	Record temperature of electrodes before each part.
4. Ambient temperature	Record.

4. **Replication:**

One part per factor combination

5. **Methods of randomization:**

Randomly select components, the order of assembling the parts, and the order of testing the parts.

6. **Design matrix:** (attach copy)

2^{5-1} (see Table 7.11.)

7. **Data collection forms:** (not shown)

8. **Planned methods of statistical analysis:**

Analyze the effects of the five factors using the design matrix, and collapse to a full factorial design if one or more factors are unimportant.

9. **Estimated cost, schedule, and other resources:**

Three person-hours administrative time to prepare the components, instructions for the operator and data collection sheets. Less than one half hour of assembly time.

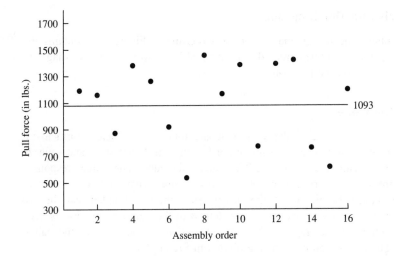

FIGURE 7.6
Run chart in order of assembly

Run Charts

Since the parts were assembled first and then tested for pull force later, the order of assembly and of test were determined using two different sets of random numbers. Therefore, two different run charts were plotted. The run chart in order of assembly is contained in Figure 7.6. The run chart in order of test is contained in Figure 7.7.

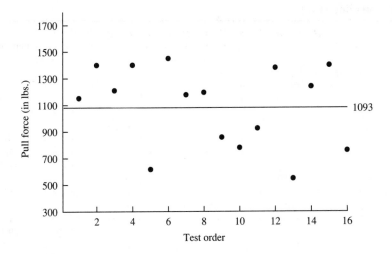

FIGURE 7.7
Run chart in order of test

Design Matrix and Dot Diagram

Table 7.11 contains the design matrix for the experiment. Figure 7.8 contains the effects of the factors computed from the design matrix as well as the dot diagram. Effects of the factors on the pull force were estimated.

Analysis of the Cube

Since a characteristic of a 2^{5-1} design is that any four or less factors form a full factorial design, the results using the important factors—heat, pressure, and squeeze time—are analyzed using a cube. Figure 7.9 contains the cube. The sixteen responses of pull force from the experiment are placed on the cube. There are two responses corresponding to each combination of the factors. Since the variation between the two responses at each factor combination is consistent with the variation previously seen on the control chart, the average of the two are used in the analysis of the paired comparisons. The paired comparisons are plotted in Figure 7.10.

Analysis of the paired comparisons indicates that the pair of measurements 871,920 corresponding to $P = 50-$, $H = 95+$, and $S = 4.5-$ are primarily responsible for the relatively large magnitude of the effect of squeeze time and its interactions with heat and pressure. If these measurements were around 1300 rather than 900, the variation in the response could be explained simply by an interaction between pressure and heat.

The process may be as complex (three important interactions) as the original analysis suggests or the apparent complexity could be a result of a special cause. To further investigate this potential special cause, checks of the measurement process and

TABLE 7.11
Design matrix for the welding study

Test	Build order	Test order	H	P	W	T	S	H P	H W	H T	H S	P W	P T	P S	W T	W S	T S	Pull force
1	16	8	−	−	−	−	+	+	+	+	−	+	+	−	+	−	−	1194
2	3	9	+	−	−	−	−	−	−	−	−	+	+	+	+	+	+	871
3	14	16	−	+	−	−	−	−	+	+	+	−	−	−	+	+	+	764
4	8	6	+	+	−	−	+	+	−	−	+	−	−	+	+	−	−	1463
5	1	3	−	−	+	−	−	+	−	+	+	−	+	+	−	−	+	1205
6	5	14	+	−	+	−	+	−	+	−	+	−	+	−	−	+	−	1256
7	15	5	−	+	+	−	+	−	−	+	−	+	−	+	−	+	−	616
8	4	12	+	+	+	−	−	+	+	−	−	+	−	−	−	−	+	1384
9	2	1	−	−	−	+	−	+	+	−	+	+	−	+	−	+	−	1152
10	12	4	+	−	−	+	+	−	−	+	+	+	−	−	−	−	+	1398
11	7	13	−	+	−	+	+	−	+	−	−	−	+	+	−	−	+	533
12	10	2	+	+	−	+	−	+	−	+	−	−	+	−	−	+	−	1382
13	9	7	−	−	+	+	+	+	−	−	−	−	−	−	+	+	+	1170
14	6	11	+	−	+	+	−	−	+	+	−	−	−	+	+	−	−	920
15	11	10	−	+	+	+	−	−	−	−	+	+	+	−	+	−	−	776
16	13	15	+	+	+	+	+	+	+	+	+	+	+	+	+	+	+	1410

Divisor = 8

Factor	Effect	Interaction	Effect
H	334	HP	403
P	−105	HW	−34
W	−3	HT	35
T	−2	HS	169
S	73	PW	14
		PT	−30
		PS	−144
		WT	−45
		WS	−31
		TS	−3

Dot diagram

FIGURE 7.8
Effect of factors on pull force

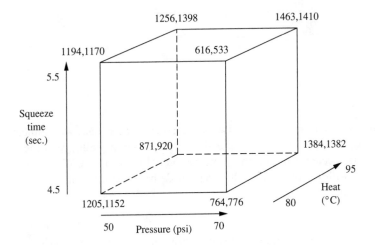

FIGURE 7.9
Cube for the welding experiment

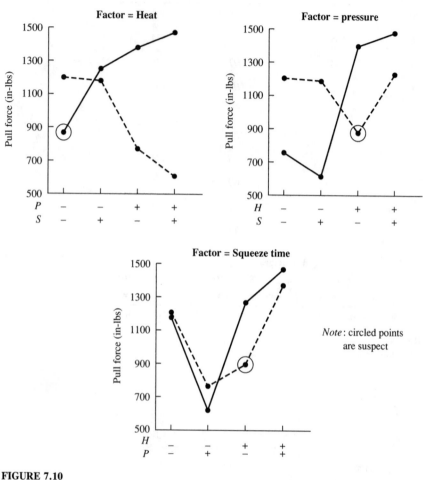

—— indicates + level; - - - indicates – level

FIGURE 7.10
Plot of paired comparisons

the assembly process should be made. The two measurements were made at a similar time (9th and 11th), and they were built at a similar time (3rd and 6th). Was there anything specific that occurred during that time period? How do the results of this experiment compare with the predictions that were made before it was run?

Response Plots

Figure 7.11 contains the response plots that summarize the results of the experiment. The validity of these plots depends on the assumption that there was no special cause of variation present during the experiment. Since the three important factors all interact in pairs, the three factors are used on the response plots to better display the relationship

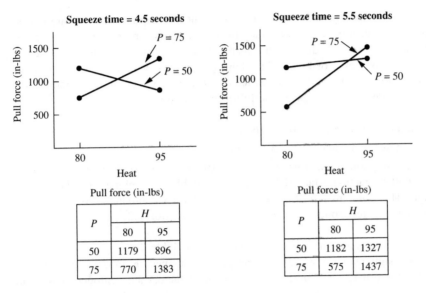

FIGURE 7.11
Response plots for the welding experiments

between them. The plots show the relationship between heat and pressure separately when squeeze time is 4.5 seconds and when squeeze time is 5.5 seconds.

Conclusions for Example 7.2

1. Three factors and three interactions between the factors were found to be important. The important factors were heat, pressure, and squeeze time.

2. The response plots displayed the interactions between the important factors. From these plots it is seen that the prediction of the average pull force that would be obtained at the current operating conditions (half way between the levels chosen for the experiment) is consistent with the 1100 in-lbs determined from the control chart. The plots also show that a substantial increase in pull force can be obtained by using high squeeze time and high heat. The combination of high heat and high pressure along with low squeeze time also results in considerable increase in pull force.

3. There is another plausible explanation of the variation in the data. There is a strong pressure-heat interaction, and there was a special cause present during the experiment. Past and future tests and knowledge of the process should be used to determine which of the two explanations provides the better model of the cause-and-effect system in the process.

7.3 FRACTIONAL FACTORIAL DESIGNS—LOW CURRENT KNOWLEDGE

A low level of knowledge is indicated by a belief that only a small proportion of the possible factors have a substantial effect on the response, and it is not known which ones they are. When a team of experimenters has a low level of knowledge, the aim of the first experiment is usually to screen out unimportant factors. Subsequent experiments can then be planned to study the important factors in more depth. In this section fractional factorial designs to screen out unimportant factors will be given.

Given a low level of knowledge about the process or product, there are three fractional factorial patterns that are particularly useful to screen out unimportant factors. These patterns are chosen primarily for three reasons:

1. They require relatively few runs (16 or less).
2. They can accommodate up to 15 factors.
3. If the assumption is incorrect that only a small number of factors is important, then another fractional factorial design is available to combine with the original design. This combination of designs will then have a confounding pattern similar to the designs for moderate knowledge given in Section 7.2.

These designs and their properties are given in Table 7.12.

The design matrix for the 2^{3-1} design is contained in Table 7.13. Any two factors form a full factorial design. The individual factors are confounded with the two-factor interactions. Because of the small number of runs in this design, the design is useful for applications to the design of a product. Since it is often expensive to build prototypes for a product, testing of alternative, possibly improved designs is sometimes not done. The 2^{3-1} design provides a relatively inexpensive way to test different product designs.

If one of the factors turns out to be unimportant, then the remaining two factors can be analyzed as a full factorial design. If all three factors turn out to be important, then four more tests can be made. These four tests can be chosen so that when they are analyzed in combination with the original four tests, the individual factors are not confounded with the two-factor interactions. With the addition of these four runs, the pattern becomes a 2^3 in two blocks. The design matrix and the properties of the 2^3 in two blocks are given in Section 7.4.

TABLE 7.13
Design matrix for a 2^{3-1} pattern
(any two factors form a full factorial pattern)

Test	1 23	2 13	3 12
1	−	−	+
2	+	−	−
3	−	+	−
4	+	+	+
Divisor = 2			

TABLE 7.12
Fractional factorial designs—low current knowledge

Experimental pattern	Number[a] of factors	Number of runs	Confounding[b]	Number of factors forming a full factorial design
2^{3-1}	3	4	Individual factors with 2fi.	2
2^{7-4}	5, 6, 7	8	Individual factors with 2fi.	Any 2 and some groups of 3.
2^{15-11}	8–15	16	Individual factors with 2fi.	Any 2 and some groups of 3 or 4.

[a] With four factors, use a 2^{4-1} design (given in Section 7.2).

[b] Only confounding of individual factors or their two-factor interactions (2fi) have been given.

The design matrix for a 2^{7-4} design is given in Table 7.14. Any two factors form a full factorial pattern. Although any arbitrary group of three factors do not form a full factorial pattern, there are some groups of three factors that do. In particular, the first three factors form a full factorial pattern. The three factors most likely to have a substantial effect on the response should be listed in the first three columns. The effects of individual factors are confounded with two-factor interactions. Using this design, the experimenters are looking for two or three factors with large effects relative to the others in the study.

If more factors than expected turn out to be important or if there is other uncertainty that needs to be resolved, eight more tests can be made. These eight tests can be chosen so that when they are analyzed in combination with the original eight tests, the individual factors are not confounded with the two-factor interactions. With the addition of these eight tests, the pattern becomes a 2^{7-3} in two blocks. The design matrix for this pattern and its properties are given in Section 7.4.

There are many alternatives for subsequent experiments after the data from a 2^{7-4} is analyzed. Table 7.15 contains some of these alternatives.

Table 7.16 contains the design matrix for a 2^{15-11} fractional factorial pattern. Any two factors and some groups of three or four factors form a full factorial pattern. In particular, the first four factors form a full factorial pattern. The four factors that the experimenters think are most likely to have a substantial effect on the response variable should be listed in the first four columns of the design matrix.

Each of the individual factors are confounded with seven 2-factor interactions. The confounding between the individual factors and the two-factor interactions can be resolved by adding another block of eight tests. The pattern then becomes a 2^{15-10} in two blocks. The design matrix and the properties of this pattern are given in Section 7.4.

TABLE 7.14
Design matrix for a 2^{7-4} pattern
(Any two factors form a full factorial design)

Test	1 24 35 67	2 14 36 57	3 15 26 47	4 12 56 37	5 13 46 27	6 23 45 17	7 34 25 16
1	−	−	−	+	+	+	−
2	+	−	−	−	−	+	+
3	−	+	−	−	+	−	+
4	+	+	−	+	−	−	−
5	−	−	+	+	−	−	+
6	+	−	+	−	+	−	−
7	−	+	+	−	−	+	−
8	+	+	+	+	+	+	+

Divisor = 4

TABLE 7.15
Follow-up to a 2^{7-4} experiment

Number of important factors[a]	Alternatives
Any 2, or any subset of 3 that forms a full factorial pattern	Analyze as a full factorial design and run other full factorial designs to improve levels or increase degree of belief.
	Run another block of eight tests to resolve uncertainty about which are the important factors (see Section 7.4).
Any 3 that do not form a full factorial pattern, or 4	Run another block of eight tests to resolve uncertainty about which are the important factors.
	Run a 2^{4-1} to study better levels.
	Run a full factorial design to study better levels and interactions.
5	Run a 2^{5-1} to study better levels or interactions.
	Run another block of eight tests to resolve uncertainty about which are the important factors.
6 or 7	Run another block of eight tests to resolve uncertainty about which are the important factors.

[a] The number of factors found to have a substantial effect on the response variable based on estimating these effects from the design matrix.

TABLE 7.16
Design matrix for a 2^{15-11} pattern (any two factors form a full factorial pattern)

Test	1	2	3	4	5	6	7	8	9	10	11	12	13	14	15
							Two-factor interactions								
1	−	−	−	−	+	+	+	+	+	+	−	−	−	−	+
2	+	−	−	−	−	−	−	+	+	+	+	+	+	−	−
3	−	+	−	−	−	+	+	−	−	+	+	+	−	+	−
4	+	+	−	−	+	−	−	−	−	+	−	−	+	+	+
5	−	−	+	−	+	−	+	−	+	−	+	−	+	+	−
6	+	−	+	−	−	+	−	−	+	−	−	+	−	+	+
7	−	+	+	−	−	−	+	+	−	−	−	+	+	−	+
8	+	+	+	−	+	+	−	+	−	−	+	−	−	−	−
9	−	−	−	+	+	+	−	+	−	−	−	+	+	+	−
10	+	−	−	+	−	−	+	+	−	−	+	−	−	+	+
11	−	+	−	+	−	+	−	−	+	−	+	−	+	−	+
12	+	+	−	+	+	−	+	−	+	−	−	+	−	−	−
13	−	−	+	+	+	−	−	−	−	+	+	+	−	−	+
14	+	−	+	+	−	+	+	−	−	+	−	−	+	−	−
15	−	+	+	+	−	−	−	+	+	+	−	−	−	+	−
16	+	+	+	+	+	+	+	+	+	+	+	+	+	+	+

Divisor = 8

TABLE 7.17
Follow-up to a 2^{15-11} experiment

Number of important factors[a]	Alternatives
Any 2, or any group of 3 or 4 that form a full factorial pattern	Analyze as full factorial design and run other full factorial designs to improve levels or increase degree of belief.
	Run another block of 16 tests to resolve uncertainty about which are the important factors (see Section 7.4).
Any 3 or 4 that do form a full factorial pattern	Run a full factorial design to study better levels and interactions.
	Run a 2^{4-1} to study better levels.
	Run another block of 16 tests to resolve uncertainty about which are the important factors.
5	Run a 2^{5-1} to study better levels or interactions.
	Run another block of 16 tests to resolve uncertainty about which are the important factors.
6 through 15	Run another block of 16 tests to resolve uncertainty about which are the important factors.

[a]The number of factors found to have a substantial effect on the response variable based on estimating these effects from the design matrix.

There are many alternatives for subsequent experiments after a 2^{15-11} design has been run. Some of these alternatives are contained in Table 7.17.

The example that follows illustrates the use of a fractional factorial design to screen out unimportant factors.

Example 7.3 2^{7-4} Design for tensile strength of rivets. An automotive part contained several components that were riveted together. The tensile strength of the part was an important quality characteristic. A study was designed by the product engineers responsible for the design of the part to determine how various configurations of the components affected tensile strength. Figure 7.12 contains the documentation of the experiment.

Run Charts

No special causes seem to be present in the run chart contained in Figure 7.13. Since the variation within each replication was small compared to the overall variation in the run chart, the four values of tensile strength at each factor combination were averaged and the averages used in the design matrix to determine the effects.

Design Matrix and Dot Diagrams

Table 7.18 contains the design matrix for the experiment and the effects of the factors. Figure 7.14 contains the dot diagram.

FIGURE 7.12
Documentation of the experiment

1. Objective:

Study the effects that seven factors have on the tensile strength of stem diaphragm plate rivets. The results will be used to determine the nominal values and tolerances for the important components.

2. Background information:

The customer's specification for tensile strength was 480 in-lbs minimum. Prototype parts built to nominal dimensions were able to meet this specification. Little information is available on the effect on tensile strength of variation in the components.

3. Experimental variables:

A.	Response variables	Measurement technique
	1. Tensile strength (in-lbs)	Pull tester

B.	Factors under study	Levels	
	1. RH = Rivet height (in.)	.015	.025
	2. PD = plate i.d. (in.)	.128	.132
	3. PT = Plate thickness (in.)	.030	.036
	4. SD = Stem rivet diameter (in.)	.123	.125
	5. SL = Stem rivet length (in.)	.200	.210
	6. BD = Bushing i.d. (in.)	.129	.133
	7. BT = Bushing thickness (in.)	.095	.105

C.	Background variables	Method of control
	1. Operators (riveters and testing)	Hold constant
	2. Time	Randomization

4. Replication:

Four for each factor combination.

5. Methods of randomization:

The order of the eight factor combinations was randomized using a random number table.

6. Design matrix: (attach copy)

2^{7-4}. see Table 7.18.

7. Data collection forms: (not shown here)

8. Planned methods of statistical analysis:

The variation within each factor combination is expected to be small relative to the effects of the factors. If so the analysis will be performed on the averages of the four tensile strength readings for each of eight factor combinations.

9. Estimated cost, schedule, and other resource considerations:

One day of administrative time is needed to organize the running of the experiment. Two hours of an operator's time will be needed to assemble the parts.

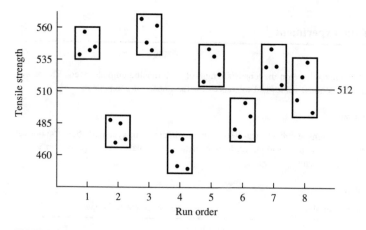

FIGURE 7.13
Run chart (2^{7-4} design)

TABLE 7.18
2^{7-4} **Design matrix for the rivet study**

		1	2	3	4	5	6	7	
		\multicolumn — 1 = RH 3 = PT 5 = SL 7 = BT							
		2 = PD 4 = SD 6 = BD							

Test	Run order	1 24 35 67	2 14 36 57	3 15 26 47	4 12 56 37	5 13 46 27	6 23 45 17	7 34 25 16	Tensile strength $\overline{X}$
1	8	−	−	−	+	+	+	−	513
2	4	+	−	−	−	−	+	+	461
3	6	−	+	−	−	+	−	+	488
4	2	+	+	−	+	−	−	−	481
5	7	−	−	+	+	−	−	+	523
6	3	+	−	+	−	+	−	−	558
7	5	−	+	+	−	−	+	−	532
8	1	+	+	+	+	+	+	+	546

Divisor = 4

Effect		−2.5	−2.0	54.0	6.0	27.0	0.5	−16.5	

FIGURE 7.14
Dot diagram (2^{7-4} design)

Analysis of the Cube

As has been indicated, a characteristic of a 2^{7-4} pattern is that there are some groups of three factors that form a full factorial pattern. In this case, the important factors, plate thickness, stem rivet length, and bushing thickness do form a full factorial pattern. Therefore, they can be analyzed on a cube. The cube is contained in Figure 7.15.

The paired comparisons are plotted in Figure 7.16. The analysis of the paired comparisons does not indicate the presence of any special causes of variation and therefore confirms the effects that were found using the design matrix.

Figure 7.15 contained the alternatives for subsequent experiments after a 2^{7-4} when three important factors formed a full factorial pattern. In this case, it was decided to follow up the 2^{7-4} by running another block of eight tests to separate the confounding between factors and interactions and, therefore, to resolve uncertainty about which are the important factors. With the addition of these eight tests, the pattern becomes a 2^{7-3} in two blocks. This pattern is discussed in Section 7.4 with further elaboration of Example 7.3.

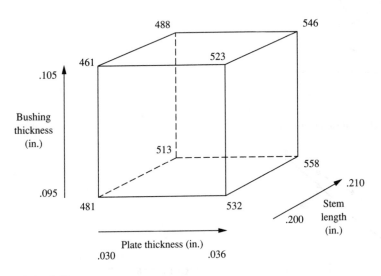

FIGURE 7.15
Cube for the Rivet Experiment (2^{7-4} design)

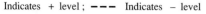

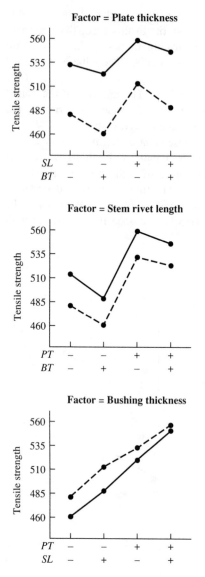

FIGURE 7.16

Plot of paired comparisons

7.4 BLOCKING IN FACTORIAL DESIGNS

Blocks are usually formed by holding background variables constant for all or part of the experimental pattern. In Chapter 5, designs for which the background variables are held constant for the entire set of combinations of factors and levels were called complete block designs. Designs for which the background variables are held constant for only a portion of the combinations of factors and levels were called incomplete block designs.

Factorial designs can be run in either complete or incomplete blocks. For example, consider a 2^3 design that was to be run in a manufacturing facility. "Shift" was identified as a major background variable. Time permitting, the experiment could be run on one shift to keep the conditions relatively constant. The background variable "shift" defines a block. The experimenters may choose to repeat the experiment on another shift to increase degree of belief in the results. If they do, the total study would consist of two replications of the full 2^3 design. Each replication would be carried out in a complete block defined by the shift. The analysis of the effects of the factors should be performed separately for each block. The results from each block can then be compared to determine if there are any interactions between the blocks and the factors.

Although it may be desirable to carry out the full 2^3 all on one shift, it may not be possible. Perhaps each test takes two hours to complete. Then only four of the eight tests could be completed on one shift. Running the remaining four tests on the next shift or waiting until the next day creates the potential to have background variables and nuisance variables confuse the results of the experiment.

In analytic studies, even if it is possible to run all of the eight tests on one shift, we may not want to do so. By running the tests on more than one shift, a wider range of conditions is included in the study. This increases the degree of belief that the results provide a basis for action. The use of incomplete blocks provides a method to reduce the effect of background variables when the factors are not studied under uniform conditions.

In this section the use of incomplete blocks with factorial or fractional factorial designs are discussed. Four particularly useful experimental patterns that incorporate incomplete blocks are given in some detail. These patterns are

- a 2^3 pattern in two blocks of four tests each,
- a 2^4 pattern in two blocks of eight tests each,
- a 2^{7-3} pattern in two blocks of eight tests each,
- a 2^{15-10} pattern in two blocks of 16 tests each.

Also, a method to develop other experimental patterns using incomplete blocks is given.

2^3 Design in Two Blocks of Size 4

If it is not possible to run all of a 2^3 factorial pattern in one block, two blocks with four tests in each block could be used. How should the eight tests be split between the two blocks? Intuitively, it seems desirable to have two tests at the low level and two tests at the high level of each factor in each block. For example, we would not want all tests at the low level of a particular factor run on shift 1 and all tests at the high level of the factor run on shift 2. If this were the case, it would be impossible to separate the effect of the factor from the effect of the shift. This discussion is similar to the discussion concerning how to chose the eight tests in a 2^{4-1} pattern. In fact, there is a direct link between fractional factorial patterns and incomplete blocks.

To divide the eight tests in a 2^3 pattern into two blocks of four tests each, a blocking variable is used. This blocking variable is usually a background variable or several background variables that are combined into a chunk. A 2^{4-1} pattern is used to accommodate the three factors and the blocking variable in eight tests. The blocking variable is used to set up the two blocks by treating it as if it were a factor. All tests that have a $+$ level of the blocking variable are included in block 1 and all tests that have a $-$ level of the blocking variable are included in block 2. The difference between the 2^{4-1} pattern and a 2^3 pattern in two blocks is in the randomization of the eight tests. The randomization is done separately in each of the two blocks. The analysis of data from a 2^3 in two blocks is carried out as if it were a 2^{4-1} pattern.

> **Example 7.3** 2^3 **Design in two blocks for the dye process.** Suppose that it was either necessary for practical reasons or desirable for increased degree of belief to run the 2^3 pattern in the dyeing process described in Example 6.2 (Chapter 6) in two blocks of four tests each. Background variables would first have to be combined into a chunk variable. Table 7.19 contains the design matrix for the 2^3 in two blocks. The randomization is done separately within the two blocks.

2^4 Design in Two Blocks of Size 8

The 2^4 factorial pattern can be separated into two blocks of eight tests each by setting up a blocking variable just as was done for the 2^3 pattern. This blocking variable becomes the fifth factor to be included in the 16 tests and is used to set up the blocks. A 2^{5-1} pattern is used to accommodate the four factors plus the blocking variable.

TABLE 7.19
2^3 Design in two-blocks—dye process (any three factors form a full factorial pattern)

Test	Run order	M	P	T	b^a	Mb PT	Pb MT	Tb PM	Response
Block 1									
1	3	−	−	+	+	−	−	+	195
2	1	+	−	−	+	+	−	−	228
3	4	−	+	−	+	−	+	−	218
4	2	+	+	+	+	+	+	+	241
Block 2									
5	6	+	+	−	−	−	−	+	259
6	7	−	+	+	−	+	−	−	238
7	8	+	−	+	−	−	+	−	200
8	5	−	−	−	−	+	+	+	189
Divisor = 4									

[a] b = Blocking variable.

Example 7.4 The solenoid experiment in two blocks. Table 7.20 contains the design matrix for the 2^4 experiment on the solenoid described in Example 6.3 (Chapter 6) as if it were run in two blocks. Figure 7.17 contains the run charts for the averages and standard deviations. The randomization for this design is done separately in each block.

The analysis of the data from a 2^4 in two blocks is carried out as if it were a 2^{5-1} pattern. Figure 7.18 contains the effects of the factors computed from the design matrix contained in Table 7.20, and it contains the dot diagram as well. A review of Figure 7.18 indicates that block *(b)* has an insignificant effect on both the average and standard deviation of flow. Since the remaining four factors form a full factorial design, the analysis can proceed in the same fashion as Example 6.3.

2^{7-3} Design in Two Blocks of Size 8

The 2^{7-3} factorial pattern can be separated into two blocks of eight tests each by setting up a blocking variable just as was done for the 2^3 and 2^4 patterns. This blocking variable becomes the eighth factor to be included in the 16 tests and is used to set up the blocks. A 2^{8-4} pattern is used to accommodate the seven factors plus the blocking variable. Table 7.21 contains the design matrix for the 2^{7-3} pattern in two blocks.

TABLE 7.20
2^4 design in two-blocks of eight—solenoid experiment (any four factors form a full factorial pattern)

Test	Run order	A	S	B	T	b^a	A S	A B	A T	A b	S B	S T	S b	B T	B b	T b	$\overline{X}$,	s
Block 1																		
1	5	−	−	−	−	+	+	+	+	−	+	+	−	+	−	−	.46,	.04
2	3	+	+	−	−	+	+	−	−	+	−	−	+	+	−	−	.45,	.10
3	7	+	−	+	−	+	−	+	−	+	−	+	−	−	+	−	.71,	.01
4	1	−	+	+	−	+	−	−	+	−	+	−	+	−	+	−	.70,	.05
5	4	+	−	−	+	+	−	−	+	+	+	−	−	−	−	+	.28,	.15
6	6	−	+	−	+	+	−	+	−	−	−	+	+	−	−	+	.60,	.07
7	8	−	−	+	+	+	+	−	−	−	−	−	−	+	+	+	.70,	.02
8	2	+	+	+	+	+	+	+	+	+	+	+	+	+	+	+	.72,	.01
Block 2																		
9	15	−	+	−	−	−	−	+	+	+	−	−	−	+	+	+	.57,	.02
10	10	−	−	+	−	−	+	−	+	+	−	+	+	−	−	+	.73,	.02
11	9	+	+	+	−	−	+	+	−	−	+	−	−	−	−	+	.70,	.01
12	13	−	−	−	+	−	+	+	−	+	+	−	+	−	+	−	.42,	.04
13	16	+	+	−	+	−	+	−	+	−	−	+	−	−	+	−	.29,	.06
14	11	+	−	+	+	−	−	+	+	−	−	−	+	+	−	−	.71,	.02
15	14	−	+	+	+	−	−	−	−	+	+	+	−	+	−	−	.72,	.02
16	12	+	−	−	−	−	−	−	−	−	+	+	+	+	+	+	.42,	.16

Divisor = 8

$^a b$ = Blocking variable.

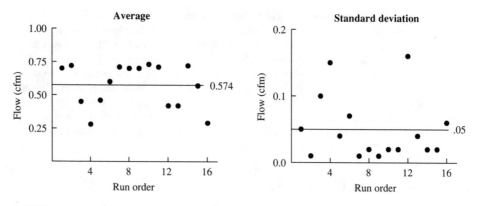

FIGURE 7.17
Run charts for the solenoid study

	Estimate (cfm)			Estimate (cfm)	
Effect	$\bar{x}$	s	Effect	$\bar{x}$	s
A	− .08	.03	Ab	.00	− .01
S	.04	− .02	SB	− .04	.02
B	.28	− .06	ST	.01	.00
T	− .04	.00	Sb	.04	.02
b	.01	.01	BT	.04	.00
AS	−.03	− .02	Bb	− .01	− .01
AB	.07	− .05	Tb	.03	.02
AT	-.03	− .01			

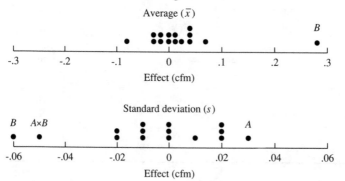

FIGURE 7.18
Effects of factors on flow

TABLE 7.21
Design matrix for a 2^{7-3} pattern in two blocks (any three
factors form a full factorial pattern)

									12	13	14	15	16	17	24
									37	27	36	26	25	23	35
									56	46	57	47	34	45	67
Test	1	2	3	4	5	6	7	b	4b	5b	2b	3b	7b	6b	1b
Block 1															
1	−	−	−	+	+	+	−	+	+	+	−	−	−	+	−
2	+	−	−	−	−	+	+	+	−	−	−	−	+	+	+
3	−	+	−	−	+	−	+	+	−	+	+	−	+	−	−
4	+	+	−	+	−	−	−	+	+	−	+	−	−	−	+
5	−	−	+	+	−	−	+	+	+	−	−	+	+	−	−
6	+	−	+	−	+	−	−	+	−	+	−	+	−	−	+
7	−	+	+	−	−	+	−	+	−	−	+	+	−	+	−
8	+	+	+	+	+	+	+	+	+	+	+	+	+	+	+
Block 2															
9	+	+	+	−	−	−	+	−	+	+	−	−	−	+	−
10	−	+	+	+	+	−	−	−	−	−	−	−	+	+	+
11	+	−	+	+	−	+	−	−	−	+	+	−	+	−	−
12	−	−	+	−	+	+	+	−	+	−	+	−	−	−	+
13	+	+	−	−	+	+	−	−	+	−	−	+	+	−	−
14	−	+	−	+	−	+	+	−	−	+	−	+	−	−	+
15	+	−	−	+	+	−	+	−	−	−	+	+	−	+	−
16	−	−	−	−	−	−	−	−	+	+	+	+	+	+	+

Divisor = 8

2^{15-10} Design in Two Blocks of Size 16

The 2^{15-10} factorial pattern can be separated into two blocks of 16 tests each by
setting up a blocking variable just as was done for the 2^3, 2^4, and 2^{7-3} patterns. This
blocking variable becomes the 16th factor to be included in the 32 tests and is used
to set up the blocks. A 2^{16-11} pattern is used to accommodate the 15 factors plus the
blocking variable. Table 7.22 contains the design matrix for the 2^{15-10} pattern in two
blocks.

> **Example 7.3 Continued Tensile strength of rivets.** To determine the factors that had
> an effect on the tensile strength of rivets, a 2^{7-4} experimental pattern was used initially.
> The design matrix for a 2^{7-4} pattern was contained in Table 7.18. After analyzing the
> results of this design, it was decided that eight more tests should be run to separate the
> confounding between the factors and interactions. The additional tests that were run were
> those in block 2 of the design matrix for the 2^{7-3} pattern in two blocks shown in Table
> 7.21.
>
> The run chart for the 16 factor combinations is shown in Figure 7.19. No special
> causes seem to be present. The design matrix for the rivet study is contained in Table
> 7.23. Figure 7.20 shows the effect of the factors on the tensile strength of stem diaphragm
> plate rivets and the dot diagram.

TABLE 7.22
Design matrix for a 2^{15-10} pattern in two blocks (any three factors form a full factorial pattern)

Test	1	2	3	4	5	6	7	8	9	10	11	12	13	14	15	b	.... Two-factor interactions ....
Block 1																	
1	−	−	−	−	−	+	+	+	+	+	−	−	−	−	+	+	
2	+	−	−	−	+	−	−	+	+	−	+	+	+	+	−	+	
3	−	+	−	−	−	−	+	−	+	+	+	+	−	+	+	+	
4	+	+	−	−	+	+	−	−	+	−	−	−	+	+	−	+	
5	−	−	+	−	−	+	+	−	−	+	+	+	+	+	+	+	
6	+	−	+	−	+	−	−	−	−	−	−	−	−	+	−	+	
7	−	+	+	−	−	−	+	+	−	+	−	−	+	+	+	+	
8	+	+	+	−	+	+	−	+	−	−	+	+	−	+	−	+	
9	−	−	−	+	−	+	+	−	+	−	+	−	+	−	+	+	
10	+	−	−	+	+	−	−	−	+	+	−	+	−	−	−	+	
11	−	+	−	+	−	−	+	+	+	−	−	+	+	−	+	+	
12	+	+	−	+	+	+	−	+	+	+	+	−	−	−	−	+	
13	−	−	+	+	−	+	+	+	−	−	−	+	−	+	+	+	
14	+	−	+	+	+	−	−	+	−	+	+	−	+	−	−	+	
15	−	+	+	+	−	−	+	−	−	−	+	−	−	+	+	+	
16	+	+	+	+	+	+	−	−	−	+	−	+	+	−	−	+	

TABLE 7.22
(continued)

Test	1	2	3	4	5	6	7	8	9	10	11	12	13	14	15	b	. . . . Two-factor interactions														
Block 2																														
17	+	+	+	+	−	−	−	−	−	−	+	+	+	+	+	−	−	−	−	+	−	+	+	−	−	+	+	−	+	−	+
18	−	+	+	+	+	+	+	−	−	−	+	−	−	+	+	−	−	−	+	+	+	+	−	−	+	+	−	−	−	+	+
19	+	+	+	+	−	−	+	−	+	−	−	−	−	−	+	−	−	+	−	+	+	−	+	−	+	+	−	+	−	−	+
20	−	−	+	+	+	+	−	+	+	+	+	+	−	−	−	−	−	+	+	−	−	+	+	+	−	+	−	+	+	+	+
21	+	+	−	+	+	+	−	+	−	+	+	−	−	−	−	−	−	+	+	+	+	−	+	+	−	+	−	+	−	+	+
22	−	+	−	+	+	+	+	+	+	+	−	+	+	−	−	−	−	+	+	−	+	+	−	+	−	+	−	−	+	−	+
23	+	−	−	+	+	+	−	−	+	+	+	−	+	+	+	−	−	+	−	+	−	+	+	−	+	−	+	+	−	+	+
24	−	−	−	+	−	−	+	−	+	+	+	+	+	+	+	−	+	+	+	−	+	+	+	+	+	−	+	+	+	+	+
25	+	+	+	−	+	+	+	−	−	−	−	+	+	+	+	−	+	−	−	−	−	−	−	−	−	−	−	−	−	−	+
26	−	−	+	−	+	+	−	+	−	−	+	−	−	+	+	−	+	+	−	−	+	−	−	+	+	−	+	−	+	+	+
27	+	+	+	−	−	−	+	+	+	+	−	−	+	+	−	−	+	−	+	−	−	+	+	+	+	−	+	+	−	−	+
28	−	−	+	−	−	−	−	−	+	+	+	+	+	−	−	−	+	+	+	+	+	−	−	+	−	+	−	+	+	+	+
29	+	+	−	−	+	+	−	−	+	+	+	−	−	+	−	−	+	−	+	+	−	+	+	+	−	+	−	+	−	+	+
30	−	+	−	−	+	+	+	+	−	−	+	+	+	−	+	−	+	+	−	−	+	+	+	−	+	−	+	−	+	−	+
31	+	−	−	−	+	+	−	−	−	−	−	+	−	+	+	−	+	−	+	−	−	−	−	+	+	−	+	−	−	+	+
32	−	−	−	−	−	−	+	+	+	+	+	−	+	+	+	−	+	+	+	+	+	+	+	+	+	+	+	+	+	+	+

Divisor = 16

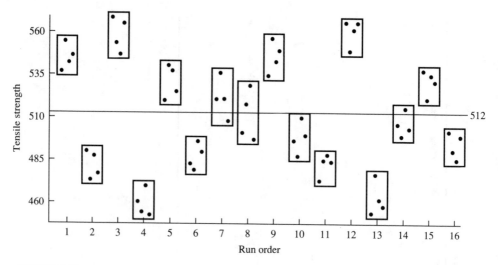

FIGURE 7.19
Run chart

TABLE 7.23
Design matrix for the rivet study

| | | 1 = RH | | 3 = PT | | 5 = SL | | 7 = BT | | | | | | | | |
| | | 2 = PD | | 4 = SD | | 6 = BD | | 8 = block | | | | | | | | |

Test	Run order	1	2	3	4	5	6	7	8	12 37 56 48	13 27 46 58	14 36 57 28	15 26 47 38	16 25 34 78	17 23 45 68	24 35 67 18	Tensile Strength $\bar{X}$
Block 1																	
1	8	−	−	−	+	+	+	−	+	+	+	−	−	−	+	−	513
2	4	+	−	−	−	−	+	+	+	−	−	−	−	+	+	+	461
3	6	−	+	−	−	+	−	+	+	−	+	+	−	+	−	−	488
4	2	+	+	−	+	−	−	−	+	+	−	+	−	−	−	+	481
5	7	−	−	+	+	−	−	+	+	+	−	−	+	+	−	−	523
6	3	+	−	+	−	+	−	−	+	−	+	−	+	−	−	+	558
7	5	−	+	+	−	−	+	−	+	−	−	+	+	−	+	−	532
8	1	+	+	+	+	+	+	+	+	+	+	+	+	+	+	+	546
Block 2																	
9	14	+	+	+	−	−	−	+	−	+	+	−	−	−	+	−	508
10	12	−	+	+	+	+	−	−	−	−	−	−	−	+	+	+	558
11	9	+	−	+	+	−	+	−	−	+	+	−	+	−	−	−	545
12	15	−	−	+	−	+	+	+	−	+	−	+	−	−	−	+	530
13	10	+	+	−	−	+	+	−	−	+	−	−	+	+	−	−	499
14	13	−	+	−	+	−	+	+	−	−	+	−	+	−	−	+	463
15	16	+	−	−	+	+	−	+	−	−	−	+	+	−	+	−	492
16	11	−	−	−	−	−	−	−	−	+	+	+	+	+	+	+	481

Divisor = 8

1 = RH	3 = PT	5 = SL	7 = BT
2 = PD	4 = SD	6 = BD	8 = block

Factor	Effect	Interactions	Effect
1 (RH)	0.3	12+37+56+48	−5.0
2 (PD)	−3.5	13+27+46+58	3.3
3 (PT)	52.8	14+36+57+28	1.5
4 (SD)	5.0	15+26+47+38	1.3
5 (SL)	23.8	16+25+34+78	3.0
6 (BD)	0.0	17+23+45+18	.5
7 (BT)	−19.8	24+35+67+18	−2.8
8 (block)	3.3		

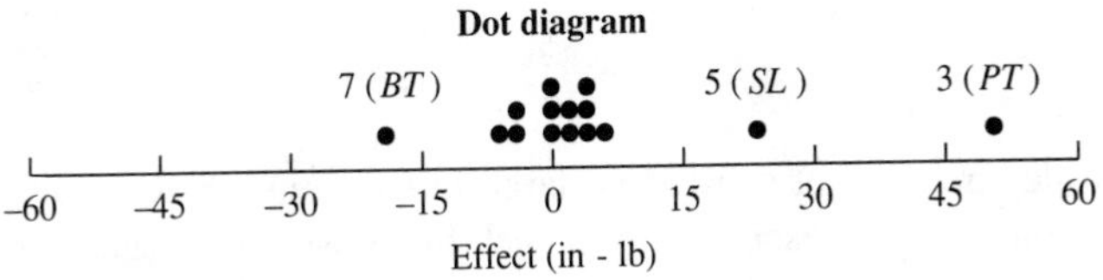

Dot diagram

FIGURE 7.20
Effects of factors on tensile strength

From Figure 7.20, it can be seen that the three factors—plate thickness, stem rivet length, and bushing thickness—have a significant effect on tensile strength. These were the same three factors found to be important using the 2^{7-4} design. This indicates that the confounded two-factor interactions were not the source of the large effects.

Analysis of the Cube

A characteristic of a 2^{7-3} in two blocks is that any three factors form a full factorial design. The three important factors were analyzed on a cube in Figure 7.21. The two values at each corner of the cube are the tensile strengths for the appropriate factor combination in each block.

The paired comparisons are plotted in Figure 7.22. The analysis of the paired comparisons does not indicate the presence of any special causes of variation and therefore confirms the effects that were found using the design matrix.

Response Plots

Figure 7.23 contains the response plots that summarize the results of the experiment. Since the factors plate thickness, stem rivet length, and bushing thickness had a

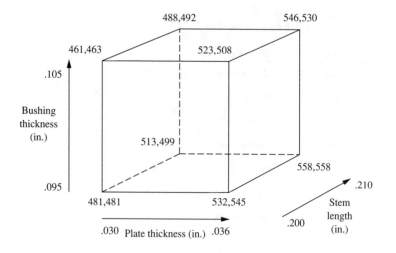

FIGURE 7.21
Cube for the rivet experiment

substantial effect on tensile strength but showed no large interactions, three separate response plots were constructed. The responses from each block were averaged since the block-to-block effect was small.

Conclusions for the Continuation of Example 7.3

1. There were no special causes of variation detected in the run chart or the analysis of the paired comparisons in the cube.
2. Plate thickness, stem rivet length, and bushing thickness had a substantial effect on the average tensile strength of the stem diaphragm plate rivets. As the plate thickness or stem length increased, the tensile strength increased; and as the bushing thickness increased, the tensile strength decreased.
3. There were no substantial interaction effects on tensile strength.
4. Very similar results were seen in each block.

Development of Other Blocking Patterns

There are many useful experimental patterns besides the three given above that arrange factorial or fractional factorial patterns in incomplete blocks. As has been seen, there is a strong connection between fractional factorial patterns and patterns using incomplete blocks. In this section, a method will be described to develop incomplete-blocking arrangements using the fractional factorial patterns discussed previously in this chapter.

Table 7.24 contains some of the useful incomplete-blocking arrangements that can be developed from the fractional factorial patterns in this chapter. Columns 1, 2, and 3 describe the number of factors, the fraction of a full factorial pattern that is to be

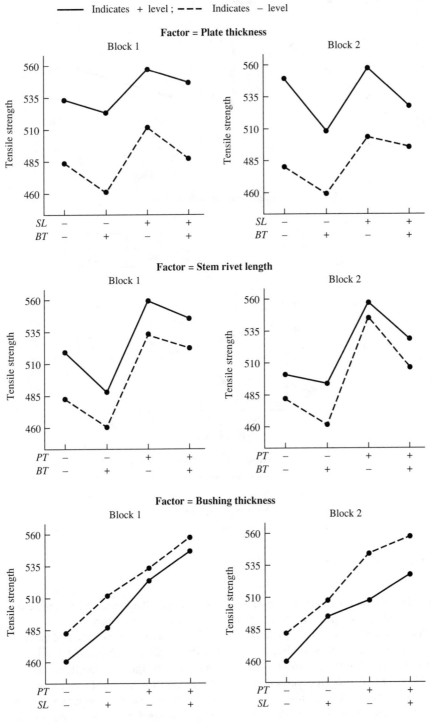

FIGURE 7.22
Plot of paired comparisons

207

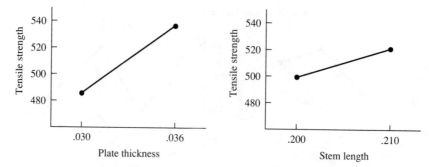

Average tensile strength (in-lbs)

PT	
.030	.036
485	538

SL	
.200	.210
499	523

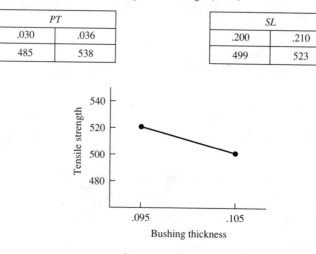

Average tensile strength (in-lbs)

BT	
.095	.105
521	501

FIGURE 7.23
Response plots for the rivet experiment

run, and the total number of tests. Columns 4 and 5 contain the blocking arrangement. Column 6 contains the method to obtain the desired experimental pattern.

The first pattern listed in Table 7.24 is a 2^3 factorial pattern arranged in two blocks of four tests. We have already seen that this pattern is obtained by using a 2^{4-1} pattern. The first three factors in the 2^{4-1} pattern correspond to the three factors in the 2^3 pattern. A blocking variable is then substituted for the fourth factor and is used to set up the blocks. All tests for which the blocking variable is negative are put in one block and all tests for which the blocking variable is positive are put in the other block.

TABLE 7.24
Incomplete blocking arrangements

Number of factors	Fraction	Number of tests	Number of blocks	Size of blocks	Development of blocks[a]
3	Full	8	2	4	1 b.v., 2^{4-1}
			4	2	2 b.v., 2^{7-4}
4	Full	16	2	8	1 b.v., 2^{5-1}
			4	4	2 b.v., 2^{8-4}
			8	2	4 b.v., 2^{8-4}
5	1/2	16	2	8	1 b.v., 2^{8-4}
			4	4	2 b.v., 2^{8-4}
	1/4	8	2	4	1 b.v., 2^{7-4}
6	1/4	16	2	8	1 b.v., 2^{8-4}
			4	4	2 b.v., 2^{8-4}
	1/8	8	2	4	1 b.v., 2^{7-4}
7	1/8	16	2	8	1 b.v., 2^{8-4}
8–13	1/16–1/512	16	2	8	1 b.v., 2^{15-11}
			4	4	2 b.v., 2^{15-11}
14	1/1024	16	2	8	1 b.v., 2^{15-11}

[a]The information in this column is the number of blocking variables (b.v.) needed to set up the desired number or blocks and the design matrix that is used to arrange the original factorial or fractional pattern into incomplete blocks.

There are times when it is desirable to arrange the factorial or fractional factorial pattern in smaller blocks, and thus more than two blocks are needed. For example, consider the 2^4 pattern arranged in four blocks of four tests each. To develop this pattern, start with a 2^{8-4} pattern. This pattern is contained in Table 7.7. Label the first four factors as the four factors in the original 2^4 pattern. Label factors 5 and 6 as blocking variables 1 and 2. Arrange the blocks so that the tests in each block have the same levels as the two blocking variables. Remembering that columns 5 and 6 in the design matrix now refer to the two blocking variables, the four blocks correspond to columns 5 and 6 being $++$, $+-$, $-+$, and $--$. For example, the first block contains tests 1, 8, 12, and 13 in Table 7.7. These four tests are in the same block because the two blocking variables (columns 5 and 6) are at the $++$ level for each of the tests.

The randomization is done separately in each of the four blocks. The data is analyzed as if it were a fractional factorial pattern with six factors in 16 tests. The confounding pattern is found as before by crossing out any interactions containing a 7 or an 8. The effects computed from columns 7 and 8 measure the impact of nuisance variables.

Table 7.25 contains the design matrix for a 2^4 pattern in four blocks of size four.

TABLE 7.25

Design matrix for a 2^4 pattern in four blocks (any three factors form a full factorial pattern)

Test[a]	1	2	3	4	5[b]	6[b]	N[c]	N[c]	12 56	13 46	14 36	15 26	16 25 34	23 45	24 35
Block 1															
1 (1)	−	−	−	+	+	+	−	+	+	+	−	−	−	+	−
2 (8)	+	+	+	+	+	+	+	+	+	+	+	+	+	+	+
3 (12)	−	−	+	−	+	+	+	−	+	−	+	−	−	−	+
4 (13)	+	+	−	−	+	+	−	−	+	−	−	+	+	−	−
Block 2															
5 (3)	−	+	−	−	+	−	+	+	−	+	+	−	+	−	−
6 (6)	+	−	+	−	+	−	−	+	−	+	−	+	−	−	+
7 (10)	−	+	+	+	+	−	−	−	−	−	−	−	+	+	+
8 (15)	+	−	−	+	+	−	+	−	−	−	+	+	−	+	−
Block 3															
9 (2)	+	−	−	−	−	+	+	+	−	−	−	−	+	+	+
10 (7)	−	+	+	−	−	+	−	+	−	−	+	+	−	+	−
11 (11)	+	−	+	+	−	+	−	−	−	+	+	−	+	−	−
12 (14)	−	+	−	+	−	+	+	−	−	+	−	+	−	−	+
Block 4															
13 (4)	+	+	−	+	−	−	−	+	+	−	+	−	−	−	+
14 (5)	−	−	+	+	−	−	+	+	+	−	−	+	+	−	−
15 (9)	+	+	+	−	−	−	+	−	+	+	−	−	−	+	−
16 (16)	−	−	−	−	−	−	−	−	+	+	+	+	+	+	+
Divisor = 8															

[a] The number in parentheses refers to the test number from the 2^{8-4} pattern contained in Table 7.7.
[b] 5 Represents the first blocking variable
  6 Represents the second blocking variable.
[c] These columns measure the effect of nuisance variables.

7.5 SUMMARY

In this chapter, a small set of fractional factorial designs have been presented. These designs, summarized in Table 7.26, have been separated into designs that are appropriate when experimenters have a moderate level of knowledge about the process and the factors involved and designs that are appropriate when there is a low level of knowledge.

There has been emphasis placed on the sequential use of fractional factorial designs. Some guidance has been given as to what to do after a particular fractional factorial design has been run. Finally, the arrangement of factorial or fractional factorial patterns in incomplete blocks has been discussed.

Fractional factorial designs provide experimenters with a powerful means of learning about their products and processes. The major advantages of these designs:

- They allow a large number of factors to be studied in relatively few tests.
- They allow each factor to be studied over a wide range of conditions.

TABLE 7.26
Summary of designs of fractional factorial patterns

Experimental pattern	Number of factors	Number of runs	Confounding	Number of factors forming a full factorial design
		Moderate Level of Knowledge		
2^{4-1}	4	8	2fi with 2fi	3
2^{5-1}	5	16	None	4
2^{8-4}	6, 7, 8	16	2fi with 2fi	Any 3 and some groups of 4
2^{16-11}	9–16	32	2fi with 2fi	Any 3 and some groups of 4 or 5
		Low Level of Knowledge		
2^{3-1}	3	4	Individual factors with 2fi	2
2^{7-4}	5, 6, 7	8	Individual factors with 2fi	Any 2 and some groups of 3
2^{15-11}	8-15	16	Individual factors with 2fi	Any 2 and some groups of 3 or 4

- They can be easily used sequentially to build knowledge.
- They take advantage of the Pareto Principle that relatively few factors have most of the influence on the response variable.

Fractional factorial designs are not without some disadvantages. The major disadvantages are:

- They are vulnerable to special causes of variation and missing values.
- It is difficult to detect the influence of special causes.
- The theory underlying them is not as easily understood by experimenters as the theory for full factorial designs.
- There is a loss of information relative to full factorial designs due to the confounding of effects.

The small set of designs included in this chapter will meet most of the needs of experimenters whose aim it is to learn more about their products or processes. There are other fractional factorial designs that have not been presented that may be better in a particular application. For more on other fractional factorial designs, how to construct fractional factorial designs, and how to add other runs to eliminate the confounding between individual factors and interactions, see the chapter reference list.

REFERENCES

Box, G., W. Hunter, and J.S. Hunter, (1978), *Statistics for Experimenters,* John Wiley & Sons, New York.
Daniel, C. (1976), *Applications of Statistics to Industrial Experimentation*, John Wiley & Sons, New York.

EXERCISES

7.1 Choose a product or process with which you are familiar. For an important quality characteristic, list potential causes that affect either the average level or the variation of the quality characteristic. (Consider use of a cause-and-effect diagram to do this.) Plan a fractional factorial experiment to determine the effects of some of the potential causes. Use the planning form.

7.2 In example 7.2, Tests 2 and 14 in Table 7.11 showed evidence of being affected by a special cause of variation. Two additional tests were made at these conditions to see if these results were repeatable. The pull force for Test 2 was 1275 and the pull force for Test 14 was 1325. Plot these results on the run chart for the original experiment. Substitute these results for the original results of 871 and 920. Analyze the data from the experiment using these two values and compare the results to the original analysis. Since the two additional tests were run after the original experiment was performed, these tests could have been influenced by special causes occurring after the original test was completed. Suggest another design to build on the knowledge gained in the first 18 tests.

7.3 Use the three entries in Table 7.24 for six factors to set up the three experimental patterns. Randomize the tests for each pattern.

7.4 The manager of an administrative group supporting the sales department of a manufacturing company was concerned about the large number of notes of credit that were sent to customers to correct errors in invoices. A preliminary investigation was done to determine the major sources of the errors. A source of data was a data base that was historically kept by the administrative group on every shipment of product. The data base contained information on customers, type of product, price, size of shipment, and the like. Also, an entry was made if a letter of credit was needed to correct the invoice.

The group identified four factors relating to a shipment of product and defined two levels for each factor:

Factor	Level	
customer (c)	minor (−)	major (+)
customer location (l)	foreign (−)	domestic (+)
type of product (t)	commodity (−)	specialty (+)
size of shipment (s)	small (−)	large (+)

The group then set up a 2^{4-1} pattern to use to sample the data base. For each of the eight cells in the pattern 100 entries in the data base were randomly selected, and the percentage of notes of credit for the 100 invoices were recorded. The results of the sampling are contained in the following table.

C	L	T	S	Percent needing notes of credit
−	−	−	−	15
+	−	−	+	18
−	+	−	+	6
+	+	−	−	2
−	−	+	+	19
+	−	+	−	23
−	+	+	−	16
+	+	+	+	21

Analyze the data to determine major sources of the notes of credit.

7.5. A chemical company had just started production of a new product in its new batch process. After some initial runs, it was found that increasing the yield and decreasing the variation in the viscosity were necessary for a successful product. The batches were run in nine day campaigns with two batches made each day. Holding tanks were available for raw materials with capacities for enough materials for 18 batches. There were four reactors in the unit that could be used for the new product.

Viscosity was measured in the laboratory using a composite sample from each batch. Yield was calculated through a material balance for each batch. The operators in the unit identified a number of factors that could affect either yield or viscosity:

Factor	Levels (maximum range for operation)
Reactor	A, B, C, or D
Reactor temperature	150 to 270 (degrees C)
Reactor pressure	130 to 180 (psi)
Reaction time	6 to 8 (hours)
Raw material supplier	Quality, A-1, Discount
Agitation rate	100 to 150 (rpm)
Distillation time	2 to 4 (hour)
Catalyst concentration	6 to 12 (percent)

Plan an experiment for this process with the following constraints:

(*1.*) Learn something about each potential factor.

(*2.*) Complete the experiment during the next campaign of the product of interest.

(*3.*) Develop the experimental pattern, blocking, replication, and randomization for the study. Complete an experimental design planning form.

CHAPTER
8

EVALUATING
SOURCES OF
VARIATION

The experimental designs studied in the last two chapters assumed that the factor combinations were interchangeable, which meant that any combination of factors and levels could be tested. Sometimes, however, the factors of interest are not interchangeable. It may not be meaningful to compare each factor at each level of the other factors. A nested or hierarchical pattern is used to accommodate these types of factors. In a nested design, levels of different factors are studied within a given level of another factor.

Figure 8.1 shows examples of nested patterns for three different studies with two, three, and four factors respectively. In each case, a factorial arrangement of the factors would not make sense. In the first example, patients are associated with a particular hospital. There is no relationship between patient 1 in hospital A and patient 1 in hospital B. It would not be reasonable or desirable to interchange the patients with the hospitals.

In the second example, cavity a in mold 1 is a physical location, not related to cavity a in the other injection mold. If instead of cavity the factor was position in the mold, then a factorial arrangement for position in the mold would be appropriate. Cavity a might be the position farthest from the center in each of the molds. In the nested experimental pattern, no commonality between the cavities in different molds is assumed.

In the third example, both of the plants have a day and night shift, but there is no relationship between the two day shifts or the two night shifts. Three operators are selected from each shift, but again there is nothing in common between the four operator 2's (O_2) in the study. Two assemblies are selected from each operator, but

A. Two factors: Hospital (3 levels)
                          Patients within a hospital (five levels)

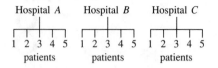

B. Three factors: Injection mold (2 levels)
                            Cavity within a mold (4 levels)
                            Parts within a cavity (3 levels)

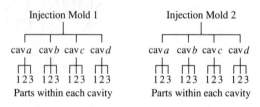

C. Four factors: Plants (2 levels)
                        Shifts within a plant (2 levels)
                        Operators (O_i) within a shift (3 levels)
                        Assemblies within an operator (2 levels)

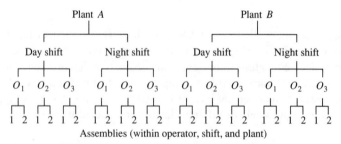

FIGURE 8.1

Three examples of nested experimental patterns

assembly 1 has no meaning beyond being the first assembly selected from a particular operator.

Nested experiments are commonly used to **identify the important sources of variation in a system.** Many nested studies involve the evaluation of sampling and testing strategies. The analysis of data from a nested design can usually be done with a dot-frequency diagram (Snee, 1983), which will be discussed in the next section. As with factorial designs, a run chart is used to evaluate the impact of nuisance variables. A method for quantifying the magnitude of effects of the factors can be used if needed.

This chapter begins by showing that a control chart can be considered as a two-factor nested design. Designs for more than two factors are then discussed. Section 8.4 summarizes the planning and analysis of a study with nested factors. Section 8.6 discusses a study with both nested and crossed factors.

8.1 THE CONTROL CHART AS A NESTED DESIGN

The Shewhart $\overline{X}$ and R control chart is an example of a nested experimental design. The two factors in the design of the control chart are *subgroups* and *within subgroups*. These are chunk-type factors that represent a number of process variables. Shewhart's concept of rational subgrouping was to organize data from the process in a way that is likely to give the greatest chance for the data in each subgroup to be alike and for the data in other subgroups to be different.

Figure 8.2 shows a schematic of an X-bar and R control chart as a nested pattern. To analyze data from the control chart, the average and range of the data from each subgroup are calculated. The range is used to evaluate the variation within a subgroup. The variation of the averages is used to evaluate the variation between subgroups. If the averages are within statistical control, then there is no important variation between subgroups in the process.

A dot-frequency diagram could be used to analyze data collected for the X-bar and R control chart. The bottom half of Figure 8.2 shows a schematic of such an analysis. The different subgroups are shown on the horizontal axis, and a scale for the measurements is on the vertical axis. A point is plotted for each measurement above the appropriate subgroup. In this example, the variation within a subgroup is the key source of variation. Each of the lines representing a subgroup overlaps all of the other lines. A control chart for these data would be in statistical control.

Figure 8.3 shows a dot-frequency diagram for a control chart that is not in statistical control. In this diagram, the variation within a subgroup is small relative to the variation between subgroups. Some of the lines representing individual subgroups

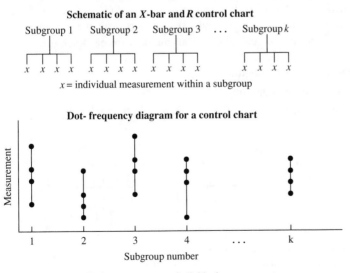

Each ● represents an individual measurement

FIGURE 8.2
A control chart as a nested design

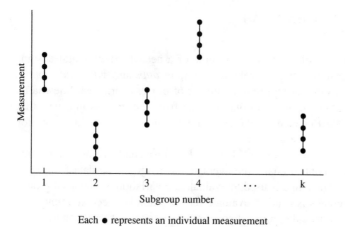

Each ● represents an individual measurement

FIGURE 8.3
Dot-frequency diagram for a control chart that is not in statistical control

do not overlap. Individual subgroup averages would be outside of the control limits calculated for these data.

8.2   NESTED DESIGN TO STUDY MEASUREMENT VARIATION

The X-bar and R control charts for a process can be modified to include an evaluation of the measurement process used to generate the data. Figure 8.4 shows a schematic of the design for this modified chart. The three factors in this study are (1) variation between subgroups, (2) variation within subgroups, and (3) variation from the measurement process. Three statistics can then be calculated from the data for each subgroup:

1. The average of the xs for each subgroup,
2. The range of the xs for each subgroup,
3. The range for the two measurements of the same sample for each subgroup.

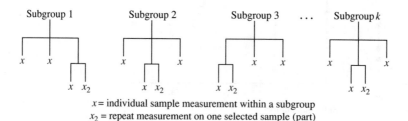

x = individual sample measurement within a subgroup
x_2 = repeat measurement on one selected sample (part)

FIGURE 8.4
Modified X-bar and R chart to evaluate measurement variation

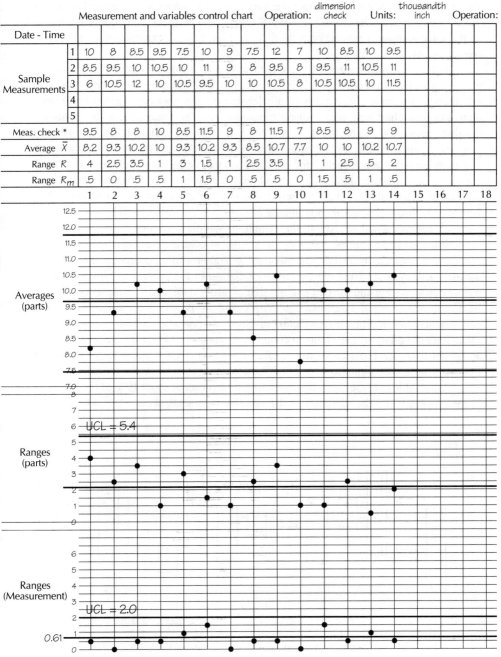

Note: This is a repeat measurement of Sample 1.

FIGURE 8.5
Example of *X*-bar, *R*, and measurement control charts

The average and range of the sample measurements (excluding the additional measurements) are plotted on the usual X-bar and R charts. The range of the two measurements on the same sample is plotted on a third chart. Figure its an example of such a set of control charts.

A dot-frequency diagram for the data in Figure 8.5 is shown in Figure 8.6. To prepare this chart, the vertical axis is scaled to include the range of all the data. The subgroup and sample (part) number are labeled on the horizontal axis. The points plotted on this chart represent a single measurement. The repeated measurements for the first sample are connected by a line. A box is drawn around the points in each subgroup. This is an incomplete or unbalanced nested design, incomplete in the sense that measurements are repeated for only one of the samples in each subgroup. The dot-frequency diagram shows that the subgroups (boxes) overlap the centerline and each other. This indicates that the most important variation is within subgroups. The length of the lines indicates the magnitude of measurement variation. The measurement variation appears small relative to the sample-to-sample variation.

Since all three control charts are in statistical control, the variation in the data can be summarized by estimating standard deviations for the process and for the measurement system. Then the percent of variation in the process attributable to the measurement system and the percent of variation attributable to the samples (parts) can be determined. Table 8.1 summarizes these calculations, which are called a variance component analysis. For the two factors in this study (variation within a subgroup and measurement variation) the variance component analysis can be viewed geometrically as a right triangle. The sum of the square of the two sides is equal to the square of the hypotenuse (see figure at bottom of Table 8.1).

From Table 8.1, the measurement variation represents about 19% of the variation in the process while the variation of the samples represents about 81%. The calcula-

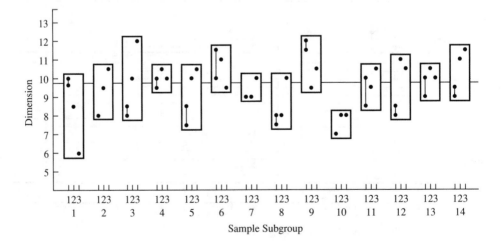

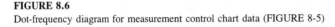

FIGURE 8.6
Dot-frequency diagram for measurement control chart data (FIGURE 8-5)

TABLE 8.1
Summary of range control charts

Variation for process

$$\bar{R} = 2.10 \qquad \delta_p = \bar{R} / d_2 = 2.10 / 1.693 = 1.24$$

Variation of measurement process

$$\bar{R}_m = 0.61 \qquad \delta_m = R_m / d_2 = 0.61 / 1.128 = 0.54$$

Variation of product

$$\delta_{\text{product}} = \sqrt{(\delta_{\text{process}})^2 - (\delta_{\text{measurement}})^2)}$$

$$\delta_{\text{product}} = \sqrt{(1.24)^2 - (0.54)^2)} = 1.12$$

Summary of variation (units = thousandth inch)

Source of variation	Standard deviation δ	Variance component δ^2	Percent of variation
Product	1.12	1.25	81.1
Measurement	0.54	0.29	18.9
Total (process)	1.24	1.54	100.0

Geometric relationship of variance components

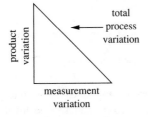

$$\hat{\sigma}^2 \text{ process} = \hat{\sigma}^2 \text{ product} + \hat{\sigma}^2 \text{ measurement}$$

$$\delta^2_{\text{process}} = \delta^2_{\text{product}} + \delta^2_{\text{measurement}}$$

tions of the percent variation due to each source of variation are based on the variance components, not the standard deviations (which are the square roots of the variance components).

8.3 A THREE-FACTOR NESTED EXPERIMENT

Figure 8.7 shows a completed planning form for a three-factor nested experiment. The objective was to evaluate the measurement system used in the process. The measurement process involved two key steps—a setup of the part to be measured and the gaging of the part using calipers.

A total of 54 measurements were made for the study (6 parts × 3 setups × 3 measurements). The data for the completed study are shown on the data collection

FIGURE 8.7
Form for documentation of a planned experiment

1. Objective:
Evaluate the variability of the measurement system and determine which part of the measurement process contributes most of the variation.

2. Background information:
This measurement system had not been evaluated since new calipers were purchased. Previous studies of similar systems had shown the setup procedure to be important.

3. Experimental variables:

A.	Response variables	Measurement technique
1.	Width of gap (millimeters)	QC standard procedure using calipers
B.	Factors under study	Levels
1.	Parts	Six parts selected.
2.	Setup within parts	Each part setup three times.
3.	Measurement within setup	Measurement made by three operators within each setup
C.	Background variables	Method of Control
1.	Time	Parts selected from an hour period when free of special causes.
2.	Operators	Three volunteers used in study.
3.	Calibration	Calipers checked at the beginning and end of study.

4. Replication:
The study has to be completed in one afternoon. Each setup takes 15 minutes, so less than 20 setups can be done. The replication is determined by the levels of each factor in the study. Only one replication of the experimental pattern will be done.

5. Methods of randomization:
The 18 setups (3 setups per part) were randomly ordered by putting the numbers 1 to 6 in a hat and drawing with replacement until 3 setups for each part were selected. The measurements by the three operators were done in the same order each time.

6. Design matrix: (see Figure 8.8)

7. Data collection forms: (see Figure 8.8)

8. Planned methods of statisical analysis:
Run chart, dot-frequency diagram, estimate of components of variation.

9. Estimated cost, schedule, and other resource considerations:
Study can be completed in 5-hour period using 3 operators.

Design matrix (experimental pattern)

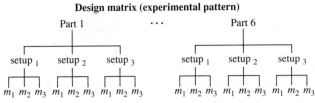

m = Measurements by three operators after each setup.
Randomize the order of the 18 setups in the study.

Data collection form (with completed data)
Measurement by three operators (gap width in millimeters)

Test order	Part	Setup	Meas 1	Meas 2	Meas 3	Notes
1	3	1	4.05	4.55	3.80	
2	2	1	4.55	4.70	4.70	
3	5	1	3.90	4.85	4.60	
4	2	2	4.30	4.20	4.50	
5	5	2	4.35	4.90	4.35	
6	4	1	2.60	2.65	2.50	verified that
7	5	3	4.10	4.00	3.95	readings were okay
8	6	1	3.95	4.10	4.00	after initial low
9	1	1	3.40	3.45	3.95	result
10	1	2	3.95	3.80	4.00	
11	3	2	4.50	4.55	4.65	took 15 minute
12	6	2	3.90	3.90	4.00	break after 12th
13	4	2	3.99	3.15	3.15	setup
14	2	3	4.75	5.15	5.20	
15	4	3	3.40	3.55	3.40	
16	6	3	4.40	4.30	4.40	
17	3	3	4.15	4.45	4.50	
18	1	3	3.95	4.00	4.15	

FIGURE 8.8
Design matrix and data collection form for a three-factor nested study

form in Figure 8.8. Figure 8.9 shows a run chart of the data. No obvious trends or other special causes are seen on the run chart. The low measurements on the sixth test were the first setup on part 4. The other two setups for part 4 (runs 13 and 15) also showed low results. Since there are no special causes, a dot-frequency diagram can be prepared to partition the variation between the three factors in the study. Figure 8.10 shows a dot-frequency diagram for this study.

The dot-frequency diagram indicates that part-to-part differences are the largest source of variation. Each box in the diagram represents a part. Part 4 has a much smaller gap than the other five parts. The variation within each part can be evaluated

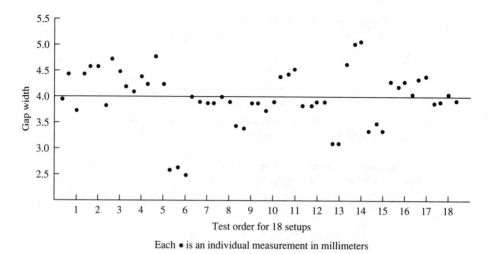

Each ● is an individual measurement in millimeters

FIGURE 8.9
Run chart for three-factor nested study

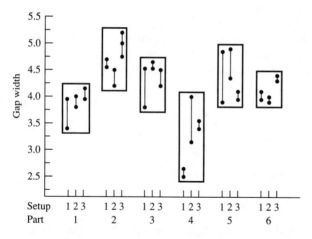

Each ● is an individual measurement in millimeters.

FIGURE 8.10
Dot-frequency diagram for three-factor nested study

by studying the lines drawn in each box. The length of the lines represents the measurement variation, while the difference between the lines represents the setup variation. For parts 1, 2, 4, and 6 the setup differences are large relative to the measurement variation.

Thus, the factors in the study can be ordered by their contribution to the variation:

1. Part—greatest source of variation.
2. Setup within a part.
3. Measurement—smallest source of variation.

Reductions in variation in the process could come from improving the setup procedure in the measurement process and then concentrating on factors that cause the part-to-part variation.

The percent of variation attributable to each of the factors can be quantified by extending the variance component analysis procedure presented in the previous example. This procedure is presented in the Appendix to this chapter. The variance component analysis verifies the visual analysis of the dot-frequency diagram:

- 72 percent of the variation is attributable to part differences.
- 18 percent is attributable to the setup.
- 10 percent is due to the measurement procedure.

8.4   PLANNING AND ANALYZING AN EXPERIMENT WITH NESTED FACTORS

The previous three sections have given examples of experimental situations where the factors were nested. The control chart and the study of measurement variation are common examples of nested studies. This section summarizes the design and analysis of experiments with nested factors.

Planning a Nested Experiment

The basic steps to plan an experiment with nested factors are no different than for a factorial experiment. Section 6.2 of Chapter 6 discussed these steps. The form "Documentation of a Planned Experiment" is useful in the planning. The objective of a nested experiment often focuses on understanding sources of variability in a process. A nested experimental design is often used as a screening study to direct focus for activities to improve a process. One of the most common applications of a nested design is to study measurement procedures and sampling strategies.

The selection of factors and factor levels is often straightforward in a nested study. In some studies, the physical layout of a process defines the factors and levels. For a study of a process with three molds and six cavities in each mold, the factors *molds* and *cavities within a mold* are obvious, with the levels fixed for both factors.

The amount of information available to evaluate the importance of each factor is another consideration in the assignment of levels. Because of the hierarchical structure, there is always more information available for factors at the lower levels of the hierarchy. In the three-factor example in Section 8.3, there were only six different parts in the study (the factor at the highest level in the hierarchy), but 54 different measurements were used to evaluate the measurement effect (the lowest factor in the hierarchy). When possible, more levels should be assigned to the highest factors in the hierarchy and fewer levels (usually two) should be used for the lowest factors in the hierarchy. This will balance the amount of information available about the contribution to variation from each factor.

Background variables such as *operator* or *raw material supplier* can be incorporated into a nested design in the same way as they are incorporated into a factorial experiment. Blocks can be established as a chunk variable using a number of background variables and then treated as a factor in the nested design. For example, the nested factor *machine within plant* might include operator, setup, configuration, and the like. Also, blocks can be set up and the nested experimental pattern can be replicated in each block.

The experimental pattern for a nested experiment displays the hierarchical nature of the factors. Figures 8.1, 8.4, and 8.8 show examples of experimental patterns for nested studies.

The amount of replication is important in a nested design. The levels of the factors in the study often dictate the replication that is done. Balanced replication (an equal number of levels for each evaluation of a nested factor) makes the analysis much easier but is not always an efficient use of test resources. The X-bar and R control charts are common examples. It is desirable to select an equal number of samples or measurements within each subgroup. A variable subgroup size can be used, but this complicates the analysis. Complete replication is not necessary in a nested experiment. For example, in the evaluation of a measurement system (Section 8.2) the measurement was repeated on only one of the three samples in the subgroup. Other forms of incomplete replication can be easily incorporated into a nested design (Bainbridge, 1965).

Randomization in a nested experiment is applicable to sampling, the order of running the tests, and the order of making measurements on the experimental units. In contrast to factorial designs, the hierarchical structure of the factors in a nested study often leads to some restrictions on the randomization. For example, once a machine is *set up*, it may be desirable to complete all of the tests within that machine before moving to the next (see the measurement example in Section 8.3). In a medical study, *hospital* and *patients within a hospital* might be two nested factors. Randomization could be used in selecting the hospitals from a list of a particular type hospital available and selecting the patients from those currently in the hospital. The order of visiting each hospital and of surveying the selected patients within each hospital could also be randomized.

Analyzing a Nested Experiment

The examples in the previous sections of this chapter contained some of the analysis procedures for nested studies. The recommended approach to the analysis is based on three steps:

1. Plot the data in run order to evaluate stability during the study. Look for obvious trends and other types of special causes in this run chart.
2. Prepare a dot-frequency diagram for the basic data. The dot-frequency diagram shows all of the observations organized by the associated nested factors.
3. Study the dot-frequency diagram and summarize the information for each factor. It may be desirable to prepare additional dot-frequency diagrams with a different ordering of factors to highlight the most important factors. It also may be desirable to prepare X-bar and R control charts to further evaluate the importance of selected factors.

STEP 1: RUN CHART. An example of a run chart for a nested experiment was given in Figure 8.9. This chart is prepared to evaluate the impact of nuisance variables in the study. Alternatively, X-bar and R charts can be prepared using the variation of the lowest factor in the hierarchy to calculate the range. Note that the X-bar chart may be out of control if any of the factors higher in the hierarchy are important.

It is important to react to any trends, runs, or cycles in the run order plot before preparing a dot-frequency diagram. If not removed, the effect of the nuisance variables that caused the nonrandom pattern will be attributed to one or more of the factors. Individual points that are affected by special causes should also be removed or adjusted prior to preparing the dot-frequency diagram, since this diagram focuses on the factors in the study. If these individual points are not removed, however, they can usually be detected in the dot-frequency diagram.

STEP 2: DOT-FREQUENCY DIAGRAM. The dot-frequency diagram is designed to partition the variation in the original data among the factors in the study. This allows a visual analysis of the data. The construction of a dot-frequency diagram is as follows:

1. Set up a vertical scale to include the highest and lowest numerical values obtained in the study.
2. Develop a horizontal scale using the factors in the study. There will be an identifier row for all of the factors except the lowest factor in the hierarchy. The lowest factor will be represented by different dots plotted directly above the other identifiers. The other factors should be ordered with the highest factor in the hierarchy at the bottom of the scale.
3. Plot the original data in the location on the plot identified by the factor levels associated with each value.
4. For the lowest factor in the hierarchy, draw a vertical line to connect the dots.
5. Draw a box around the lines for the second lowest factor in the hierarchy. Additional boxes can be drawn to identify other factors in the hierarchy.

STEP 3: ANALYSIS OF THE DOT-FREQUENCY DIAGRAM. Figure 8.11 contains illustrations of dot-frequency diagrams for a three-factor nested experiment (factors A, B within A, and C within B and A). Four different diagrams are shown illustrating the following situations:

1. Factor A most important,
2. Factor B most important,
3. Factor C most important,
4. All factors equally important.

Special causes can impact the interpretation of a dot-frequency diagram. Figure 8.12 illustrates dot-frequency diagrams with two types of special causes:

1. The special cause affects a single value.
2. The special cause is a trend across all values.

The impact of the trend will depend on the amount of randomization done. The trend will be associated with the lowest factor in the hierarchy for which the order was randomized. In the example in Figure 8.12, the response increases linearly with run order. Since factor B is the lowest factor in the hierarchy that was randomized, the trend makes factor B appear to have a large effect.

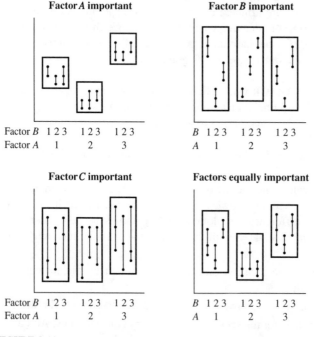

FIGURE 8.11
Illustrations of dot-frequency diagrams

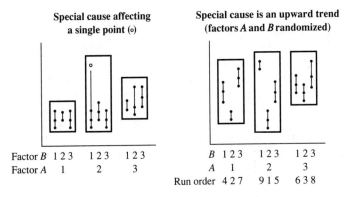

Special cause affecting a single point (o)

Special cause is an upward trend (factors A and B randomized)

Factor B 1 2 3   1 2 3   1 2 3
Factor A   1   2   3

B 1 2 3   1 2 3   1 2 3
A   1   2   3
Run order 4 2 7   9 1 5   6 3 8

FIGURE 8.12
Special causes on a dot-frequency diagram

Sometimes it is desirable to prepare a dot-diagram to summarize the key factors in the study. The factors can be reordered to highlight the most important factors, and factors not important can be eliminated from the diagram. The example in Section 8.5 can be used to illustrate this point. In foundry 2, both the day effect and the part effect are insignificant. A dot-frequency diagram based on the six heats (ignoring days) and the nine measurements of hardness within each heat (ignoring parts) would highlight the important factors.

X-bar and R control charts can be prepared to verify observations on the dot-frequency diagram. The data can be organized in rational subgroups with the least important factors included within subgroups and the important factors included between subgroups. In the three-factor example in Section 8.3, a control chart could be prepared with subgroups defined by the parts, and the averages of the three measurements on each setup within a part could be used to calculate the average and range. The subgroup size (three) would be used to calculate control limits. The control chart would verify the importance of the variation between parts by showing part averages outside of the control limits.

For some nested studies, it may be desirable to quantify the magnitude of each of the factors. A procedure called variance component analysis is available to estimate the component of variation associated with each factor. The measurement study examples in Section 8.2 and Section 8.3 illustrated this analysis (see Table 8.1). This analysis is further described in the Appendix of this chapter.

8.5 MORE THAN THREE FACTORS IN A NESTED DESIGN

The procedures used to evaluate the measurement process in Sections 8.2 and 8.3 can be extended to more than three factors. A study was done to determine the most important factors affecting hardness of parts. The specification for hardness was 30 to 40 Rockwell units. The following factors were considered in the study:

1. Foundries (2 different suppliers of parts).
2. Days within foundries (3 days selected for each foundry).

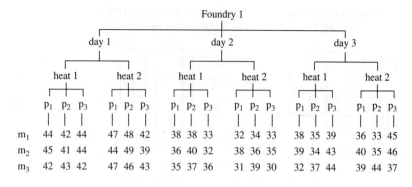

Foundry 1

	day 1				day 2				day 3	
heat 1			heat 2			heat 1			heat 2		

	P_1 P_2 P_3	P_1 P_2 P_3	P_1 P_2 P_3	P_1 P_2 P_3	P_1 P_2 P_3	P_1 P_2 P_3
m_1	44 42 44	47 48 42	38 38 33	32 34 33	38 35 39	36 33 45
m_2	45 41 44	44 49 39	36 40 32	38 36 35	39 34 43	40 35 46
m_3	42 43 42	47 46 43	35 37 36	31 39 30	32 37 44	39 44 37

Foundry 2

	P_1 P_2 P_3	P_1 P_2 P_3	P_1 P_2 P_3	P_1 P_2 P_3	P_1 P_2 P_3	P_1 P_2 P_3
m_1	25 24 28	35 31 29	38 43 34	35 30 32	25 26 31	33 30 34
m_2	26 25 23	33 32 34	39 37 33	34 32 36	24 29 30	32 36 38
m_3	22 29 23	28 30 36	42 38 37	34 32 34	27 23 28	28 34 36

P_1, P_2, P_3 = three parts within each heat
m_1, m_2, m_3 = three measurements of each part
Note: Measurements are in Rockwell hardness scale.

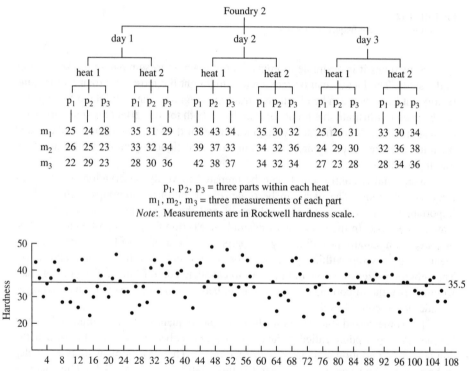

FIGURE 8.13
Design matrix and run chart for a five-factor nested design

3. Heat within a day (2 heats selected for each day).
4. Parts within a heat (3 parts selected from each heat).
5. Measurements within a part (hardness measured 3 times).

A complete nested design was conducted with a resultant 108 ($2 \times 3 \times 2 \times 3 \times 3$) data points. The design matrix is shown in Figure 8.13. Each part was collected and labeled, and then the hardness tests were conducted in random order. A run chart indicating the test order is also shown in Figure 8.13. The run chart does not indicate any special causes. A dot-frequency diagram for the study is shown in Figure 8.14. Each line on the dot-frequency diagram represents an individual part. Each box shown on the graph represents a particular heat of parts. Additional boxes can be drawn to represent each day and each foundry.

The dot-frequency diagram allows one to visually assign the variation in the response variable to each of the factors. The length of the lines drawn represents the measurement variation. The length of the lines ranges from 2 to 12 units and "averages" about 4 or 5 units. The height of the boxes (and the differences in the lines within a box) represents the part-to- part variation. Note that the "average" height of the boxes is greater than the specification width of 10 units.

The difference between the consecutive pairs of boxes indicates the heat-to-heat variation within a day. For foundry 1 there is almost no difference in these pairs of boxes, while for foundry 2 the pairs are very different. The variation due to heats is insignificant in foundry 1, but it is a very important source of variation in foundry 2.

The day-to-day variation within a foundry can best be seen by drawing a box around the two heats in each day (Figure 8.15*a*). Note that there is some day-to-day variation in foundry 1, but the day-to-day variation in foundry 2 is no greater than the heat-to-heat variation.

The variation due to foundry differences can be seen by comparing all of the measurements on the left half of the dot-frequency diagram with those on the right half (Figure 8.15*b*). The measurements average near the upper specification of 40 for foundry 1 and near the lower specification of 30 for foundry 2. Significant improvements would result if foundry 1 could target hardness 5 units less and foundry 2 could target hardness 5 units more.

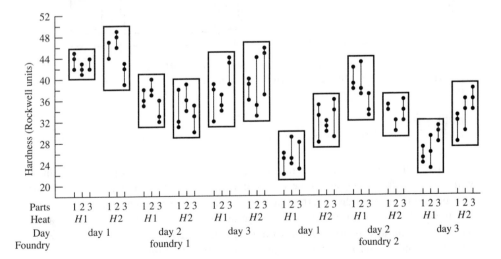

FIGURE 8.14
Dot-frequency diagram for five-factors affecting hardness of parts

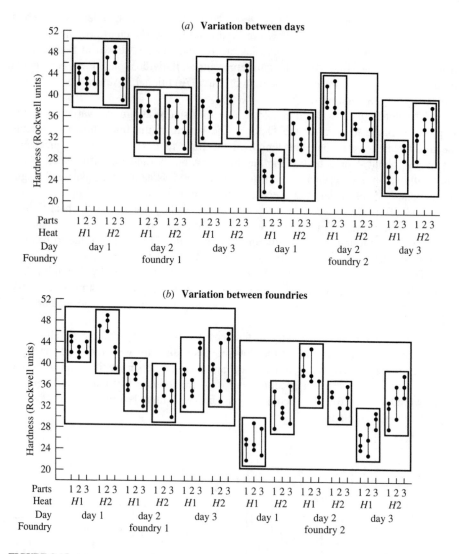

FIGURE 8.15

Dot-frequency diagram (hardness of parts) for five-factor nested design with different factors emphasized

To summarize the dot-frequency diagrams, reduction in the variation in hardness will require different actions in the two foundries. For foundry 1, the process should first be changed to produce parts that average about five hardness units less. Next, the focus should be on the day-to-day variation. The procedure used to measure hardness should also be reviewed for improvement. The heat-to-heat and the part-to-part variations are not important sources of hardness variation.

For foundry 2, the process should first be changed to produce parts that average about five hardness units more. Then the focus should be on the variation from heat to heat. The measurement process is also a significant source of variation for the parts from foundry 2. The part-to-part variation and the day-to-day variation are not important sources of hardness variation.

The amount of variation due to each of the factors in the study can be quantified by a variance component analysis. This analysis is given in the Appendix of this chapter.

The summary of the variance components supports the conclusions from the dot-frequency diagrams. The action required for improvement is different in the two foundries. In both foundries, the measurement process represents about 21% of the total variation. In foundry 1 the variation between days accounts for 67% of the variation, while in foundry 2 the variation between heats within a day accounts for 69% of the total variation.

8.6   A STUDY WITH NESTED AND CROSSED FACTORS

Sometimes a study will contain some factors that are crossed (a factorial pattern) and some factors that are nested (a nested pattern). The analysis of such a study combines graphical tools from both factorial and nested designs. The following example illustrates this.

An interlaboratory study was done to evaluate a new analytical method to determine particle size, a method that would be used in three laboratories within one company. The response variables were the weights of materials that passed through various size screens. The key response variable was the amount of material not passing through a 200-mesh screen. The following seven factors were included in the study:

1. Alignment of screens (stack A and stack B).
2. Sample weights (50 gram and 100 gram).
3. Flow aids (1 and 2).
4. Laboratories (A, B, C).
5. Analyst within laboratory (two within each lab).
6. Days within analyst (two days for each analyst).
7. Measurement within days (two measurements each day).

The first three factors are across the entire study and thus form a factorial design with each factor at two levels. The last four factors form a nested design. The design matrix for the entire experiment is shown in Figure 8.16. The factors are organized on the design matrix to highlight the focus of the study. The three factorial factors represent different ways to run the test, while the three nested factors represent different conditions under which a given test method will be run in the future. One of the objectives of the study was to choose a test condition that would minimize the measurement variation across different laboratories and analysts.

Table 8.2 contains the data obtained from the study. Run charts were prepared for the data from each laboratory. Excessive variation was noted in laboratory C for analyst B. Figure 8.17 shows eight dot-frequency diagrams for the four nested factors. There is one diagram for each of the eight factorial combinations.

Factorial design matrix for three crossed factors

	Alignment - stack A		Alignment - stack B	
	Flow aid 1	Flow aid 2	Flow aid 1	Flow aid 2
Sample size = 50 g	22 tests (see design below)	22 tests	22 tests	22 tests
Sample size = 100 g	22 tests	22 tests	22 tests	22 tests

**Experimental pattern in each of the eight cells
in the factorial design matrix**

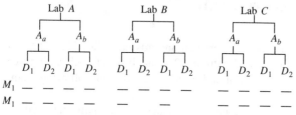

Lab: laboratory in which tests will be run.
A_i: Analyst a or analyst b in each laboratory.
D_i: First day of tests or second day of tests by each analyst in each lab.
M_i: First or second measurement each day by each analyst.

FIGURE 8.16
Design matrix for nested/factorial study to evaluate new method of determining particle size

The dot-frequency diagrams indicate that laboratory C has much more variable results than the other laboratories. The repeat tests by the same analyst on the same day are usually consistent, except for 5 of the 56 cases. The variation between days (repeated tests on different days) can be studied by comparing the lines within each box. These lines overlap in all but 7 of the 32 cases in laboratories A and B. Variation between analysts in the same lab (comparison between consecutive pairs of boxes) is usually small. The second-day results for analyst A in laboratory A for the conditions stack A, flow aid 2, and the 50-gram sample are inconsistent with other data and were not used in further analysis. In summary, for the four nested factors:

1. Laboratories A and B show good agreement on most tests. The test results for laboratory C show much more variation and are not used in further analysis.
2. Analysts within a laboratory show generally good agreement. The differences within laboratories are not consistent with any one analyst.
3. The variation from day to day is not much greater than the variation from repeated tests. Since the variation between labs A and B is small, all the results can be pooled together in further analysis.

TABLE 8.2
Data from interlaboratory study

				Analyst A				Analyst B			
				Day 1		Day 2		Day 1		Day 2	
Stk	Fa	Sample size	Lab	M_1	M_2	M_1	M_2	M_1	M_2	M_1	M_2
A	1	50	A	29.4	28.2	27.3	31.6	30.2	26.0	28.0	27.5
			B	26.6	26.7	26.4	—	27.3	29.4	27.2	—
			C	29.	34.	50.	52.	25.	30.	42.	50.
A	1	100	A	31.3	32.0	29.0	27.5	29.7	32.3	36.5	35.2
			B	35.4	43.6	44.0	—	27.7	28.7	28.0	—
			C	31.	31.	34.	32.	57.	43.	46.	48.
A	2	50	A	33.7	32.3	44.8	41.7	27.4	29.0	30.4	31.4
			B	32.8	30.5	33.0	—	31.4	27.3	29.2	—
			C	31.	31.	30.	30.	43.	49.	30.	30.
A	2	100	A	27.8	27.8	34.4	36.3	36.8	37.1	37.7	37.4
			B	40.5	40.3	37.8	—	39.8	42.1	40.3	—
			C	30.	27.	36.	35.	50.	51.	35.	34.
B	1	50	A	31.7	27.5	27.3	31.6	31.2	27.2	27.4	27.4
			B	26.3	26.6	27.1	—	26.3	26.6	26.8	—
			C	35.	34.	33.	34.	27.	27.	27.	28.
B	1	100	A	28.0	27.9	29.0	28.2	32.4	30.7	28.8	29.6
			B	34.0	34.2	32.5	—	35.0	28.5	29.6	—
			C	39.	40.	31.	32.	43.	33.	31.	29.
B	2	50	A	29.7	32.1	32.4	31.9	27.8	29.8	28.8	28.9
			B	29.6	28.5	29.3	—	27.1	27.7	27.1	—
			C	36.	34.	30.	51.	30.	29.	28.	29.
B	2	100	A	45.1	42.2	34.5	35.5	32.6	36.1	35.6	34.7
			B	35.2	38.4	29.2	—	32.0	33.1	33.1	—
			C	56.	56.	32.	36.	44.	50.	33.	38.

Notes:
Stk: Stack *A* or stack *B* (alignment of screens).
Fa: Flow aids 1 or 2.
Size: Sample size of 50 or 100 grams.
Lab: Laboratory in which tests were done (*A*, *B*, or *C*).
M_1 Individual measurements of percent over 200-mesh screen. (No repeat analysis in laboratory *B*
    on the second day; data from laboratory *C* reported in whole numbers.)

To analyze the crossed factors, the data is first summarized. Only data from laboratories *A* and *B* was used in calculating the summary statistics shown in Figure 8.18.

The effects of the factors on both the average and standard deviation are important in this study. The standard deviation is a measure of the consistency of the test method on different days, by different analysts, and in different laboratories. Figure 8.19 shows these statistics displayed on cubes (see Chapter 6).

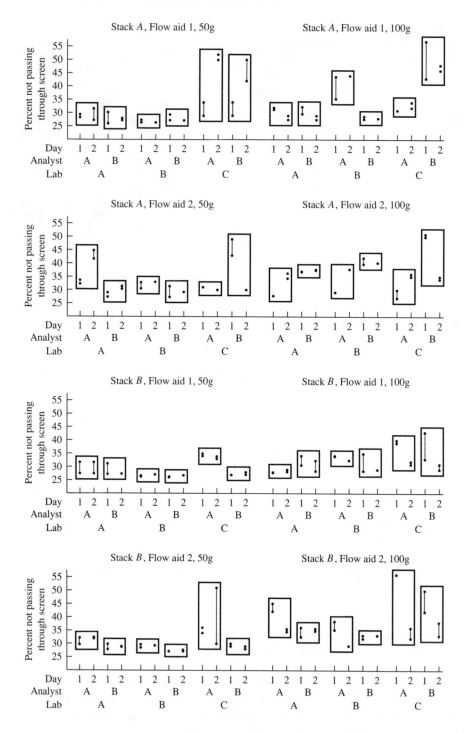

FIGURE 8.17
Dot-frequency diagram for nested factors in nested/factorial study

	Alignment - stack A		Alignment - stack B	
	Flow aid 1	Flow aid 2	Flow aid 1	Flow aid 2
Sample size = 50 g	$\bar{x} = 27.99$ $s = 1.61$	$\bar{x} = 30.70*$ $s = 2.13*$	$\bar{x} = 27.93$ $s = 1.98$	$\bar{x} = 29.34$ $s = 1.76$
Sample size = 100 g	$\bar{x} = 32.92$ $s = 5.47$	$\bar{x} = 36.86$ $s = 4.35$	$\bar{x} = 30.60$ $s = 2.52$	$\bar{x} = 35.52$ $s = 4.11$

x = percent of material not through 200-mesh screen.
$\bar{x}$ = average of 14 test results from laboratories A and B.
s = standard deviation of 14 test results from labs A and B.
*Data for lab A, analyst A, day 2 not used.

FIGURE 8.18
Statistical summary of data for labs A and B

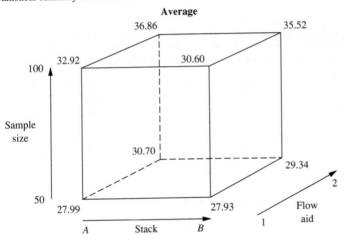

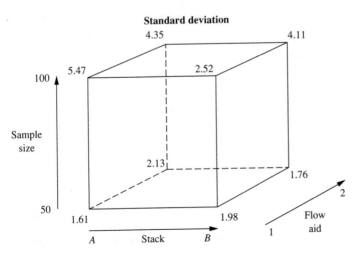

FIGURE 8.19
Cube analysis for summary statistics

Table 8.3 contains estimates of the effects for the factors and interactions and also contains response plots for the important factors.

The following conclusions can be made for this study:

1. Laboratory C reported inconsistent results in the study. The data was only reported to whole numbers. Large differences between analysts and between days under the same test conditions were observed. Training in the test method should be done in laboratory C.

2. The agreement between laboratories A and B was good. Analyst and day variation were also small in these laboratories.

3. The sample size (50 or 100 grams) was the most important factor in the study. Increasing the sample size from 50 to 100 grams increased the percent of material not passing through a 200-mesh screen by about 5%. The standard deviation of

TABLE 8.3
Estimates of factor effects and interactions (interlaboratory study)

Factors (interactions)	Average effects (percent)	Standard deviation effects (percent)
Sample size (S)	4.98	2.24
Flow-aid (F)	3.25	0.19
Screen alignment (A)	−1.27	−0.08
$S \times F$	1.19	0.04
$S \times A$	−0.56	−0.80
$F \times A$	−0.08	0.49
$S \times F \times A$	−0.57	0.86

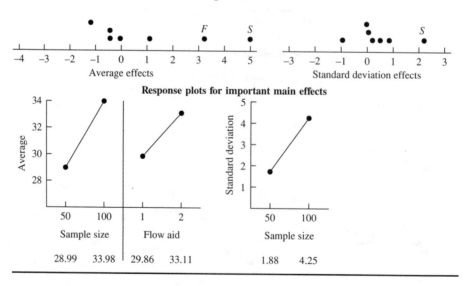

Dot diagrams for effects

Average effects

Standard deviation effects

Response plots for important main effects

28.99   33.98 | 29.86   33.11   1.88   4.25

repeated tests (different laboratories, days, and analysts) more than doubles from 1.88 to 4.25 percent.

4. Using flow aid 2 results in about a 4.2% increase over flow aid 1 in material not passing through a 200-mesh screen.

In other nested/factorial experiments, it may be appropriate to analyze the crossed factors first and then the nested factors.

8.7   SUMMARY

This chapter has discussed planned experimentation for nested factors used to evaluate the important sources of variation. Physical and location constraints and sampling and testing considerations often lead to nested factors. The Shewhart control chart can be viewed as a two-factor nested design. Many of the examples in this chapter dealt with measurement processes.

Planning a nested study is similar to planning a factorial design, but one difference is how the randomization is performed. The dot-frequency diagram is the primary analysis tool to partition the variation in the response variables among the factors. Components of variance to quantify the amount of variation associated with each factor can be estimated. Appendix 8A contains details on these calculations.

Nested factors can be mixed with crossed factors to form more complex experimental patterns.

APPENDIX 8A:
CALCULATION OF VARIANCE COMPONENTS

Appendix 8A presents methods for quantifying the importance of the factors in a nested study. These procedures are known as a *variance component analysis*. This analysis is only meaningful for a stable process; that is, a process with a constant cause system. With an unstable process, special causes are present, so the analysis can give meaningless results.

The basic steps in the procedure are given, and the analysis is illustrated for two of the examples in Chapter 8. The variance component analysis is an extension of the procedure used to separate measurement and product variation from total process variation (Table 8.1) presented in Section 8.2.

Estimates of the variation due to each factor are called variance components. Ranges can be calculated from the data and from averages of the data to obtain estimates of these variance components. But the ranges calculated from the averages will also include variation from the sources that were averaged. It is therefore necessary to understand exactly which components of variance are estimated from each quantity and then solve for the appropriate variance components. The procedure given here to estimate the components of variance for each factor has been described by Box, Hunter, and Hunter (1978) and Snee (1983).

The following steps summarize the calculation of the variance components:

1. Calculate ranges and averages for the factor in the lowest hierarchy. (*Note*: The standard deviation can be used instead of the range at each step in this analysis. For factors with greater than five levels the standard deviation is preferred.)

2. Using the averages from Step 1, calculate ranges and averages for the second-lowest factor in the nested design.

3. Continue calculating ranges and averages for each factor in the design, using the averages from the previous step.

4. For the highest factor in the hierarchy, the range will be based on only one subgroup of averages.

5. Calculate the average range and an upper control limit for each set of ranges calculated in the first four steps. Any ranges greater than the upper control limit should be investigated for special causes. Corrections should be made or the range eliminated and the average range recalculated.

6. Estimate the standard deviation for each average range using the tabled values of d_2^* from the table in Appendix B. (*Note*: If the standard deviation is used instead of the range, the pooled standard deviation would be calculated in place of the average range. No factor would be needed.)

7. Square the standard deviation to obtain a variance associated with each factor. Write out the quantity estimated from each calculated variance. In general, each variance will estimate the variance component associated with the averages from which the range was calculated and some fraction of the variance component for all factors lower in the hierarchy. The fraction will be one over the total number of measurements of the factor in the average used to calculate the range.

8. Use the form of the calculated variance to calculate each variance component estimate. Begin with the lowest factor in the hierarchy and substitute the calculated components in determining the component for the next factor in the hierarchy. It is possible to obtain a negative value for a variance component using this procedure. If a negative value is calculated, an estimate of zero should be used for that component.

9. Summarize the variance components by calculating the percent of the total variation associated with each nested factor.

The following two examples illustrate these steps. In the first example, the standard deviation is used instead of the range. The second example uses the range.

MEASUREMENT STUDY EXAMPLE (SECTION 8.3)

In Section 8.3, a three-factor nested study of a measurement process was described. The three factors studied were parts, setups within parts, and measurement within a setup. Table 8.4 shows the summary calculations required to do a variance component analysis for the data from this study. First, averages and standard deviations of the three measurements are calculated. Then the average and standard deviation are calculated

TABLE 8.4
Analysis of three-factor nested study

		Measurements			Summary statistics			
Part	Setup	1	2	3	$\bar{X}_1$	S_1	$\bar{X}_2$	S_2
1	1	3.40	3.45	3.70	3.52	.16		
	2	3.95	3.80	4.00	3.92	.10		
	3	3.95	4.00	4.15	4.03	.10	3.82	.268
2	1	4.55	4.70	4.70	4.65	.09		
	2	4.30	4.20	4.50	4.33	.15		
	3	4.75	5.15	5.20	5.03	.25	4.67	.350
3	1	4.30	4.70	3.85	4.28	.43		
	2	4.50	4.55	4.65	4.57	.08		
	3	4.15	4.45	4.50	4.37	.19	4.41	.148
4	1	2.60	2.65	2.50	2.58	.08		
	2	3.00	3.15	3.15	3.10	.09		
	3	3.40	3.55	3.40	3.45	.09	3.04	.438
5	1	3.90	4.85	4.60	4.45	.49		
	2	4.35	4.90	4.35	4.53	.32		
	3	4.10	4.00	3.95	4.02	.08	4.33	.274
6	1	3.95	4.10	4.00	4.02	.08		
	2	3.90	3.90	4.00	3.93	.06		
	3	4.40	4.30	4.40	4.37	.06	4.11	.232

$$\bar{\bar{X}} = 4.06 \qquad \bar{S}_1 = .204 \qquad\qquad \bar{S}_2 = .299$$
$$\text{UCL} = .521 \qquad S_3 = .578 \quad \text{UCL} = .77$$

Notes
$\bar{X}_1$ = Average of the three measurements for each setup.
S_1 = Standard deviation of the three measurements for each setup.
$\bar{X}_2$ = Average of the three setup averages for each part.
S_2 = Standard deviation of the three setup averages for each part.
$\bar{\bar{X}}$ = Overall average of data.
$\bar{S}_1$ = Pooled standard deviation of measurements.
$\bar{S}_2$ = Pooled standard deviation of averages for each setup.
S_3 = Standard deviation of the averages for each part.
UCL = Upper control limits for the individual standard deviations used to compute the pooled
     values.

for each set of three setup averages. Control limits for the standard deviations are
calculated to evaluate stability of the individual estimates, and pooled averages of the
individual standard deviations are calculated. Finally, the standard deviation of the six
part averages is calculated.

Table 8.5 shows the calculation to partition the variation among the factors. The
three standard deviations obtained from Table 8.4 are used to obtain the estimates of
the variance components. An expression is written for each standard deviation and

TABLE 8.5
Calculation of components of variation

Summary statistics from Table 8.4 (units = millimeters)

$\bar{S}_1 = .203 =$ Pooled standard deviation of measurements.
$\bar{S}_2 = .299 =$ Pooled standard deviation of setup averages.
$S_3 = .578 =$ Standard deviation of the averages for each part.

Standard deviation of measurement ($\hat{\sigma}_m$)

$$\hat{\sigma}_m = \bar{S}_1 = .203$$

Standard deviation of setup ($\hat{\sigma}_s$)

$$\bar{s}_2^2 = \hat{\sigma}_s^2 + \sigma_m^2/3$$

$$\hat{\sigma}_s = \sqrt{\bar{s}_2^2 - \hat{\sigma}_m^2/3}$$

$$\hat{\sigma}_s = \sqrt{.299^2 - .203\,^2/3}$$
$$\hat{\sigma}_2 = .275$$

Standard deviation of parts ($\hat{\sigma}_p$)

$$s_3^2 = \sigma_p^2 + \sigma_s^2/3 + \sigma_m^2/9$$

$$\hat{\sigma}_p = \sqrt{s_3^2 - \hat{\sigma}_s^2/3 - \hat{\sigma}_m^2/9}$$

$$\hat{\sigma}_p = \sqrt{.578^2 - .275^2/3 - .203\,^2/9}$$
$$\hat{\sigma}_p = .552$$

Summary of sources of variation (units = millimeters)

Source of variation	Standard deviation	Variance componenet	Percent of variation
Part to part	.552	.3043	72.2
Setup within a part	.275	.0758	18.0
Measurement	.203	.0412	9.8
Total	.649	.4213	100.0

then solved to compute the components. This variance component analysis verifies the visual analysis of the dot-frequency diagram given in Section 8.3:

- 72 % of the variation is attributable to part differences,
- 18 % is attributable to the setup, and
- 10 % is due to the measurement procedure.

FOUNDRY EXAMPLE (SECTION 8.5)

In Section 8.5, a five-factor nested study for hardness (in Rockwell units) of parts from two foundries was described. The amount of variation due to each of the factors in the study can be quantified by a variance component analysis. Tables 8.6 through 8.9 contain these calculations for the five-factor nested study.

TABLE 8.6
Analysis of hardness study data—foundry 1

Day	Heat	Part	Measurements 1	2	3	$\overline{X}_a$	R_a	$\overline{X}_b$	R_b	$\overline{X}_c$	R_c
1	1	1	44	45	42	43.6	3				
		2	42	42	43	42.0	2				
		3	44	44	42	43.3	2	43.0	1.6		
	2	1	47	44	47	46.0	3				
		2	48	49	46	47.7	3				
		3	42	39	43	41.3	4	45.0	6.4	44.0	2.0
	1	1	38	36	35	36.3	3				
		2	38	40	37	38.3	3				
		3	33	32	36	36.7	4	37.1	2.0		
	2	1	32	38	31	33.7	7				
		2	34	36	39	36.3	5				
		3	33	35	30	32.7	5	34.2	3.6	35.7	1.6
3	1	1	38	39	32	36.3	7				
		2	35	34	37	35.3	3				
		3	39	43	44	42.0	5	38.0	6.7		
	2	1	36	40	39	38.3	4				
		2	33	35	44	37.3	11				
		3	45	46	37	42.7	9	39.4	5.4	38.7	1.4

$$\overline{\overline{X}} = 39.5 \quad \overline{R}_a = 4.61 \quad \overline{R}_b = 4.28 \quad \overline{R}_c = 1.68 \quad \overline{R}_d = 8.35$$
$$\text{UCL} = 11.8 \quad \text{UCL} = 11.0 \quad \text{UCL} = 5.5$$

	S_a^2	S_b^2	S_c^2	S_d^2
Number of subgroups (k)	18	6	3	1
Measurements/subgroup (n)	3	3	2	3
Standard deviation factor d_2^*	1.69	1.73	1.23	1.91
Standard deviation (S)	2.72	2.53	1.49	4.93
Variance (S^2)	7.40	6.40	2.21	24.32

$\overline{X}_a$ = Part averages, average of the three measurements.
R_a = Range of the three measurements.
$\overline{X}_b$ = Heat averages, average of the three part averages.
R_b = Range of the three part averages.
$\overline{X}_c$ = Day averages, average of the two heat averages.
R_c = Range of the two heat averages.
$\overline{\overline{X}}$ = Foundry average, average of the three day averages.
R_d = Range of the three day averages.
S_a^2 = Variance based on measurements.
S_b^2 = Variance based on part averages.
S_c^2 = Variance based on heat averages.
S_d^2 = Variance based on day averages.
UCL = Upper control limit for the ranges in this column.
d_2^* = Standard deviation factor (from Appendix 8B).

TABLE 8.7
Analysis of hardness study data—foundry 2

Day	Heat	Part	Measurements 1	2	3	$\overline{X}_a$	R_a	$\overline{X}_b$	R_b	$\overline{X}_c$	R_c
1	1	1	25	26	22	24.3	4				
		2	24	25	29	26.0	5				
		3	28	23	23	24.7	5	25.0	1.7		
	2	1	35	33	28	32.0	7				
		2	31	32	30	31.0	2				
		3	29	34	36	33.0	7	32.0	2.0	28.5	7.0
2	1	1	38	39	42	39.7	4				
		2	43	37	38	39.3	6				
		3	34	33	37	34.7	4	37.9	5.0		
	2	1	35	34	34	34.3	1				
		2	30	31	32	31.0	2				
		3	32	36	34	34.0	4	33.1	3.3	35.5	4.8
3	1	1	25	24	27	25.3	3				
		2	26	29	23	26.0	6				
		3	31	30	28	29.7	3	27.0	4.7		
	2	1	33	32	28	31.0	5				
		2	30	36	34	33.3	6				
		3	34	38	36	36.0	4	30.2	5.0	30.2	6.4

$$\overline{X} = 31.4 \qquad \overline{R}_a = 4.39 \qquad \overline{R}_b = 3.62 \qquad \overline{R}_c = 6.07 \qquad R_d = 7.00$$
$$\text{UCL} = 11.3 \qquad \text{UCL} = 9.3 \qquad \text{UCL} = 19.8$$

Number of subgroups (k)	18	6	3	1
Measurements/subgroup (n)	3	3	2	3
Standard deviation factor d_2*	1.69	1.73	1.23	1.91
Standard deviation (S)	2.59	1.88	4.93	3.66
Variance (S^2)	6.72	4.38	24.35	13.43
	S_a^2	S_b^2	S_c^2	S_d^2

$\overline{X}_a$ = Part averages, average of the three masurements

R_a = Range of the three measurements.

$\overline{X}_b$ = Heat averages, average of the three part averages.

R_b = Range of the three part averages.

$\overline{X}_c$ = Day averages, average of the two heat averages.

R_c = Range of the two heat averages.

$\overline{X}$ = Foundry average, average of the three day averages.

R_d = Range of the three day averages.

S_a^2 = Variance based on measurements.

S_b^2 = Variance based on part averages.

S_c^2 = Variance based on heat averages.

S_d = Variance based on day averages.

UCL = Upper control limit for the ranges in this column.

d_2* = Standard deviation factor (from Appendix 8B).

TABLE 8.8
Calculation of variance components—five-factor hardness data study

Variance	Foundry 1	Foundry 2	Form of the calculated variance
S_a^2	7.40	6.72	$\hat{\sigma}_m^2$
S_b^2	6.40	4.38	$\hat{\sigma}_m^2/3$
S_c^2	2.21	24.35	$\hat{\sigma}_m^2/9 + \hat{\sigma}_p^2/3$
S_d^2	24.32	13.42	$\hat{\sigma}_m^2/18 + \hat{\sigma}_p^2/6 + \hat{\sigma}_h^2/2 + \hat{\sigma}_d^2$

Calculation of variance components for foundry 1

$$\hat{\sigma}^2 = S_a^2$$
$$\hat{\sigma}^2 = 7.40$$
$$\hat{\sigma} = 2.72$$

$$\hat{\sigma}_p^2 = S_b^2 - \hat{\sigma}_m^2/3$$
$$\hat{\sigma}_p^2 = 6.40 - 7.40/3$$
$$\hat{\sigma}_p^2 = 3.93$$
$$\hat{\sigma}_p = 1.98$$

$$\hat{\sigma}_h^2 = S_c^2 - \hat{\sigma}_p^2/3 - \hat{\sigma}_m^2/9$$
$$\hat{\sigma}_d^2 = 2.21 - 3.93/3 - 7.40/9$$
$$\hat{\sigma}_h = 0.08$$
$$\hat{\sigma}_h = 0.28$$

$$\hat{\sigma}_d^2 = S_d^2 - \hat{\sigma}_c^2/2 - \hat{\sigma}_p^2/6 - \hat{\sigma}_m^2/18$$
$$\hat{\sigma}_d^2 = 24.32 - 0.08/2 - 3.88/6 - 7.40/18$$

$$\hat{\sigma}_d^2 = 23.11$$
$$\hat{\sigma}_d = 4.80$$

Calculation of variance components for foundry 2

$$\hat{\sigma}^2 = S_a^2$$
$$\hat{\sigma}^2 = 6.72$$
$$\hat{\sigma} = 2.59$$

$$\hat{\sigma}_p^2 = S_b^2 - \hat{\sigma}_m^2/3$$
$$\hat{\sigma}_p^2 = 4.38 - 6.72/3$$
$$\hat{\sigma}_p^2 = 2.14$$
$$\hat{\sigma}_p = 1.46$$

$$\hat{\sigma}_h^2 = S_c^2 - \hat{\sigma}_p^2/3 - \hat{\sigma}_m^2/9$$
$$\hat{\sigma}_d^2 = 24.35 - 2.14/3 - 6.72/9$$
$$\hat{\sigma}_h = 22.89$$
$$\hat{\sigma}_h = 4.78$$

$$\hat{\sigma}_d^2 = S_d^2 - \hat{\sigma}_c^2/2 - \hat{\sigma}_p^2/6 - \hat{\sigma}_m^2/18$$
$$\hat{\sigma}_d^2 = 13.43 - 22.89/2 - 2.14/6 - 6.72/18$$
$$\hat{\sigma}_d^2 = 1.26$$
$$\hat{\sigma}_d = 1.12$$

TABLE 8.9
Summary of variance components for hardness data

Source of variation	Standard deviation $\hat{\sigma}$	Variance component $\hat{\sigma}^2$	Percent of variation
Variance components for foundry 1			
Days within foundry	4.80	23.11	66.9
Heats within days	0.28	0.08	0.3
Parts within heats	1.98	3.93	11.4
Measurement within parts	2.72	7.40	21.4
Total		34.52	100.0
Variance components for foundry 2			
Days within foundry	2.59	1.26	3.8
Heats within days	1.46	22.89	69.3
Parts within heats	4.78	2.14	6.5
Measurement within parts	1.12	6.72	20.4
Total		33.01	100.0

The dot-frequency diagrams in Section 8.5 indicated that the important factors were different in the two foundries. Therefore, the variance component analysis is done separately for each foundry. The variance components for each foundry (Table 8.9) are compared to the visual analysis of the dot-frequency diagrams in Section 8.5.

APPENDIX 8B:
CALCULATING AND COMBINING STATISTICS ($\overline{X}$, R, or S)

In chapter 8, statistics were frequently combined to obtain an overall statistic. Appendix 8B summarizes the formula for combining averages, ranges, and standard deviations. A table of factors for estimating the standard deviation from the range is also included.

When n is equal for all k values of the statistic:

$$\overline{R} = \Sigma R/k$$

$$\overline{\overline{x}} = \Sigma \overline{x}/k$$

$$\overline{S} = \sqrt{\Sigma S^2/k}$$

When n is different for some of the statistics:
(the range should not be used with unequal groups of data)

$$\overline{\overline{x}} = \frac{n_1 \overline{x}_1 + n_2 \overline{x}_2 + \ldots + n_k \overline{x}_k}{n_1 + n_2 + \ldots + n_k}$$

$$\overline{s}^2 = \frac{(n_1 - 1)s_1^2 + (n_2 - 1)s_2^2 + \ldots + (n_k - 1)s_k^2}{(n_1 - 1) + (n_2 - 1) + \ldots + (n_k - 1)}$$

$$\sigma = \overline{R}/d_2$$

(use d_2^* when less than 10 subgroups)

Table of d_2 and d_2^* values

n	d_2	$k = 10$	$k = 8$	$k = 6$	$k = 5$	$k = 4$	$k = 3$	$k = 2$	$k = 1$
2	1.128	1.16	1.17	1.18	1.19	1.21	1.23	1.28	1.41
3	1.693	1.72	1.72	1.83	1.74	1.75	1.77	1.81	1.91
4	2.059	2.08	2.08	2.09	2.10	2.11	2.12	2.15	2.24
5	2.326	2.34	2.35	2.35	2.36	2.37	2.38	2.40	2.48
6	2.534	2.55	2.55	2.56	2.56	2.57	2.58	2.60	2.67
7	2.704	2.72	2.72	2.73	2.73	2.74	2.75	2.77	2.83
8	2.847								
9	2.970				(use d_2)				
10	3.078								

REFERENCES

Bainbridge, T. R. (1965): "Staggered, Nested Designs for Estimating Variance Components," *Industrial Quality Control*, no. 22, pp. 12–20.

Box, G. E. P., W.G.Hunter, and J.S. Hunter (1978): *Statistics for Experimenters*, Wiley-Interscience, New York, Chapter 17.

Snee, R. D. (1983): "Graphical Analysis of Process Variation Studies," *Journal of Quality Technology*, vol. 15, no. 2, April, pp. 76–88.

Trout, R. (1985): "Design and Analysis of Experiments to Estimate Components of Variation—Two Case Studies," *Experiments in Industry*, Chemical and Process Industries Division, American Society for Quality Control. Quality Press, 230 West Wells Street, Milwaukee, WI 53203

EXERCISES

8.1. Develop a design matrix for a three-factor nested design with the following number of levels for each factor:

factor	levels
A	4
B within A	3
C within B	2

8.2. Select a control chart that you are familiar with. Describe the control chart as a nested design. What do each of the factors represent?

8.3. How many runs are required for a four-factor factorial design with each factor at two levels? How many runs are required for a four-factor nested design with each factor at two levels? What are the key differences in designing these two studies? What are important differences in analyzing the results of the two types of studies?

8.4. Describe a process where the physical layout would require a nested design to study the process. What are the factors, and how many levels does each factor have? Develop a design matrix for a study of the process.

8.5. Consider a four-factor nested design with each factor at three levels. How many runs are required for a complete replication of the design? Sketch a dot-frequency diagram from such a study where the second factor in the hierarchy (e.g., B within A) is the dominant factor. The other three factors are of very little importance.

8.6. Develop a design matrix for a three-factor nested design with each factor at two levels. Assume the study was completed but the order of testing was not randomized. The chemical tests to measure the response variable were done in the order given on the design matrix. None of the three factors were important, but the test apparatus had a steady drift during the study that was undetected. Sketch a dot-frequency diagram for the results of this study.

8.7. Consider the measurement study example in Section 7.3. If four measurements of each production part could be made, how many setups should be made to obtain the most precise average measurement of the part?

8.8. *Delays in expense reports*. The Accounting Department Quality Improvement Team is studying problems in handling expense reports. Many employees have complained about long delays in receiving reimbursement of expenses. A number of factors were identified that might cause varying processing times. The following three factors were selected as the most likely causes:

1. Reports originated in the field versus headquarters.

2. Processing group—two groups handled field reports and two other groups handled reports from headquarters.

3. Variation in the clerks processing the expense reports.

Data was collected for the 18 expense reports processed during the next week to investigate the problem.

Analyze this data as a three-factor nested design. Prepare a run order plot and a dot-frequency diagram. (*Note*: This example was developed by Rob Stiratelli of Rohm & Haas Company).

Expense report	Originating location	Processing group	Clerk	Processing time (days)
1	field	B	2	6
2	headquarters	C	1	4
3	headquarters	C	2	2
4	headquarters	D	2	9
5	field	A	2	6
6	field	A	1	2
7	field	B	1	9
8	field	B	2	8
9	headquarters	D	1	1
10	headquarters	D	2	7
11	field	B	3	5
12	field	B	3	8
13	headquarters	C	1	2
14	headquarters	C	2	5
15	field	A	2	3
16	headquarters	D	1	4
17	field	B	1	5
18	field	A	1	5

8.9. *Measurement variation study.* A study was done to evaluate the measurement of the flushness of a fitted assembly. The measurement process simulated the assembly by clamping the key part into a fixture constructed for the purpose. Then a micrometer was used to measure the flushness (the gap at a critical location between the part and the fixture). Measurements were in fractions of a millimeter.

Ten parts were selected from the process during a period when the process was stable. Each part was placed in the fixture (a setup) two different times, and the micrometer measurement was performed twice after each setup. The 20 setups were done in a random order. The following data was obtained.

			Measurement	
Run order	Part	Setup	1st	2nd
1	6	1	0.75	0.80
2	2	1	0.50	0.55
3	10	1	0.65	0.70
4	3	1	0.40	0.50
5	6	2	0.60	0.60
6	9	1	1.95	1.80
7	3	2	1.80	1.90
8	9	2	1.90	1.75
9	1	1	1.70	1.70
10	4	1	1.65	1.75
11	5	1	1.50	1.40
12	7	1	1.30	1.40
13	2	2	1.15	0.95
14	4	2	0.95	1.00
15	5	2	1.00	0.90
16	10	2	0.85	1.80
17	7	2	0.70	0.85
18	8	1	0.70	0.65
19	1	2	0.65	0.65
20	8	2	0.50	0.55

Analyze this data as a three-factor nested design. Prepare a run chart and a dot-frequency diagram. What is the biggest contributor to variation of the measurement procedure?

8.10. *Variation in a milling process.* A milling process was used to obtain the correct width of a metal part. Both the width and the microfinish (smoothness) of the part were important quality characteristics. The process was run by two operators, each responsible for two mills. Each mill had two spindles and there were two part positions (left and right) for each spindle. Thus, each machine could mill four parts at a time (two spindles with two positions each).

To study the important sources of variation in this process, the process supervisor selected two parts from each position during a one-hour period. The pairs of parts were labeled by location and then randomly ordered for measurement of width (inches) and microfinish (micro units). The following data was obtained:

Order	Operator	Mill	Spindle	Position	Width (inches) Part 1	Width (inches) Part 2	Microfinish Part 1	Microfinish Part 2
1	A	101	1	L	2.003	1.998	110	122
2	A	123	1	L	2.011	2.007	112	115
3	B	220	2	L	1.991	1.989	123	113
4	A	123	2	R	2.009	2.008	130	126
5	B	220	1	L	2.001	1.998	138	130
6	B	285	1	R	2.015	2.012	155	120
7	B	285	1	R	2.001	1.999	148	125
8	A	101	1	L	1.996	2.001	160	146
9	B	220	2	R	1.900	1.988	158	136
10	A	123	2	R	2.012	2.010	171	155
11	A	101	2	R	2.001	2.004	182	164
12	B	285	2	L	2.003	1.999	168	160
13	B	220	2	L	1.999	2.000	180	172
14	A	123	2	L	2.009	2.007	182	166
15	A	101	1	R	1.997	1.999	198	183
16	B	285	1	R	2.000	2.002	186	192

Develop a design matrix for this study. Show the results of the study on the design matrix.

Analyze the data for each quality characteristic with run charts and dot-frequency diagrams. For each characteristic, list the factors in order of contribution to variation.

What action is needed to reduce variation in this process?

CHAPTER
9

EXPERIMENTS
FOR SPECIAL
SITUATIONS

This chapter presents experimental designs for some special situations that occur in activities to improve quality. The focus of the chapter is on factorial patterns with more than two levels and designs for assemblies with interchangeable parts. An overview of designs for mixtures will be given in Section 9.5. A brief description of other special designs is included in Section 9.6.

9.1 FACTORIAL DESIGNS WITH MORE THAN TWO LEVELS

Chapter 6 discussed factorial designs at two levels. Usually, these designs can be used in sequence to understand the effects of factors, but sometimes it is desirable to study a factor in an experiment at more than two levels. For example, if the effect of a raw material supplier on the process yield is to be evaluated, and there are three raw material suppliers, it often is desirable to include all three in a study. Thus there are three levels of the factor *suppliers*.

If the factor is a quantitative variable, such as temperature or speed, then an infinite number of levels are possible. The motivation for more than two levels of a quantitative factor is usually to study nonlinear effects of the factor.

Sometimes the levels of a factor appear to be qualitative but have some underlying continuum. For example, three different types of chemicals would usually be considered as three levels of a qualitative factor. But if the primary difference between the chemicals was the concentration of an inhibitor, then the factor could be thought of as quantitative. Another example is three different operators in a study. The operators could be considered as a factor *job experience* studied at three different levels of number of years experience. It is usually better to treat the factor as quantitative when it is possible to do so.

Designs for qualitative factors with more than two levels are discussed in this section. Designs for quantitative factors with more than two levels are discussed in Section 9.2 and 9.3.

The following example describes a study with qualitative variables studied at more than two levels.

Example 9.1. A study was conducted to help set tolerances for the microfinish of a precision part. Needs of the customer suggested that the microfinish should be specified less than 30 units. Three factors thought by the operators to affect microfinish were studied. One of the factors, *tool type*, was studied at four levels since four different types of tools could be used in the process. The second factor, *supplier*, was studied at three levels since castings could be purchased from three sources. The third factor, *coolant level*, was quantitative and was studied at two levels.

Figure 9.1 shows the design matrix (a $4 \times 3 \times 2$ factorial design) and the data collected for this study. A run chart is also shown. The experimental design was run

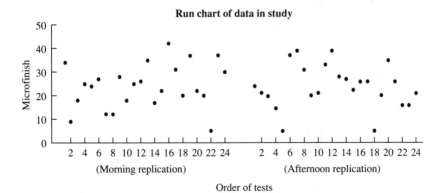

FIGURE 9.1
Design matrix for $4 \times 3 \times 2$ factorial

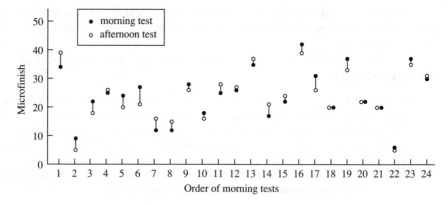

FIGURE 9.2
Run chart rearranged to compare morning to afternoon test results

once in the morning and then repeated in the afternoon. A random order of the 24 tests was done each time. The run chart indicates no special causes or problems with the data. Microfinish results ranged from a low of 5 units up to a high of 43 units.

To better study the importance of the common cause variation in the study, the run chart was next rearranged to compare the morning and afternoon results for each combination of factor levels. Figure 9.2 contains this plot.

No systematic differences between the morning and afternoon data is apparent. The differences between the morning and afternoon test results (the common cause variation) are small relative to the effects of the factors. No special causes are obvious in the data. Because of the consistent results, the morning and afternoon test results can be averaged to study the factor effects. Figure 9.3 shows these averages.

				Tool type			
				T1	*T2*	*T3*	*T4*
Coolant	Low	Supplier	*A*	5.5	13.5	14.0	35.0
			B	19.0	22.0	20.0	25.5
			C	23.0	30.5	27.0	40.5
	High	Supplier	*A*	7.0	17.0	20.0	24.0
			B	22.0	26.5	26.5	20.0
			C	28.5	36.5	36.0	36.0

FIGURE 9.3
Statistical summaries of microfinish data

If the between morning and afternoon variability had been greater, it might have been of interest to study the effect of the factors on this variability. The ranges between morning and afternoon could have been studied as a response variable.

It is possible to compute estimates of the factor effects from a design matrix when there are more than two levels for a factor. The calculation procedure is, however, beyond the scope of this text. We will primarily use response plots to study the effects of factors at more than two levels. Often, with qualitative variables the average response at a particular setting is more interesting than the factor effect.

Response plots for the averages were next developed to study factor effects and two-factor interactions. These response plots are shown in Figure 9.4. The following observations were made from these plots:

1. The effects of both tool type and supplier are large. The coolant effect is smaller.
2. There are numerous interactions between the three factors that can be exploited to minimize microfinish.
3. The parts made from supplier C material have consistently higher microfinish. Supplier A had the best performance except for tool type 4 where the parts from supplier A averaged 29.5 units and the parts from supplier B averaged 23 units. If tool 4 could be eliminated, supplier A could be a single supplier.

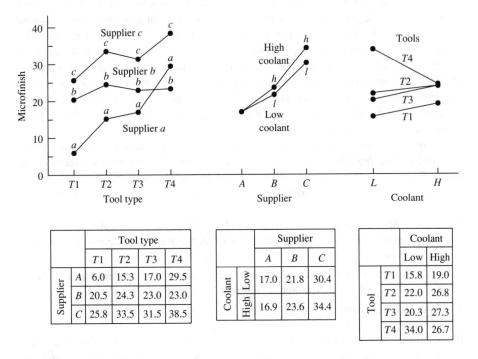

		Tool type			
		T1	T2	T3	T4
Supplier	A	6.0	15.3	17.0	29.5
	B	20.5	24.3	23.0	23.0
	C	25.8	33.5	31.5	38.5

		Supplier		
		A	B	C
Coolant	Low	17.0	21.8	30.4
	High	16.9	23.6	34.4

		Coolant	
		Low	High
Tool	T1	15.8	19.0
	T2	22.0	26.8
	T3	20.3	27.3
	T4	34.0	26.7

FIGURE 9.4
Response plots for microfinish averages

4. Tool type 1 generally resulted in lower microfinish readings than the other tools.

5. If it is necessary to use all four tools, the following conditions would give the lowest expected microfinish:

tool type	casting supplier	coolant level
T1	A or B	low
T2	A	low
T3	A or B	low
T4	B	high

6. To obtain the lowest microfinish readings, use tool 1, supplier A, and either coolant level. These conditions should be run and control charts developed to confirm the improved process capability for microfinish.

The response plot will be the most important graphical representation of the data in studies with more than two levels. Replication is important in this type of study so that effects of nuisance variables can be evaluated prior to studying the response plots. Replication also makes it possible to study the effect of the factors on the variation of the response variable due to nuisance variables.

If replication is not practical, it is more difficult to assess the impact of nuisance variables using the run charts or a rearranged run chart. In this case, more detailed response plots can be prepared to evaluate the consistency of the factor effects. One way to do this is to prepare a response plot for each two-factor interaction at each level of the other factors. If the plots are similar for each level, then nuisance variables are relatively unimportant. If the response plots at each level of the other factors are not similar, then either nuisance variables are important or there is a complex system of interactions among the factors. In either case, replication or some other type of confirmation will be necessary.

Figure 9.5 shows response plots for the morning replication of the microfinish experiment (data from Figure 9.1). Three plots are prepared, one for each factor in the study. The data from the experiment is ordered by the levels of the two other factors in the study. The following conclusions can be made from a study of these plots:

1. *Supplier Effect:* The lowest microfinish is for parts from supplier A, except for tool type 4. Supplier C is consistently the worst (eight out of eight cases). The supplier effect is not much different for the two coolant levels.

2. *Tool Type Effect:* Tool 1 is consistently better, and Tool 4 consistently the worst. The tool type effect is different at different coolant levels.

3. *Coolant Effect:* Low coolant is slightly better, except for tool type 4.

Based on these observations, summary response plots of the tool type/supplier interaction and the tool/coolant interaction would be prepared to summarize the findings. These response plots would be similar to those in Figure 9.4.

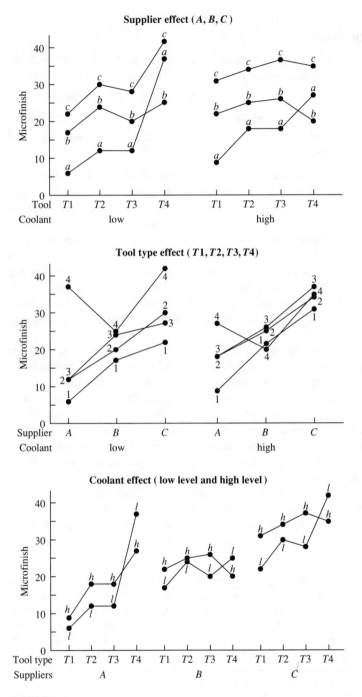

FIGURE 9.5
Detailed response plots for first replication of the microfinish study

9.2 AUGMENTING 2^k FACTORIAL DESIGNS WITH CENTER POINTS

Sometimes it is desirable to study quantitative factors at more than two levels. Some examples of situations that would lead to three or more levels are the following:

1. The effect of the factor is expected to be nonlinear in the range of interest.
2. The purpose of the study is to find the optimum level at which to set the factor.
3. It is important to include the current levels of the factors in the study as well as the high and low levels.

A possible approach in these cases is the use of three-level factorial designs (3^k factorials). With three or more factors, however, the 3^k factorials become very large (three factors, 27 tests; four factors, 81 tests; five factors, 243 tests). Designs based on fractions of the 3^k have been developed to make these designs somewhat more manageable. Section 9.3 discusses 3^k factorial and fractional factorial designs.

An alternative to three-level factorial designs is a type of experimental pattern called composite designs (Box and Wilson, 1951). The simplest composite design is a 2^k factorial design with a center point. Example 9.2 shows how to analyze this design. Other types of composite designs can be developed by adding additional runs to the 2^k or 2^{k-p} factorial designs. These designs will be discussed briefly at the end of this section.

When centerpoints are added to a factorial design, the importance of curvature or the presence of nonlinear effects can be evaluated. A center point is located at the mid-level for each of the factors in the factorial design. The curvature cannot be assigned to a particular factor, but the center point gives an indication of whether it is appropriate to interpolate between the factorial points. The magnitude of the curvature effect can be directly compared to the factor effects for relative importance if the number of center points is equal to one-half the number of factorial points. Thus, two center points would be required for a two-factor design, four for a three-factor design, and eight for a four-factor design.

If fewer runs than required for direct comparison are run, the nonlinear effect can still be subjectively evaluated. The responses at the center points can be compared to the other responses to evaluate gross departures from linearity. For example, if the center point results are close to the lowest or highest response, the nonlinearity is probably important.

Example 9.2 The operators in the filling process wanted to study variables in the process that affected the variation in the fill weights of their products. A number of factors that affected the average fill weight had been identified using control charts, but identification of factors that affected the range had been inconsistent. The operators suspected that interactions between some of the factors might be the key to reducing variation of the fill weights. They proposed to run a 2^3 factorial design on the following three factors:

Factor	Current level	Low level	High level	Units
Product temperature	200	180	220	degrees Fahrenheit
Line speed	350	300	400	Cans per minute
Product consistency	15	10	20	Viscosity (coded)

FIGURE 9.6
Design matrix for fill weight variation study

Factor	low level (−)	high level (+)	current level (0)
Temperature (T)	180	220	200
Speed (S)	300	400	350
Consistency (C)	10	20	15

Test	Run order	T	S	C	TS	TC	SC	TSC	Range of five cans (grams) 1st	2nd	$\bar{R}$
1	5	−	−	−	+	+	+	−	2.7	3.5	3.10
2	7	+	−	−	−	−	+	+	2.0	4.1	3.05
3	2	−	+	−	−	+	−	+	6.8	5.4	6.10
4	11	+	+	−	+	−	−	−	5.6	6.9	6.25
5	10	−	−	+	+	−	−	+	5.4	5.8	5.60
6	3	+	−	+	−	+	−	−	2.2	3.4	2.80
7	6	−	+	+	−	+	−	−	8.6	9.4	9.00
8	9	+	+	+	+	+	+	+	6.5	5.8	6.15
9	1	0	0	0	0	0	0	0	3.4	4.3	3.85
10	4	0	0	0	0	0	0	0	3.0	5.2	4.10
11	8	0	0	0	0	0	0	0	4.6	3.2	3.90
12	12	0	0	0	0	0	0	0	4.1	4.2	4.15

The levels of the factor were selected by adding and subtracting an equal amount from the current levels of the process. In addition to the eight runs in the factorial pattern, four tests were run at the levels of the factors currently used in the process. Figure 9.6 shows the design matrix and the results of the study. The ranges for two five-can samples collected at each test condition were calculated. The response variable used was the average of these two ranges.

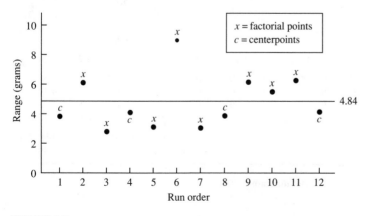

FIGURE 9.7
Run chart of range of fill weights

A run chart of the average range is shown in Figure 9.7. Note that the center points were spread throughout the study, while the order of the other runs was randomized. The repeated centerpoints give a measure of the importance of nuisance variables in the study. From the run chart, the effect of nuisance variables appears small relative to the factor effects.

Estimates of the effects were made and plotted on a dot-diagram (Figure 9.8). The curvature effect is the difference between the average of the four tests at the centerpoint and the average of the eight factorial points. Since four centerpoints were run (one-half the number of factorial points), the curvature effect (N) can be directly compared to the other effects on the dot-diagram. The important effects are the line speed, the temperature/consistency interaction, and the curvature.

The average ranges are shown on a cube in Figure 9.9. The average of the centerpoints is shown in the center of the cube. The important effects are also shown on response plots in Figure 9.9. The response plot for the speed effect indicates less variation in the fill weights at the lower line speed. But the centerpoint indicates lit-

FIGURE 9.8
Analysis of fill weight variation data

Factor	low level ($-$)	high level ($+$)	current level (0)
Temperature (T)	180	220	200
Speed (S)	300	400	350
Consistency (C)	10	20	15

Test	Run order	T	S	C	TS	TC	SC	TSC	1st	2nd	$\bar{R}$
1	5	$-$	$-$	$-$	$+$	$+$	$+$	$-$	2.7	3.5	3.10
2	7	$+$	$-$	$-$	$-$	$-$	$+$	$+$	2.0	4.1	3.05
3	2	$-$	$+$	$-$	$-$	$+$	$-$	$+$	6.8	5.4	6.10
4	11	$+$	$+$	$-$	$+$	$-$	$-$	$-$	5.6	6.9	6.25
5	10	$-$	$-$	$+$	$+$	$-$	$-$	$+$	5.4	5.8	5.60
6	3	$+$	$-$	$+$	$-$	$+$	$-$	$-$	2.2	3.4	2.80
7	6	$-$	$+$	$+$	$-$	$+$	$-$	$-$	8.6	9.4	9.00
8	9	$+$	$+$	$+$	$+$	$+$	$+$	$+$	6.5	5.8	6.15
Effects		-1.39	3.24	1.26	.04	-1.43	.14	$-.06$		Average $\bar{R}$ =	5.26
9	1	0	0	0	0	0	0	0	3.4	4.3	3.85
10	4	0	0	0	0	0	0	0	3.0	5.2	4.10
11	8	0	0	0	0	0	0	0	4.6	3.2	3.90
12	12	0	0	0	0	0	0	0	4.1	4.2	4.15

Range of five cans (grams) — column headers: 1st, 2nd, $\bar{R}$

Average of center points = 4.00
non-linear effect (N): 5.26 $-$ 4.00 = 1.26

Dot-diagram of effects

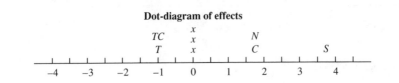

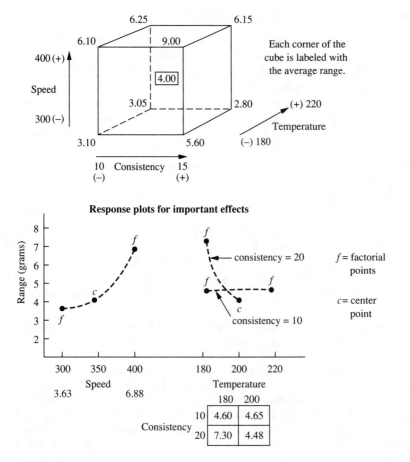

FIGURE 9.9
Analysis of fill weight variation

tle increase in variation from 300 to 350 cans per minute. The response plot for the temperature and consistency interaction indicates that there is no temperature effect for the low-consistency product but that variation is much greater at the low temperature for the high-consistency product. Again, the results at the center point indicate that the temperature/consistency effects may not be linear. Additional tests would be required to describe the important nonlinear effects.

If nonlinear effects are important, there are a number of alternative ways to design future studies. The three-level factorial designs discussed in the next section are one alternative.

More complex composite designs can be used to study factors with nonlinear effects. *Response surface methodology* (Box and Draper, 1987) comprises designs of this type. In addition to the factorial points and centerpoints, additional runs are made at other levels of the factors. The complete analysis of these designs requires the use of a statistical model. The book by Box and Draper (1987) present the design and analysis of these composite designs.

9.3   THREE-LEVEL FACTORIAL DESIGNS

As discussed in Section 9.2, factorial designs with factors studied at three levels are of limited use because of the large size of the experiments. It is also not possible to develop small fractions of most of these designs that enable main effects to be separated from two-factor interactions. It may be desirable to study three levels of a factor if previous two-level factorials have indicated a nonlinear effect on the response variable and if the factor interacts with other important factors. Three-level factorial designs could also be used to study a process when it is known that the selected factors are important and when their effects are expected to be nonlinear.

Table 9.1 summarizes recommended experimental designs when it is desirable to study factors at three levels. The designs can incorporate from two to five factors. If more than five factors are being considered, two-level fractional-factorial designs (Chapter 8) should be used, augmented with center points to evaluate the importance of nonlinearity. Each of these designs can be used to evaluate the main effects and two-factor interactions of each of the factors (except where noted in Figure 9.10). The design matrix for each of these designs is shown in Figure 9.10. Cochran and Cox (1957) contain more information on these designs. Run charts (or control charts) and response plots of interactions or main effects can be used to analyze results of these designs.

The following example illustrates fractional factorial design for four factors, each at three levels. The analysis of the other 3^k designs is similar.

> **Example 9.3** A study in a chemical process was done to understand the effects of four factors in the region of the current operating conditions. Previous studies had shown that each of these factors had an effect on the response (yield), and the effects had been nonlinear in some of the studies. One of the factors, *catalyst concentration,* did not interact with the other factors in the previous studies. The one-third replicate of a 3^4 factorial was used for the study (see Figure 9.10). Figure 9.11 shows the design matrix for this study.

The factor levels were chosen by making the current process level the middle level. The high and low levels were then determined by increasing or decreasing this level by an amount that would not upset the process. The factor *catalyst* was designated the *A* factor since the previous studies had indicated it was unlikely to interact with the other factors.

TABLE 9.1
Recommended three-level designs

Number of factors	Experimental pattern	Number of runs
2	3^2	9
3	3^3	27
4	3^{4-1}	27
5	3^{5-1}	81
>5	use 2^{k-p} with centerpoints	

FIGURE 9.10
Design matrices for three-level factorials

Code	Interpretation
−	Low level of factor
0	Middle level of factor
+	High level of factor

Two-factor design (3^2)

Test	A	B
1	−	−
2	0	−
3	+	−
4	−	0
5	0	0
6	+	0
7	−	+
8	0	+
9	+	+

Three-factor design (3^3)

Test	A	B	C
1	−	−	−
2	0	−	−
3	+	−	−
4	−	0	−
5	0	0	−
6	+	0	−
7	−	+	−
8	0	+	−
9	+	+	−
10	−	−	0
11	0	−	0
12	+	−	0
13	−	0	0
14	0	0	0
15	+	0	0
16	−	+	0
17	0	+	0
18	+	+	0
19	−	−	+
20	0	−	+
21	+	−	+
22	−	0	+
23	0	0	+
24	+	0	+
25	−	+	+
26	0	+	+
27	+	+	+

Four-factor design (3^{4-1})

Test	A	B	C	D
1	−	−	−	−
2	+	+	0	−
3	0	0	+	−
4	0	+	−	−
5	−	0	0	−
6	+	−	+	−
7	+	0	−	−
8	0	−	0	−
9	−	+	+	−
10	+	+	−	0
11	0	0	0	0
12	−	−	+	0
13	−	0	−	0
14	+	−	0	0
15	0	+	+	0
16	0	−	−	0
17	−	+	0	0
18	+	0	+	0
19	0	0	−	+
20	−	−	0	+
21	+	+	+	+
22	+	−	−	+
23	0	+	0	+
24	−	0	+	+
25	−	+	−	+
26	+	0	0	+
27	0	−	+	+

Note: For the 3^{4-1} design, most of the information on 2-factor interactions are clear. The following effects have some confounding:

$$AB \text{ with } CD$$
$$AC \text{ with } BD$$
$$AD \text{ with } BC.$$

If the interactions effects with factor **A** can be considered negligible, then the remaining interactions are clear.

FIGURE 9.10 (*continued*)
Design matrices for three-level factorials

Code	Interpretation
−	Low level of factor
0	Middle level of factor
+	High level of factor

Five-factor design (3^{5-1})

Test	A	B	C	D	E	Test	A	B	C	D	E	Test	A	B	C	D	E
1	−	−	−	−	−	28	0	0	−	−	0	55	+	+	−	−	+
2	+	−	0	−	−	29	−	0	0	−	0	56	0	+	0	−	+
3	0	−	+	−	−	30	+	0	+	−	0	57	−	+	+	−	+
4	0	+	−	−	−	31	+	−	−	−	0	58	−	0	−	−	+
5	−	+	0	−	−	32	0	−	0	−	0	59	+	0	0	−	+
6	+	+	+	−	−	33	−	−	+	−	0	60	0	0	+	−	+
7	0	0	0	−	−	34	+	+	0	−	0	61	−	−	0	−	+
8	−	0	+	−	−	35	0	+	+	−	0	62	+	−	+	−	+
9	+	0	−	−	−	36	−	+	−	−	0	63	0	−	−	−	+
10	0	+	+	0	−	37	+	−	+	0	0	64	−	0	+	0	+
11	−	+	−	0	−	38	0	−	−	0	0	65	+	0	−	0	+
12	+	+	0	0	−	39	−	−	0	0	0	66	0	0	0	0	+
13	+	0	+	0	−	40	−	+	+	0	0	67	0	−	+	0	+
14	−	0	−	0	−	41	+	+	−	0	0	68	−	−	−	0	+
15	0	0	0	0	−	42	0	+	0	0	0	69	+	−	0	0	+
16	−	−	−	0	−	43	−	0	−	0	0	70	0	+	−	0	+
17	0	−	0	0	−	44	+	0	0	0	0	71	−	+	0	0	+
18	+	−	+	0	−	45	0	0	+	0	0	72	+	+	+	0	+
19	+	0	0	+	−	46	−	+	0	+	0	73	0	−	0	+	+
20	0	0	+	+	−	47	+	+	+	+	0	74	−	−	+	+	+
21	−	0	−	+	−	48	0	+	−	+	0	75	+	−	−	+	+
22	−	−	0	+	−	49	0	0	0	+	0	76	+	0	0	+	+
23	+	−	+	+	−	50	−	0	+	+	0	77	0	+	+	+	+
24	0	−	−	+	−	51	+	0	−	+	0	78	−	+	−	+	+
25	−	+	+	+	−	52	0	−	+	+	0	79	+	0	+	+	+
26	+	+	−	+	−	53	−	−	−	+	0	80	0	0	−	+	+
27	0	+	0	+	−	54	+	−	0	+	0	81	−	0	0	+	+

Note: All two-factor interactions are clear in this design

The order of the 27 tests was randomized. The four factors were set to the required level, and then the process was allowed to stabilize (from 1 to 4 hours). The yield of the process based on online measurements was then determined for a 20-hour period. Figure 9.12 shows a run chart for the percent yield for each test. The yields did not show any trends or special causes during the month-long study. The average yield of 95.6% was slightly lower than the process average before the study (96%). A few of the tests had yields greater than 98%.

Response plots of the three factors that were expected to interact were made. Figure 9.13 shows these plots. The temperature/pressure interaction was important. The effect of the inhibitor level was independent of temperature and pressure.

FIGURE 9.11
Design matrix for 3^{4-1} factorial study

Factor	Code	Low (−)	Middle (0)	High ()
Temperature (°F)	Temp	190	200	210
Pressure (psi)	Pres	300	350	400
Inhibitor level (ppm)	Inhb	40	50	60
Catalyst concentration (lbs.)	Catl	8	10	12

Four-factor design (3^{4-1})

Test	Catl	Inhb	Pres	Temp	Run Order	Yield Percentage
1	−	−	−	−	17	93.1
2	+	+	0	−	12	94.3
3	0	0	+	−	7	96.8
4	0	+	−	−	15	91.9
5	−	0	0	−	16	93.6
6	+	−	+	−	1	98.5
7	+	0	−	−	5	92.2
8	0	−	0	−	19	97.7
9	−	+	+	−	27	94.3
10	+	+	−	0	10	93.6
11	0	0	0	0	22	96.4
12	−	−	+	0	9	97.1
13	−	0	−	0	25	92.7
14	+	−	0	0	18	97.8
15	0	+	+	0	13	96.0
16	0	−	−	0	23	97.4
17	−	+	0	0	2	93.5
18	+	0	+	0	26	96.6
19	0	0	−	+	4	97.0
20	−	−	0	+	11	96.8
21	+	+	+	+	21	95.5
22	+	−	−	+	14	98.4
23	0	+	0	+	3	96.1
24	−	0	+	+	8	95.0
25	−	+	−	+	20	93.7
26	+	0	0	+	6	96.7
27	0	−	+	+	24	98.8

The effects of each of the factors are nonlinear over the range studied. The pressure effect is about 4% at the low temperature, but it is insignificant at the high temperature. At low pressure the temperature is also important, but at high pressure a variation in the temperature has no impact. Hence, yield can be increased slightly by setting temperature to the high level, but there is no increase if both temperature and pressure are increased. The increased temperature will also result in less variation in yield if pressure changes inadvertently.

Higher levels of inhibitor decrease the yield, but the decrease in yield as a result of inhibitor levels raised from 50 to 60 ppm is much less than for inhibitor

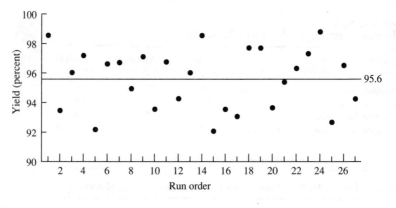

FIGURE 9.12
Run chart for percent yield

levels raised from 40 to 50. Decreasing the catalyst charge from the current level of 12 pounds to 8 pounds lowers the yield by about three percent. There is no gain in yield by increasing the catalyst charge.

Response plots to summarize the important effects are shown in Figure 9.14. Response plots are shown for the temperature/pressure interaction and the inhibitor and catalyst main effects. Note that the interaction plot for temperature and pressure is the same as in Figure 9.13, but the pressure effect has been placed on the horizontal axis to get a second display of the interaction.

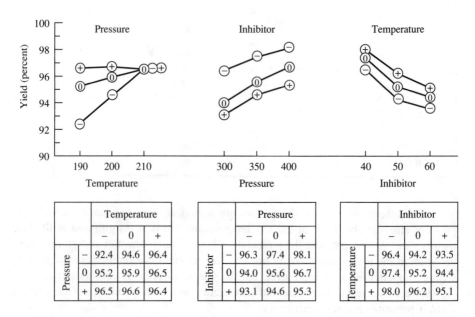

		Temperature		
		−	0	+
Pressure	−	92.4	94.6	96.4
	0	95.2	95.9	96.5
	+	96.5	96.6	96.4

		Pressure		
		−	0	+
Inhibitor	−	96.3	97.4	98.1
	0	94.0	95.6	96.7
	+	93.1	94.6	95.3

		Inhibitor		
		−	0	+
Temperature	−	96.4	94.2	93.5
	0	97.4	95.2	94.4
	+	98.0	96.2	95.1

FIGURE 9.13
Response plots for two-factor interactions

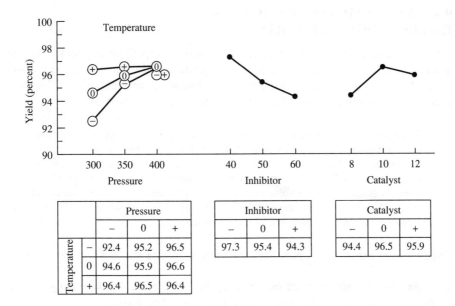

	Pressure		
	–	0	+
–	92.4	95.2	96.5
0	94.6	95.9	96.6
+	96.4	96.5	96.4

(row label: Temperature)

Inhibitor		
–	0	+
97.3	95.4	94.3

Catalyst		
–	0	+
94.4	96.5	95.9

FIGURE 9.14
Response plots for important effects

The process should be evaluated for improvement of yield after setting the temperature or pressure (or both if advantageous) at the high level in the study and reducing the level of the inhibitor.

This example illustrates the analysis of a three-level factorial design. In this case, previous studies had indicated that nuisance variables were relatively unimportant. If this is not the case, some replication should be included in the design. The run chart and the response plots of main effects and two-factor interactions are the methods of analysis. Response plots summarizing the important effects should be prepared for presentation of results.

Mixed Two- and Three-Level Factorial Designs

Factorial designs can be run with some of the factors at two levels and some of the factors at three levels. Table 9.2 describes some possible designs. Response plots can be used to analyze the results of these designs in a manner similar to the last example.

Section 9.4 discusses a special application of two-level factorial designs.

TABLE 9.2
Mixed two- and three-level factorial designs

Design	Factors at two levels	Factors at three levels	Number of tests required
$2^1 3^1$	1	1	6
$2^2 3^1$	2	1	12
$2^3 3^1$	3	1	24
$2^1 3^2$	1	2	18
$2^2 3^2$	2	2	36

9.4   EXPERIMENTAL DESIGN FOR INTERCHANGEABLE PARTS

Ott (1975) described the use of factorial designs in an experimental plan to study an assembled product that can be readily disassembled and reassembled. The plan uses the performance of the assembled product to determine the high and low levels of each factor in the study. The following example illustrates this design.

> **Example 9.4** A manufacturer of automobiles developed a method of measuring noise in their engines. The method was calibrated to human responses. Engines that measured greater than 7.0 were considered noisy, and engines that measured less than 6.0 were considered quiet. Control charts on this measurement indicated that the engines currently being produced had greater noise levels than those that had been assembled in the previous month. The cause of the increase in noise was suspected to be found in the crank or the balancer since both of these components contained gears.

The following experimental plan was developed to isolate the source of the noise:

1. Eight engines were obtained for the study. Four of the engines were noisy (greater than 7.0) and four of the engines were quiet (less than 6.0).
2. The eight engines were disassembled into three components:

   the balancer

   the crankshaft, and

   the rest of the engine (the block).

   The three components thus become the three factors in the study. Each factor was studied at two levels, *quiet* or *noisy*. The level was based on whether the component came from a noisy or quiet engine. For each component, four were labeled *noisy* and four were labeled *quiet*.
3. Eight engines were reassembled using the 2^3 factorial design shown in Table 9.3. Each of the three components of the engines was randomly selected from the noisy or quiet group according to the design. Each engine was then tested for noise. The results are shown in Table 9.3.

TABLE 9.3
2^3 **Factorial design for noisy engines**

Test	Test order	Balancer	Crankshaft	Block	Measured noise
1	3	quiet	quiet	quiet	8.5
2	5	noisy	quiet	quiet	9.0
3	1	quiet	noisy	quiet	5.9
4	6	noisy	noisy	quiet	5.2
5	2	quiet	quiet	noisy	10.3
6	8	noisy	quiet	noisy	10.5
7	4	quiet	noisy	noisy	7.2
8	7	noisy	noisy	noisy	5.7

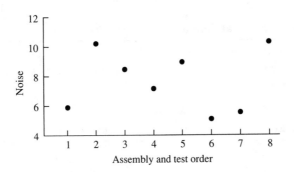

FIGURE 9.15
Run chart of 2^3 design for noisy engines.

Figure 9.15 shows a run chart for the measured noise of the assembled engines. The run chart indicated no trends or special causes.

Figure 9.16 shows the analysis of the factor effects. The main effects and interactions were estimated from the design matrix in the usual manner (see Chapter 6).

A main effect in this study is the change in noise when going from a component labeled quiet to a component labeled noisy. An important interaction effect indicates that the main effect depends on whether one of the other components in the engine is labeled quiet or noisy.

The dot diagram of the effects shows that the effect of the balancer is more important than the other factors. None of the interactions appeared to be important. These results confirmed predictions made by the engineers involved. The balancer was a new component to the engine.

FIGURE 9.16
Analysis of 2^3 design for noisy engines

Test	run order	Noisy (−)			Quiet (+)				Measured noise
		C	**B**	**E**	**CB**	**CE**	**BE**	**CBE**	
1	3	−	−	−	+	+	+	−	8.5
2	5	+	−	−	−	−	+	+	9.0
3	1	−	+	−	−	+	−	+	5.9
4	6	+	+	−	+	−	−	−	5.2
5	2	−	−	+	+	−	−	+	10.3
6	8	+	−	+	−	+	−	−	10.5
7	4	−	+	+	−	−	+	−	7.2
8	7	+	+	+	+	+	+	+	5.7
Effects		−0.38	**3.58**	1.28	−.73	−.28	−.38	−.13	**Average** 7.79

Dot-diagram of effects

```
                    x
            x       x                                    B
            x       x           x                        x
    |───┬───┬───┬───┬───┬───┬───┬───┬───|
   −3      −2      −1       0       1       2       3       4
```

B = Balancer
C = Crankshaft
E = Engine

TABLE 9.4
2^{7-4} Fractional factorial design for noise study

	Factors in study	Code	Factors in study	Code	
	RH Gear	RG	RH Shaft	RS	Noisy (−)
	Hoursing	HO	LH Shaft	LH	Quit (+)
	LH Gear	LG	Cover	CO	
	GE Rotor	GR			

Test	Run order	RG	HO	LG	RS	LS	CO	GR	Measured noise
1	2	−	−	−	+	+	+	−	10.0
2	6	+	−	−	−	−	+	+	8.2
3	4	−	+	−	−	+	−	+	9.3
4	8	+	+	−	+	−	−	−	6.5
5	1	−	−	+	+	−	−	+	6.8
6	5	+	−	+	−	+	−	−	6.3
7	7	−	+	+	−	+	−	−	7.6
8	3	+	+	+	+	+	+	+	5.8
	Effects	−1.73	−.53	−1.88	−.58	.58	.68	−.08	Avg. = 7.56

A second study was done to understand what part of the balancer was causing the noise. The consensus was that the noise was somehow related to the two gears in the balancer. Seven components of the balancer were identified that might contribute to the noise: the right and left gear, the right and left shaft, the housing, the cover, and the rotor. These seven components were then used as factors in a 2^{7-4} fractional factorial design.

The four balancers from the original noisy engines were disassembled and each of the seven components (factors) were labeled *noisy*. Similarly, the four balancers from the quiet engines were disassembled and the components labeled *quiet*. Eight balancers were then reassembled using a component specified as appropriate by the fractional factorial design. The eight balancers were then assembled in an engine for testing.

Table 9.4 shows the design matrix and the results of the study. The main effect of each factor is estimated using the design matrix.

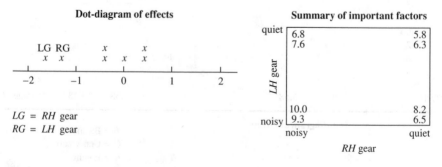

FIGURE 9.17
Analysis of 2^{7-4} design

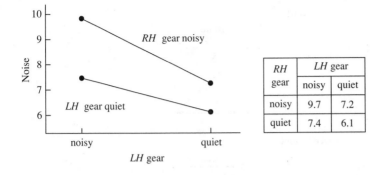

FIGURE 9.18
Response plot for important factors

Figure 9.17 shows a dot diagram of the effects. The two important effects are also shown on a square.

As predicted, the two gears had the biggest effect on the noise. The effect of each type of gear on the noise was about the same. Figure 9.18 shows a response plot for the interaction of these two factors.

Follow-up studies focused on understanding the difference between noisy and quiet gears. The preparation of the materials in the gears turned out to be the cause of the noise.

This example illustrates an application of two-level factorial designs that is effective and quick for improving the performance of products with multiple components. This type of study should be considered whenever an assembly of components is involved.

9.5 EXPERIMENTS FOR FORMULATIONS OR MIXTURES

An experiment for a formulation or a mixture is a special type of study in which the factors are ingredients that are mixed together. The response variables are thought to depend on the relative proportions of the components rather than the absolute amounts.

Examples of situations in which a mixture experiment is appropriate are

a study of viscosity of oil blends with different concentrations of the additives,

a study of the hardness of steel with different mixtures of alloys, and

a test of taste preference of a drink with different levels of key ingredients.

In each of these cases, the levels of one factor cannot be changed independently of the levels of the other factors. The sum of the percentages of each factor in the mixture must add to 100%. Therefore, a factorial design can be used for testing mixtures only in special circumstances. Other experimental patterns will be used for this type of experiment. Figure 9.19 compares the possible combinations of the factors for a factorial design to those for a mixture experiment. This section provides an introduction to two of these patterns; simplex and extreme vertices. For a more in-depth treatment of designing experiments for mixtures, see Cornell (1983).

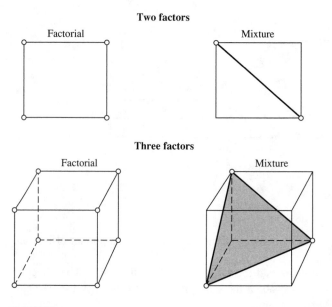

FIGURE 9.19
Allowable Combinations of a Factorial Design versus a Mixture Design

Using Factorial Designs for Mixture Experiments

In some situations factorial designs can be used to study formulations or mixtures even though the components must add to 100%. Some of these situations include

- One of the components makes up the majority (more than 80%) of the mixture.
- One of the components is an inert filler.
- One or more of the components are used to modify the properties of another component and the ratio of the modifiers to the primary component is thought to be important.

In the first two situations the components are changed independently as they normally would be in a factorial design. The major component or the filler is then adjusted so that the percentage of all the components add up to 100%. The components that are adjusted to make the sum 100% are called *slack components*. Table 9.5 contains a test for a four-component mixture with one of the components making up the majority of the mixture. The experiment is designed by changing three of the components according to a 2^3 factorial design and adjusting the slack component to make the sum of the four components equal 100%.

The third situation mentioned above occurs when the ratios of components are important rather than the actual proportions of each in the mixture. This is frequently the case when some of the components are included to modify the properties of a primary component. Table 9.6 contains the pattern for an experiment to deter-

TABLE 9.5
Use of a 2^3 design for a four-component mixture

Factor (component)	Levels (%)
A	0–1
B	1–3
C	2–4
D (slack)	adjusted to 100%

Test	A	B	C	D
1	0	1	2	97
2	1	1	2	96
3	0	3	2	95
4	1	3	2	94
5	0	1	4	95
6	1	1	4	94
7	0	3	4	93
8	1	3	4	92

TABLE 9.6
A 2^{4-1} Experiment to study the properties of rubber compounds

Factor	Levels(Amount/amount of rubber)
1	.2–.3
2	.05–.1
3	.0–.02
4	.02–.05

Test	1	2	3	4
1	.20	.05	.00	.02
2	.30	.05	.00	.05
3	.20	.10	.00	.05
4	.30	.10	.00	.02
5	.20	.05	.02	.05
6	.30	.05	.02	.02
7	.20	.10	.02	.02
8	.30	.10	.02	.05

mine the formulation that achieved the desired properties of a rubber compound. Five components were studied; rubber and four components that were included in the formulation to modify the properties of the rubber. The amount of rubber was held constant. The levels of the four modifiers were designated as ratio of the amount of the modifier to the amount of the rubber. A 2^{4-1} fractional factorial design was used to run the experiment.

In these examples the circumstances were such that factorial or fractional factorial patterns could be used to design a study for mixtures. The data obtained from these studies are analyzed in the usual way for factorial or fractional factorial patterns. There will be many other circumstances in which factorial patterns are not appropriate and other patterns will be needed.

Simplex Patterns When All Components Have No Constraints

One of the situations in which factorial patterns cannot be used for mixture experiments is when all the components can be varied anywhere between 0 and 100%. In these situations simplex experimental patterns are used. A simplex pattern for a three-component mixture is most easily displayed using triangular graph paper. Figures 9.20 to 9.24 illustrate the scaling of triangular graph paper.

A simplex pattern for an experiment with three factors is developed by spreading the tests over the allowable experimental region. A simplex pattern using six tests is composed of the three pure mixtures and three mixtures with two components at 50%. This pattern is listed in Table 9.7. For simplex patterns with four or more factors, see Connell(1983).

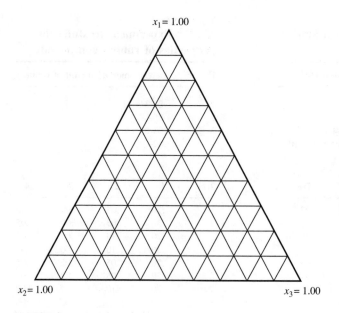

FIGURE 9.20
Triangular graph paper.

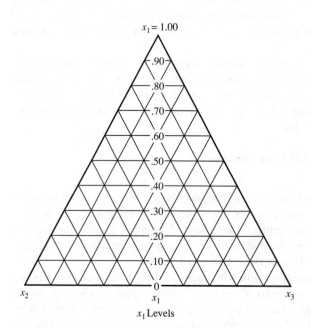

FIGURE 9.21
Scale for component 1.

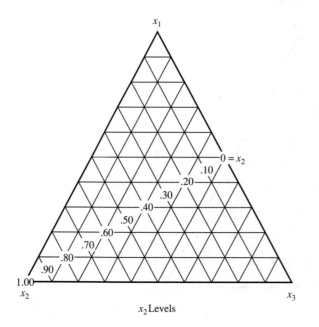

FIGURE 9.22
Scale for component 2.

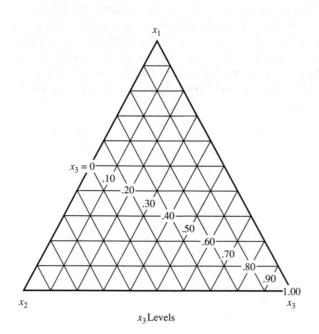

FIGURE 9.23
Scale for component 3.

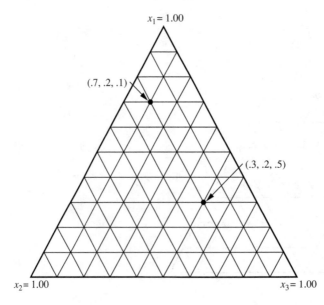

FIGURE 9.24
Plotting points on triangular graph paper.

By adding the center point, all factors at 1/3 of the mixture, the interior of the experimental region can be studied. It is advisable to include a center point and replicate the point to have a measure of the effect of nuisance variables. In addition the center point it is useful to include other points interior to the triangle if it is suspected that none of the factors will be excluded from the final mixture. Table 9.8 contains a simplex pattern using 10 tests, and Figure 9.25 contains a display of the pattern on triangular graph paper.

TABLE 9.7
Simplex pattern using six tests

	Factor		
Test	1	2	3
1	100	0	0
2	0	100	0
3	0	0	100
4	50	50	0
5	50	0	50
6	0	50	50

TABLE 9.8
**Simplex pattern using ten
tests**

Test	Factor		
	1	2	3
1	100	0	0
2	0	100	0
3	0	0	100
4	50	50	0
5	50	0	50
6	0	50	50
7	33.3	33.3	33.3
8	66.7	16.6	16.6
9	16.6	66.7	16.6
10	16.6	16.6	66.7

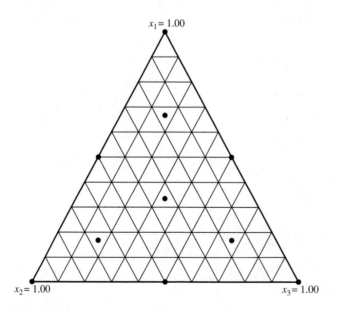

FIGURE 9.25
Display of a simplex design using 10 tests.

Example 9.5   Improving the Density of Ceiling Tile.   One of the critical properties of ceiling tile is the density of the tile. The ceiling tiles perform best if the density of the tile is 0.9 lb/sq. ft. One of the primary ingredients in ceiling tile is recycled material. There are three sources of the recycled material: A, B, and C. Source A is semifinished or finished tile that is scrapped in production. Source B is dust that is generated and collected at the sanding operation. Source C is sludge that is collected at the clarifier. Each of the different sources of recycled material has different properties that affect the density of the tile. An experiment was performed to determine what percentage of each type of material should be used in the formulation for the tile. The total amount of recycled material in the formulation was kept approximately constant. The experiment consisted of a simplex design with 10 different mixtures. Two replications of each mixture were made. The run chart indicated that the effect of nuisance variables was small relative to the variation between different mixtures. Figure 9.26 contains the average density for each of the mixtures in the experiment.

The data indicate that density increases as the percentages of dust and sludge are increased. A mixture of 70% scrap, 15% sludge, and 15% dust results in a density of approximately 0.9 lb/sq. ft. This mixture was tested in production over a period of a week, and the measurements of density were plotted on a control chart to confirm the results of the experiment.

Mixture Experiments When the Components Are Constrained

It is often the case that all the components cannot be varied from 0 to 100%. One or more may be constrained to a range, for example 10 to 20% of the mixture. In these situations the simplex patterns discussed above are not appropriate. Before a pattern can

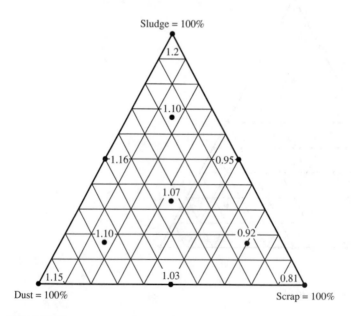

FIGURE 9.26
Data from the ceiling tile study.

be determined, the allowable region for experimentation (based on the constraints) must be determined. Then a pattern that consists of mixtures that cover the allowable region is developed. One type of pattern that is useful in this type of application is called an *extreme vertices pattern*. The extreme vertices design will be discussed through an example. For more information on extreme vertical patterns see Connell (1983).

> **Example 9.6: Fruit Punch Study.** Juices from watermelon, pineapple, and orange were combined to make a fruit punch. The proportions for the different fruits were constrained as follows:
>
> 40% < watermelon < 80%   10% < pineapple < 50%   10% < orange < 30%

Since there are three factors, the allowable region is obtained by drawing in the constraints on triangular graph paper. Figure 9.27 contains the allowable region for this example.

Once the allowable region is determined, a pattern that covers this region is developed. In developing the region, first consideration should be given to the vertices. In this example there are four vertices, and all four were included in the study. If two or more vertices are close together, some can be left out. It is usually desirable to include the center of the region. The experimental pattern chosen for this experiment is contained in Table 9.9. Also to be considered for inclusion in the pattern are other interior points and points along the edges.

Six children were asked to rank the drinks from best to worst. A rank of 1 was given to the best drink and a rank of 5 was given to the worst drink. The median ranks of the six children are displayed in Figure 9.28.

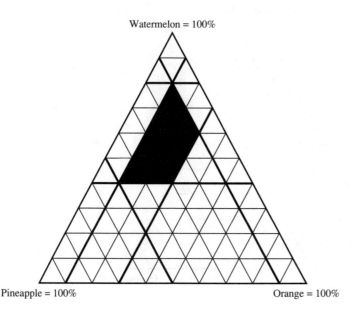

Watermelon = 100%

Pineapple = 100%   Orange = 100%

FIGURE 9.27
Allowable region for the juice experiment.

TABLE 9.9
Pattern for the
juice experiment

Test	W	P	O
1	40	50	10
2	80	10	10
3	60	10	30
4	40	30	30
5	55	25	20

The data indicate that the children preferred the drinks with higher concentrations of orange and pineapple juice. It was decided to make the drink with a combination of 30% orange juice, 30% pineapple juice, and 40% watermelon juice, test number 4 with a 1.5 average response.

This section provides only a brief introduction to mixture experiments. For a more in depth study, see Connell (1983).

9.6 EVOLUTIONARY OPERATION (EVOP)

Evolutionary operation (EVOP) is a strategy developed by Box and Draper (1969) to run experiments on a process while in operation. The EVOP studies were designed to be run by operators on a full-scale manufacturing process while continuing to produce a product of satisfactory quality.

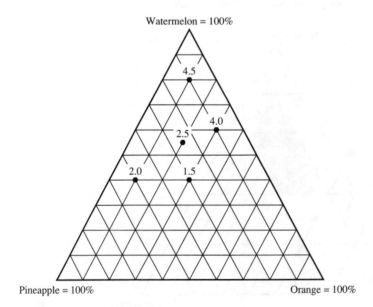

FIGURE 9.28
Data from the juice experiment.

The important features of the EVOP strategy include the following:

1. A series of sequential experiments.
2. Simple factorial designs with two or three factors that do not interrupt operations.
3. Small changes to the factor levels so that radical changes in product quality or efficiency do not occur.
4. A large number of runs required per factor combination.
5. Simple graphical analysis.

The concepts and methods presented in the first eight chapters of this book provide all the elements required to use an EVOP strategy. These elements include:

1. The model for improving quality (Chapter 1).
2. The importance of stability of the process response variables (Chapter 2, control charts, and Chapter 5, run charts).
3. Factorial designs with the selection of factor levels to meet different objectives (Chapter 6).
4. Replication (Chapter 4 and Chapter 6).
5. Centerpoints to evaluate nonlinearity of effects (Chapter 9).
6. Confirmation of new quality or efficiency levels (Chapter 2, control charts).
7. Selection of the next experiment; that is, continuous improvement (Chapter 1, the model).

The model for improving quality presented in Chapter 1 provides the road map to implementing the EVOP concept in an operating process. The model emphasizes the importance of stability of the process as part of the current knowledge before beginning to experiment on the process. The four-phase improvement cycle provides the structure to plan and conduct a series of sequential experiments on a process. After each experiment, the current knowledge is first updated and the next experiment is planned. Chapter 6 on factorial experiments provides the experimental designs to be used in a series of sequential studies.

Chapter 11 includes a case study that illustrates the application of the model for improvement of quality to a manufacturing process.

9.7 SUMMARY

This chapter presented experimental patterns for special situations and applications. Cochran and Cox (1957) have described the designs with factors at more than two levels presented here, as well as many others. The key ideas in this chapter include:

1. There are situations in the improvement of quality that require factorial designs at more than two levels of both qualitative and quantitative factors.
2. Response plots are the primary tool to study nonlinear effects of factors.

3. Centerpoints can be added to 2^k factorial designs to evaluate the importance of nonlinear effects.
4. Three-level factorials can be used in situations where factor interactions and non-linear effects are expected.
5. Special experimental patterns may be required when the factors are components of a mixture.
6. The model for improving quality presented in Chapter 1 and the experimental patterns in other chapters provide a framework similar to EVOP for improvement of processes while in operation.

REFERENCES

Box, G. E. P., and N. R. Draper (1969): *Evolutionary Operation: A Statistical Method for Process Improvement*, John Wiley & Sons, New York.

Box, G. E. P., and N. R. Draper (1987): *Empirical Model-Building and Response Surfaces*, John Wiley & Sons, New York.

Box, G. E. P., and K. B. Wilson (1951): "On the Experimental Attainment of Optimum Conditions," *Journal of the Royal Statistical Society B*, no. 13, pp.1–45.

Cochran W. G., and G. M. Cox (1957): *Experimental Design*, John Wiley & Sons, Inc., New York.

Cornell, John A. (1983): *How to Run Mixture Experiments for Product Quality*, American Society for Quality Control, Milwaukee.

Ott, Ellis R. (1975): *Process Quality Control*, McGraw-Hill Book Company, New York, Chapter 6.

EXERCISES

9.1. Describe a situation where it would be desirable to include more than two levels of a factor in a study. Give an example for a qualitative factor and a quantitative factor.

9.2. Refer to the 2×2 Plasticizer example in Chapter 6. In addition to the eight runs in the study, two runs were made at the current process levels of 45% ingredient X and 182 °C. One run was made at the beginning of the study (reaction time = 6.5 hours) and the second run was made at the end of the study (reaction time = 6 hours). Analyze the example including these two additional runs. Are the conclusions of the study affected by the new data?

9.3. Refer to the 2^3 Dye Process example in Chapter 6. Four additional tests were made at a mid-level oxidation temperature and oven pressure. Two of the runs were done with material quality A, and two of runs were done with material quality B. The following results were obtained:

Run order	Material	Pressure	Temperature	Response shade
1	A	mid-level	mid-level	191
5	B	mid-level	mid-level	207
8	A	mid-level	mid-level	189
12	B	mid-level	mid-level	210

Re-analyze this example using the additional four runs. Plot the new runs on the cube. Estimate the nonlinear effect. Is the effect important relative to the factor effects? Plot the mid-level points on the response plots.

9.4. Describe a potential study in which the concept of interchangeable parts (Section 9.4) would be appropriate. What is the response variable? What are potential factors? How would the levels of the factors be established?

9.5. Describe a study to improve quality where it is appropriate to use a mixture design.

9.6. An in-process scale was used in preparing additives for an industrial process. The quality team had made significant improvements in the accuracy of the weightings done on the scale by standardizing operations and using control charts for calibration decisions. Environmental factors (wind and temperature) were identified as possible causes of weight variations. The scale was supposed to compensate for ambient temperature fluctuations, but special causes on the control chart indicated that sometimes temperature changes appeared to have an effect.

A study was designed to evaluate the effect of wind speed and ambient temperature on scale accuracy. The accuracy was measured by loading a standard weight on the scale and recording the difference from the standard value. The team conjectured that possible interactions and nonlinear effects of these factors made it difficult to understand their effects using the control charts. A 3×3 factorial experiment was planned. Three levels of each factor were established from previous wind and temperature readings. The order of the tests was determined by the occurrence of the proper weather conditions. The standard weight was measured three times during each test. The following data was obtained:

	Test			Measurement − standard (pounds)		
Test	Order	Wind	Temperature	Meas1	Meas2	Meas3
1	3	low	low	.4	−.8	.6
2	8	low	mid	−.7	.5	.3
3	4	low	high	2.6	3.2	2.8
4	6	mid	low	−1.0	.8	−.7
5	1	mid	mid	−.5	1.3	.6
6	5	mid	high	3.6	2.5	3.5
7	2	high	low	2.1	−1.6	−.8
8	7	high	mid	−1.3	.5	1.6
9	9	high	high	1.5	4.3	2.6

Analyze the data from this study.

1. Prepare run charts of the average (to evaluate the bias of the weighing process) and the range (to evaluate the precision of the weighing process) of the weighings.
2. Prepare response plots for the average and range.
3. Do either of the environmental factors affect the accuracy of the scale?
4. What additional study should be done? What are recommendations for action?

CHAPTER
10

QUALITY
BY
DESIGN

10.1 INTRODUCTION

Today's technological innovations and changing markets make the improvement of existing products and the development of new products and services competitive necessities. Competition for an existing product will increase with time as new ways are found to better meet the needs of the customer.

Product development programs are an integral part of the business strategy of organizations. A telling study by Harvard professor Kim B. Clark (1989) compared product development programs of 20 U.S., European, and Japanese auto companies between 1982 and 1987. Clark found that the average lead time at U.S. and European auto companies is about 60 months, compared to 46 months for the Japanese. Moreover, the U.S. companies introduced one-third fewer new products than the Japanese companies in that time period.

What are the important areas of concern in designing a new product or service?

- Understanding the future needs of the customer.
- Designing the products and services that meet those needs.
- Designing the processes to produce the products and services.

The system to translate the needs of the customer to a product was illustrated in Chapter 1 (Figure 1.1) with Deming's view of production as a system. This system provides the framework for matching products and services to a need through the design and redesign stage of the system.

Questions from Chapter 1 that need to be addressed at the design and redesign stage by engineers, scientists, or managers are:

- How do we make products out of concepts that have potential to meet needs of the customer?
- How do we select the best concept to meet the needs?
- How do we select from hundreds of choices the vital few factors for design?
- How do we design a product that will work under the wide range of conditions that will be encountered during production and subsequent use by the customer?
- How do we choose the best operating conditions for a manufacturing process?

Previous chapters have discussed methods of planned experimentation that are useful in the design and redesign of products and processes to improve the matching. The aim of this chapter is to link these methods to the design and redesign stage of Deming's system to achieve quality by design.

Application of the Model for Improving Quality at the Design and Redesign Stage

Chapter 1 introduced the four basic activities to improve quality: (1) design a new product; (2) redesign an existing product; (3) design a new process (including service); and (4) redesign an existing process. Figure 1.5 introduced the model for improving quality. In this chapter, the focus is on improving quality at the design stage for new products and processes and at the redesign stage for products. As a result, the second component of the model, current knowledge, was modified as illustrated in Figure 10.1. Products and processes are brought closer to the needs of the customer by application of the model to increase knowledge of the processes in the design and redesign stage. The key tasks of these processes in the design and redesign stage include:

1. Defining quality using quality characteristics. This is the outcome of the consumer research in production viewed as a system.
2. Setting targets for these quality characteristics based on the consumer research.
3. Designing the products or processes that are close to the targets under a wide range of conditions.

In the next chapter, two case studies, redesigning a process and redesigning a product, illustrate these ideas.

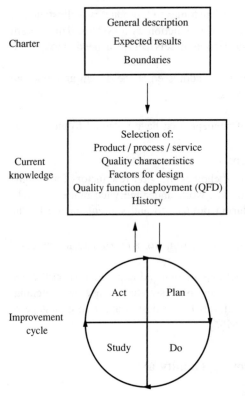

Charter — General description / Expected results / Boundaries

Current knowledge — Selection of: Product / process / service / Quality characteristics / Factors for design / Quality function deployment (QFD) / History

Improvement cycle — Act / Plan / Study / Do

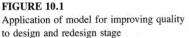

FIGURE 10.1
Application of model for improving quality to design and redesign stage

Processes in the Design and Redesign Stage

The major processes in the design of new products are:

1. Definition of quality.
2. Generation of product or process concepts.
3. Design of product.
4. Design of production process.

Figure 10.2 illustrates these major processes and the key outcomes of each. The next four sections of this chapter will assume a new product is being designed. However, the concepts described in these sections also apply to redesign of a product and to design and redesign of a process.

Each of these processes overlaps the previous one. Each of these processes provides information necessary for the next process, but each of these processes begins before the previous process is completed. Because of the interrelations of these four processes, teamwork among the marketing, product planning, R&D, engineering, and manufacturing divisions is essential during the design of a new product. After the production process is designed and the product is successfully manufactured in

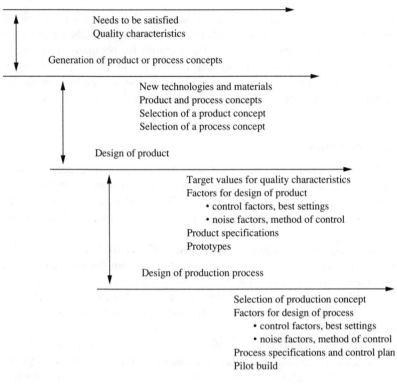

Definition of quality

Needs to be satisfied
Quality characteristics

Generation of product or process concepts

New technologies and materials
Product and process concepts
Selection of a product concept
Selection of a process concept

Design of product

Target values for quality characteristics
Factors for design of product
• control factors, best settings
• noise factors, method of control
Product specifications
Prototypes

Design of production process

Selection of production concept
Factors for design of process
• control factors, best settings
• noise factors, method of control
Process specifications and control plan
Pilot build

FIGURE 10.2
Processes and their outcomes in the design of new products

a pilot run, production begins. After production begins, additional processes include sales and delivery and customer service.

How can the key outcomes for each of these four processes in the design of new products be improved? The strategy is to apply the model for improving quality given in Figure 10.1 to the four processes given in Figure 10.2. This strategy includes methods of planned experimentation developed in the previous chapters, as well as the methods of quality-function deployment and the Taguchi method described in this chapter.

Sections 10.2, 10.3, 10.4, and 10.5 of this chapter present the application of the strategy. Section headings are the four major processes. Each outcome of the process is a subheading within the section. Section 10.6 applies the strategy to designing a new process for service rather than for a new product.

10.2 DEFINITION OF QUALITY

The first process in Figure 10.2 is to define quality. The needs of the customer must be understood and embodied in a definition of quality using quality characteristics. The process is usually led by marketing or a product-planning team.

Needs to be Satisfied

The statement of purpose of an organization should define the needs the organization intends to meet. Customers must be identified. Information must be gathered to match new products to the needs of the customers. The methods for obtaining information from customers include written surveys, personal interviews, focus groups, or simply observation of customers' uses of your current products (or your competitor's product). Recent marketing environments and trends should be included in the analysis.

Quality Characteristics

The needs to be satisfied must be understood, and then quality can be defined using a set of quality characteristics that then serve as response variables for experiments conducted during the design of a product or process. This translation of customer needs into quality characteristics is facilitated by a diagram called a quality characteristic diagram.

The quality characteristic diagram is illustrated in Figure 10.3. The needs of the customer are listed at the top. The first column of the diagram defines quality at a primary level in the language of the customer. Examples are: easy to service, easy to close, does not rattle, proper size, lasts a long time, comfortable to use, doesn't skip, or easy to read. This primary level states how the product will give the customer what he or she wants.

The primary level of quality characteristics must be refined through more detailed steps, as illustrated by the additional columns (secondary and tertiary) of the diagram. Additional quality characteristics may be added by reviewing the dimensions of quality given in Chapter 1. The details of the quality characteristic should be measurable for comparison with competitive products. More detail can be attained by subdividing a quality characteristic into two or more subcharacteristics.

A weighting factor of some type for each quality characteristic (e.g., most important to least important; or high, medium, and low) might provide useful information for potential trade-offs later. Customer surveys may be necessary to justify the weightings.

Quality characteristics should be:

- Continuous variables, as far as possible.
- Measurable.
- A family (vector) of measurements that provides a definition of quality sufficient for the product.
- Specific enough to be useful for design of the product or process.

The quality characteristic diagram defines quality and provides a communication link for marketing, R&D, and engineering. It provides these departments major input into the design of the product or process.

Needs of the customer:			
Quality characteristic[1]			
Primary[2]	Secondary[3]	Tertiary[3]	#

[1] Do not include design factors in this list. (*Test*: You should *not* be able to set the levels of these quality characteristics.)

[2] Express in the language of the customer.

[3] To add more detail, subdivide into two or more quality characteristics.

FIGURE 10.3
Quality characteristic diagram

Example 10.1   Solenoid Valve. A manufacturer of components for automobile engines is designing a valve to turn a pollution control device on and off (see Chapter 6, Example 6.3). How should the manufacturer define quality? What are the important quality characteristics? The quality characteristics selected are given by the quality characteristic diagram in Figure 10.4.

Needs of the customer:	Low pollution efficient engine		
	Quality characteristic[1]		
Primary[2]	**Secondary**[3]	**Tertiary**[3]	**#**
performs well	airflow	airflow	1
	overload protection	overload protection	2
is reliable	time until replacement	time until replacement	3
	thermal stability	thermal stability	4
	impact resistance	impact resistance	5
can be serviced	easy to remove	time to remove	6
		effort to remove	7
	easy to install	time to install	8
		effort to install	9

[1] Do not include design factors in this list. (*Test*: You should *not* be able to set the levels of these quality characteristics.)
[2] Express in the language of the customer.
[3] To add more detail, subdivide into two or more quality characteristics.

FIGURE 10.4
Quality characteristic diagram for designing a solenoid valve

Example 10.2   Wallpaper.   A manufacturer of wallpaper is redesigning an existing wallpaper product. The quality characteristic diagram is given in Figure 10.5. Which quality characteristics are most important to you?

10.3   GENERATION OF PRODUCT OR PROCESS CONCEPTS

The second process in Figure 10.2 is the generation of product or process concepts given a selected technology and materials. The activities of this ongoing process are usually performed by people in research. Outcomes of this process include an increased understanding of new technologies and materials and concepts for new products based on these technologies.

Needs of the customer:	Decorative walls		
Quality characteristics [1]			
Primary [2]	**Secondary** [3]	**Tertiary** [3]	#
looks good	consistent with fashion trends		1
	attractive		2
	lack of visible seams		3
easy to clean	time to clean		4
	effort to clean		5
wears well	resistance to scratches		6
	stain resistance		7
	gloss retention		8

[1] Do not include design factors in this list. (*Test*: You should *not* be able to set the levels of these quality characteristics.)
[2] Express in the language of the customer.
[3] To add more detail, subdivide into two or more quality characteristics.

FIGURE 10.5
Quality characteristic diagram for redesigning wallpaper

New Technologies and Materials

People in R&D must have a deep understanding of natural phenomena and an independent drive for creativity. Theirs must be an environment that allows risk taking and failure as a learning experience. The improvement cycle and planned experimentation will be important methods to maximize learning in this environment. New materials, new technologies, and new production techniques should be considered. Sometimes a new technology leads to the expansion of old concepts for new products. Miniaturization of the microchip is an example. New technology is generated by a combination of needs, concepts, and hardware. There must be sufficient activity addressing all major strategic needs of the customer.

Product and Process Concepts

Methods for generating new product and process ideas include (1) direct search of opportunities involving marketing, R&D, and engineering; (2) exploratory consumer studies; (3) technology information (patents and inventions); (4) individual effort; and (5) creative group methods. It is important that several concepts are considered for a product.

Sometimes a new product or process idea does not require or warrant new technology. The capacity to generate technology is important, but the challenge is to bring the technology to the marketplace with lower cost and higher quality.

What are the product possibilities? How well can they be manufactured? What is their potential in meeting the needs of the customer? The table given in Figure 10.6 captures important information for new and current concepts. As a goal for the process of generating product concepts, the concepts selected must be capable of meeting the customer's needs through product possibilities that can be designed (with few parts) for ease of manufacture and assembly. Assessment of process concepts can be done by the table as well.

Selection of Product Concept

The first task is to select the product concept. Identify several conceptual designs and the form or structure the product might take. Avoid selecting one concept without serious consideration of others. Screen product concepts while increasing current knowledge by running planned experiments to reduce the number of conceptual designs. An example of a planning form using a randomized block design for such selection of a product concept is given in Figure 10.7.

Product concept	Product form/ structure	Ease in manufacturing assembly	Needs addressed	Potential risk in meeting needs low/medium/high
(current)				
(new)				

FIGURE 10.6
Assessment of product concepts

FIGURE 10.7
Example of a planning form for selecting a product concept

1. Objective:
Select a product concept from available candidates.

2. Background information:
Information on each product concept that is in contention. (Figure 10.6)

3. Experimental variables:

A. Response variables	Measurement technique
1. Quality characteristics	(may be subjective rankings)
2. Manufacturability measures	
• assembly time	
• number of parts, steps	
• manufacturing costs	
3. Life in service	

B. Factors under study	Levels
1. Product concept	Current Concept 1 Concept 2

C. Background variables	Method of control
1. Wide range of conditions (production and customer)	(consider chunk-type block variables)

Selection of Process Concept

The selection of concepts to be used for manufacturing the new product should follow the same process as the selection of the product concept. Consideration should be given to modular design, number of assembly steps, assembly time, and automation.

10.4   DESIGN OF THE PRODUCT

Having selected a product concept, the third process is to design components and systems for the new product. Product specifications must be completed and prototypes demonstrated. The objective of this process is to design the new product to meet the needs of the customer as defined by the quality characteristic diagram (Figure 10.3). The key tasks are to set targets for the quality characteristics and to design a product that stays close to the targets under a wide range of conditions (see the analytic study from Chapter 3).

Taguchi's Contributions

Some concepts from Taguchi are needed for the remaining sections in this chapter. Taguchi (1986) has made many contributions to designing quality into products and processes. Three of these contributions are:

Customer loss function,

Reduction of variation from noise factors, and

Parameter design and tolerance design.

CUSTOMER LOSS FUNCTION.   Taguchi defines loss to the customer as a quadratic function of variation from a target of some quality characteristic. As the measurement of the characteristic moves away from the target, the loss increases regardless of where the specifications are. The traditional interpretation of being within specification implies a loss function that is zero while inside the limits and constant while outside the limits. Figure 10.8 illustrates these two interpretations of customer loss. Taguchi's loss function, represented on the left, is usually an accurate description of nature; moreover, it provides the driving force for the philosophy of continuous improvement.

The loss function attains its minimum at the target value and increases as the quality characteristic deviates from the target value. There are three cases of the target value (t). They are:

1. smaller is better ($t = 0$): wear, shrinkage, deterioration.
2. bigger is better ($t =$ infinity): strength, life, fuel efficiency.
3. target value is best ($t = t_0$): dimension, clearance, weight, viscosity.

Although the loss function in Figure 10.8 is symmetrical around the target value, there are many situations that will not yield a symmetrical graph. See Ross (1988, Chapter 1) for examples and more discussion on the economics of reducing variation using the three types of loss function.

REDUCTION OF VARIATION FROM NOISE FACTORS.   Taguchi focuses on reduction of noise factors that cause variation in the actual performance of product. These noise factors are:

1. *External factors:* the environment in which the product is used or distributed.
2. *Internal factors:* product deterioration with age or use.
3. *Unit-to-unit factors:* variations in the manufacturing process.

These sources of product performance variation must be considered during the design phase of the product. Figure 10.9 illustrates where this variation can be reduced (based on Kackar, 1985).

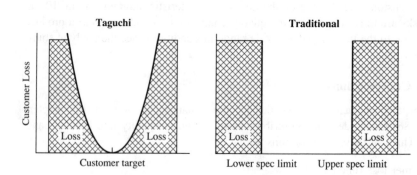

FIGURE 10.8
Taguchi's loss function vs. specification loss function

Process/noise:	External	Internal	Unit-to-unit
Design of product	high	high	medium
Design of production process	low	medium	high
Production	low	low	medium
Sales/service	medium	low	low

FIGURE 10.9
Leverage on noise factors to reduce variability

The greatest leverage in improvement of quality will come during the design phases of the product and the production processes. The leverage for improvements during these phases in the product cycle is many times greater than for improvements made during actual production of that product. When an engineer calls for changes early in product development, there is time to make improvements at the lowest cost. The majority of the production costs, including levels of scrap, are determined at the beginning during the design of the product and the design of the production process.

PARAMETER DESIGN AND TOLERANCE DESIGN. After the selection of the concept for a product or process, the next step should include parameter design and tolerance design. *Parameter design* is the selection of levels for the control factors to obtain the targets for the quality characteristics and to reduce the effect of noise factors. The next step is the *tolerance design,* during which acceptable levels of variation for control factors (considering costs) of product or process are set.

The activities for parameter design and tolerance design are illustrated as a process in Figure 10.10. The efficiency of these two sets of activities can frequently be raised by planned experimentation. Taguchi's approach combines planned experimentation with engineering knowledge to design products and production processes that are less sensitive to noise factors (external, internal, and unit-to-unit).

The input required for this process for reducing variability includes clear definitions of the needs of the customer and related quality characteristics as determined by

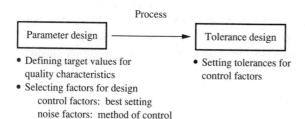

FIGURE 10.10
Process for reducing variability of product design and production design

the quality characteristic diagram. Additional input includes the selected concepts for product or process. The outcomes of this process include identification of product and process specifications, key factors and their control, and best operating conditions. Throughout the product's design, prototypes are tested. Pilots are done to test the design of the production process. Methods of planned experimentation are used to gain maximum knowledge about product or process. Additional reading on Taguchi's approach can be found in Taguchi (1986, 1987) and Ross (1988).

Target Values for Quality Characteristics

Quality characteristics are used as response variables in planned experimentation to increase product and process knowledge. When there are multiple measurements of the quality characteristic at the same combination of levels of the factors, it is often useful to form several statistics. The overriding factor in selection of a quality characteristic statistic is its impact on customer loss and its relationship to customer need. Taguchi's loss function could be used as a response variable. Following is a discussion of some commonly used statistics:

AVERAGE. The average of the observations in the experiment reduces the magnitude of variation due to nuisance variables. The average is widely used as a response variable when replication is included in the design.

STANDARD DEVIATION. The standard deviation of the observations in a replicated experiment represents the magnitude of the nuisance or background variables within the experimental pattern. Depending on the design, this magnitude could represent any or all of Taguchi's three noise factors. Ranges could be used instead of standard deviation if the number of observations is small (less than 10) and constant for each factor combination.

AVERAGE VS. VARIATION. Both average and measures of variation such as standard deviations could be used as response variables (see Example 6.3). Choice of one depends on the relative impact of each on customer loss. The objective is to minimize the loss to customer.

Taguchi's signal-to-noise ratio has no advantage, as a response variable, over looking at average and variation separately, and it is more difficult to interpret.(See Box [1988] for more discussion on signal-to-noise ratios.)

Once the statistic of the quality characteristic is defined, the target value for each statistic is determined. This target is set at minimum loss for the Taguchi loss function. Figure 10.11 gives examples of needs, quality characteristics, selected statistics of quality characteristics, and target values from examples presented in earlier chapters of this book.

Factors for Design of Product

The next step in the process of designing a new product is to select factors for design that have an effect on the quality characteristics. A methodology to relate the factors for design to the quality characteristics is *quality function deployment* (QFD).

FIGURE 10.11
Examples of quality characteristics and targets

Need of the Customer	Quality characteristic	Statistic of the quality characteristics	Target values	Example in book
Increased tool-life	wear rate	slope of line	smaller	5.3
Labels on cans	labels loose or not	percent cans with loose labels	smaller	5.5
Desired viscosity	reaction time	average	8 hr.	6.1
Color	shade	shade reading	220	6.2
Low-pollution efficient engine	air flow of solenoid	average, standard deviations	.60 cfm smaller	6.3
Smoothness of finished part	microfinish	average	smaller	9.1

QFD is a concept and methodology that originated in Japan in 1972 at Mitsubishi's Kobe shipyard site. The aim of QFD is to deploy the needs of the customer throughout the evolution of a product, during both design of the product and design of its production process. This deployment ensures that the needs of the customer guide the work in all phases of the process. By disciplining the designer of the product to focus on the needs of the customer, this emphasis increases the likelihood that the new product will be perceived as high quality when it comes to the marketplace.

In viewing production as a system, QFD also connects customer research to the design and redesign stage. It links the four major processes in the design of a new product. Constant interaction between marketing, research, engineering, manufacturing, and sales/service is needed to translate the needs of the customer into a new or improved product that will better meet those needs. QFD promotes this interaction and the breaking down of barriers between departments. See Sullivan (1986) and Hauser and Clausing (1988) for more discussion on QFD.

A diagram combining the "house of quality" of QFD and Taguchi's parameter design is illustrated in Figure 10.12. The QFD relation diagram provides a method for relating quality characteristics to factors that should be addressed in the design of a new product or component. This diagram helps define the current knowledge for a product and is analogous to the cause-and-effect diagram for current knowledge of a process.

The extreme lefthand column of the QFD relation diagram lists the quality characteristics that have been identified using the quality characteristic diagram given in Figure 10. 3. The next column lists the target value. The remaining columns are factors that may have an effect on the quality characteristic. These factors are classified as control factors and noise factors and are defined as follows:

> *Control factors:* Factors that can be assigned at specific levels. These factors are set by those designing the product; they are not directly changed by the customer.

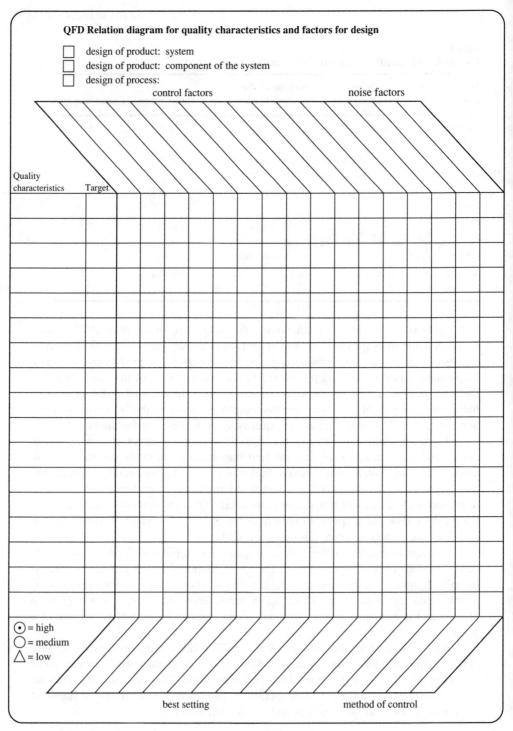

QFD Relation diagram for quality characteristics and factors for design

☐ design of product: system
☐ design of product: component of the system
☐ design of process:

control factors   noise factors

Quality characteristics   Target

⊙ = high
◯ = medium
△ = low

best setting   method of control

What planned experiments need to be run to test this diagram?

FIGURE 10.12
QFD relation diagram

Noise factors: These factors can potentially affect the quality characteristic but cannot be controlled at the design phase. They may be external, internal, or unit-to-unit factors.

An example of control factors when designing, say, a new tennis racket, would be the shape of the frame and the type of composite material selected for the frame. Examples of noise factors include temperature or humidity (external), wear or warp of the frame (internal), and the effect of variation in shape of different frames on accuracy of the hit (unit-to-unit).

At the intersections of the diagram, symbols are used to represent the degree of relationship between the quality characteristics and the factors. Planned experiments may be required to test the degree of relationship and to determine the best settings for each control factor. A method of control must also be determined for all noise factors.

If there is an interaction between the control factor and noise factor, the best setting for the control factor may reduce the effect of the noise factor (contribution of Taguchi's parameter design).

Example 10.1 (continued) QFD Relation Diagram of a Solenoid Valve. A planned experiment was run to determine the best settings for some important components in the design of a solenoid valve for a pollution control device on an automotive engine. This experiment is described in Chapter 6 (Example 6.3). Airflow was the quality characteristic chosen as the response variable in a particular experiment. Other experiments were run to study the other quality characteristics listed in Figure 10.4.

The partially completed QFD relation diagram of the solenoid valve based on this planned experiment is given in Figure 10.13. Best settings were chosen for two of the four control factors based on the response plots given in Chapter 6 (Figure 6.22). The remaining two could be set to impact the other quality characteristics favorably.

The QFD relation diagram is useful in planning the types of experiments that are needed to decide the best settings for control factors to ensure that the quality characteristics will be close to the target (minimum loss) and have minimum sensitivity to the noise factor.

The terms used in planning an experiment and the terms used in selecting factors for design are usually related in the following way:

$$Response\ variables\ =\ Quality\ characteristic$$

$$Factors\ under\ study\ =\ Control\ factors\ and\ noise\ factors$$

$$Background\ variables\ =\ Noise\ factors$$

Although the above relationships are the general rule, there will be some exceptions. For example, inner noise factors such as wear may be used as response variables. As more knowledge is gained from planned experiments, the QFD relation diagram should be updated for relationships between factors and quality characteristics, best settings for control factors, and method of control of noise factors.

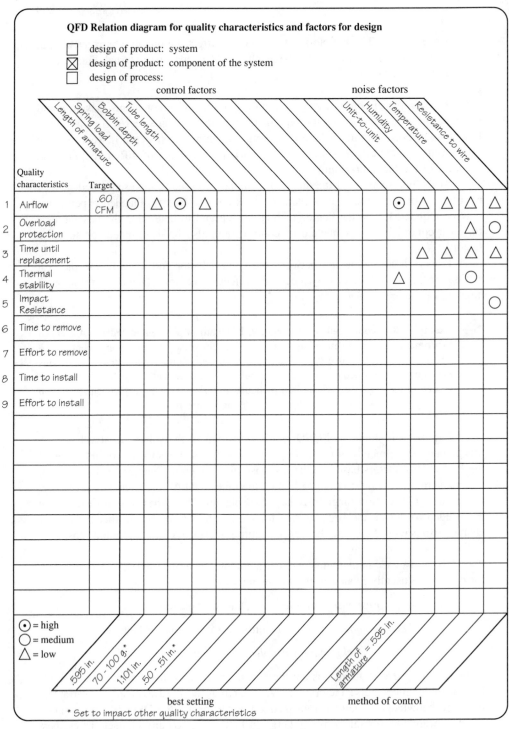

What planned experiments need to be run to test this diagram?

FIGURE 10.13
QFD relation diagram for designing a solenoid valve (partially completed)

The overall strategy of parameter design is setting control factors to desensitize the product or process to noise factors. Quality by design requires a change from thinking about designing a product or process as a search for something that works to recognizing the importance of understanding the cause system to ensure that the product or process is close to targets under a variety of conditions. The following example demonstrates this strategy and thinking.

Example 10.2 Setting of Control Factors for New Wallpaper. An engineering group from a manufacturer of wallpaper is identifying the factors for design of a vinyl-coated wallpaper product. Figure 10.14 provides the QFD relation diagram.

The group is at a moderate level of knowledge, with the new design being more attractive. Previous tests have identified a seam problem because of shrinkage of the wallpaper (probably due to temperature or humidity during installation). When two sheets were put together, the seam tended to show. When the width of the seam was greater than 0.2 mils, it called attention to the seams and detracted from the appearance of the wall.

The group decided to determine the effect that installation factors have on seams and appearance. It was hoped that by setting control factors, noise factors would be reduced. Figure 10.15 contains the planning form for the study.

Three control factors and two installation (noise) factors were studied. Four measurements were made on each seam with averages and standard deviations calculated. A 2^{5-1} design was chosen because there is no confounding.

After the design was carried out, the usual analysis of appearance, averages, and standard deviations for visible seams was made. Appearance of the new design remained good throughout the 16 runs of the experiment. Laboratory testing for scratches, stains, and gloss retention confirmed that no deterioration took place for these quality characteristics.

Figure 10.16 contains the response plots for standard deviations of visible seams; they summarize the important results of the experiment. The first plot shows a strong interaction between a control factor and a noise factor. Thickness of wear level interacts with the temperature/humidity noise factor. The effect of this noise factor on visible seams is less with the 2.0 mil wear layer. The thinner wear level (1.0 mil) was affected by a high temperature/humidity combination.

The second plot shows a strong interaction between the other two control factors. Backing material and the formulation of the intermediate layer interacted with respect to seams. Both backing materials are needed for different applications. Thus, material *A* is chosen with formula 1, and material *B* is chosen with formula 2.

The third plot demonstrates a strong relationship between an installation factor and visible seams. By laying out the roll for 10 minutes before installation, the seams will have less variation. This adds to the installation time.

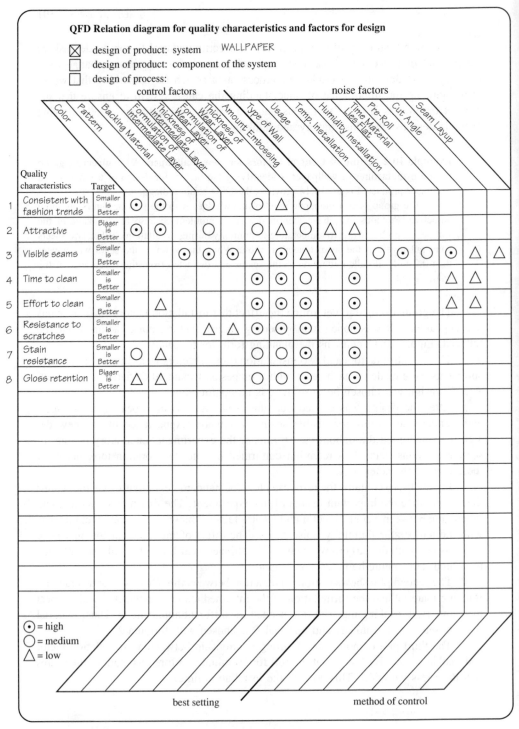

What planned experiments need to be run to test this diagram?

FIGURE 10.14
QFD relation diagram for redesigning wallpaper

FIGURE 10.15
Documentation of the wallpaper experiment

1. **Objective:** Find best settings of the control factors to minimize visible seams while preserving the appearance.

2. **Background information:** The redesign has a new color and pattern that has a beautiful appearance. There has been a problem with seams. Probable cause is temperature or humidity at the installation site.

3. **Experimental variables:**

A.

Response variables	Measurement technique
1. Visible seams	gage (mils)
2. Appearance	subjective scoring

B.

Factors under study	Levels	
1. Backing material	Material *A*	Material *B*
2. Formulation of intermediate layer	Formula 1	Formula 2
3. Thickness of wear layer	1.0 mils	2.0 mils
4. Temperature/humidity	50°/40%	90°/80%
5. Pre-roll	0 min.	10 min.

C.

Background variables	Method of control
1. Installer	One person, standard instructions
2. Type of adhesive	Use standard for all applications
3. Subwall	Use plywood (porous material)
4. Thickness intermediate layer	4.0 mils
5. Formulation of wear layer	Same as old design
6. Time material lays flat	4 minutes
7. Cut angle	standard 90°
8. Seam layup	set at low

4. **Replication:** Four measurments for visible seams per piece.

5. **Methods of randomization:** Order of the 16 runs was randomized using a random permutation table.

6. **Design matrix:** 2^{5-1} factorial design.

7. **Data collection forms:** (not shown here)

8. **Planned methods of statistical analysis:** Compute average and standard deviation of four readings per piece.

9. **Estimated cost, schedule, and other resource considerations:** Evaluated in the test room with temperature and humidity controls. Four days are required to complete. Appearance scoring done by appearance team.

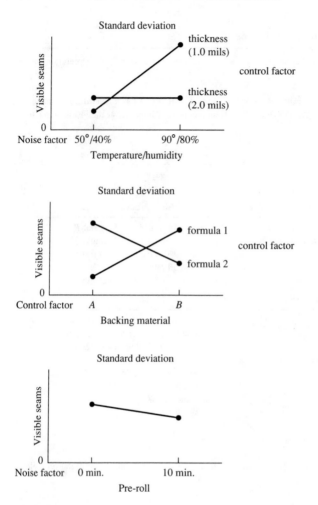

FIGURE 10.16
Response plots for the wallpaper experiment

    Based on these analyses and the current knowledge, the following conclusions (and updates for Figure 10.14) are:

Control factors	Best setting
Thickness of wear layer	2.0 mils
Backing material	A with Formula 1
	B with Formula 2

Noise factors	Method of control
Pre-roll	Have installation instructions include a pre-roll at 10 minutes
Temperature/humidity	Desensitized by setting control factor, thickness of wear layer, at 2.0 mils.

Product Specifications

Having reduced the effect of all noise factors, the next step is to set tolerances for control factors. In Example 10.1 the bobbin depth was set at 1.101 in. to obtain an average flow of .6 cfm. The armature length was set at .595 in. to reduce variation in flow. Since flow is sensitive to variations in bobbin depth, there must be specifications on bobbin depth to aid in the design of the production process. Because variation has been reduced by setting armature length at .595 in., the specification on bobbin depth can be loosened while not increasing the overall variation in flow.

It is usually more cost effective to reduce variation through parameter design than through the setting of tight specifications.

Prototypes

The objectives of testing prototypes are

- to confirm that the criteria for function and performance are built into the design,
- to detect likely causes of quality characteristic variation around targets under a variety of conditions, and
- to reevaluate costs.

Knowledge of reliability and accelerated testing as well as failure analysis on existing products is essential. The philosophy of testing prototypes should be to challenge the design rather than pamper it. The number of improvement cycles used to test prototypes need not be large if sufficient knowledge is gained with each cycle.

Example 10.1 (continued) Testing Prototype Solenoid Valves. The quality characteristic of air flow in Chapter 6 (Example 6.3) is important in the design of a solenoid valve for a pollution control device on an automotive engine. The 2^4 experiment (based on a moderate level of knowledge) showed that the control factor *bobbin depth* can be used to increase air flow. Armature length was recommended to be set at .595 inches to minimize variation of air flow.

The best settings for a redesign of the solenoid are:

Control factors	Best setting
bobbin depth	1.101 inches
length of armature	.595 inches
spring load	(set both based on other quality characteristics)
length of the tube.	

Thirty prototypes of the new design were developed. Is the prototype design better than the current design in the field? Will this design work well under the wide range of conditions that will be encountered during actual production and use by customers?

Figure 10.17 contains the planning form for this new experiment. The experiment chosen is a paired-comparison block design with the noise factors arranged in blocks (chunk variables). Four background variables that potentially affect the per-

FIGURE 10.17
Documentation of the solenoid experiment

1. **Objective:** Determine if a redesign of a solenoid valve holds up under various noise factors. This information will be used to determine if the current solenoid valve should be replaced by the new design.

2. **Background information:** Previous experiments using fractional factorial and factorial designs lead to redesigning the current valve. Noise factors are suspect in affecting variation of the current design.

3. **Experimental variables:**

A.

Response variables	Measurement technique
1. Average and standard deviation of flow (cfm)	Flow meter 65c

B.

Factors under study	Levels	
1. Solenoid design	Current	Redesign

C.

Background variables	Levels	
1. Pressure (lb/sq. in.)	30	45
2. Humidity (percent):	80	95
3. Temperature (F)	40	60
4. Resistance of wire (ohms)	50	100

Method of control: Treat the four background variables as noise factors. Create five blocks made up of different treatment combinations of the background variables.

4. **Replication:** Five replications (blocks as noise factors) will be made using six current and six redesigned solenoids within each block.

5. **Methods of randomization:** Flip a coin to determine order (heads: current design; tails: redesign).

6. **Design matrix: (attach copy)** See Table 10.1.

7. **Data collection forms: (attach copy)** See Table 10.1.

8. **Planned methods of statistical analysis:** Compute the average and standard deviation for the six flow readings for current and redesigned solenoids. Plot run charts for average and standard deviation of flow and run charts for average and standard deviation of flow adjusted for block effect.

9. **Estimated cost, schedule, and other resource considerations:** Study can be completed during a 5-day period.

---

formance of the solenoid are pressure, humidity, temperature, and resistance. These variables are treated as noise factors arranged in blocks.

The blocks were chosen by judgment to create extreme conditions. This experiment will increase understanding of the performance of the new design with respect to potential problems. This allows for testing our current and new designs under the widely varying conditions that could be expected in the field.

Table 10.1 lists the design matrix and the data form for the study.

The results of the study are given in Table 10.2. Run charts for average and

TABLE 10.1
Design matrix and data collection form for example 10.1

Noise factor	Block 1	Block 2	Block 3	Block 4	Block 5
Pressure	45	30	30	45	45
Humidity	80	80	95	80	95
Temp.	40	40	60	40	60
Resistance	50	100	50	100	100
Response: average and standard deviation of six solenoids for flow (cfm)					
current c	(1)	(1)	(2)	(2)	(1)
avg.	—	—	—	—	—
std. dev.	—	—	—	—	—
redesign r	(2)	(2)	(1)	(1)	(2)
avg.	—	—	—	—	—
std. dev.	—	—	—	—	—

NOTE: The (1) or (2) in parenthesis indicates the order in which the values were tested.

standard deviation of flow and run charts for average and standard deviation of flow adjusted for block effect are given in Figure 10.18.

Based on the analysis of the prototype study, the conclusions are as follows:

1. Average flow for current (.594) and redesign (.598) are both close to the target value (.600). The sensitivity (block effect) of both designs was similar for average flow.
2. Standard deviations for the redesign were smaller than for the current design. The sensitivity (block effect) for the redesign was much better.

It was felt that these results will hold up in the future and that it would be best to accept the redesign and run a pilot of the redesigned solenoid.

An alternative design for Example 10.1 would use four background variables as factors and run a 2^{5-1} to test the interaction of the background variables with the solenoid design. Taguchi (1986, 1987) uses the "outer-array" approach to accommodate background variables as noise factors.

10.5   DESIGN OF THE PRODUCTION PROCESS

The strategy of parameter design discussed in Section 10.4 applies directly to designing a production process. The leverage for improvement of quality is in reducing variation of the quality characteristics of the product due to unit-to-unit noise (variations in the manufacturing process). It is often the case that the control factors for the product become the quality characteristics for the manufacturing process. For example two important control factors for the solenoid in Example 10.1 were bobbin depth and length of armature. The targets were 1.101 inches for bobbin depth and 0.595 inches for length of armature. These two factors are control factors for the product design but become quality characteristics for the process design.

TABLE 10.2
Test results for example 10.1

Design	Block 1	Block 2	Block 3	Block 4	Block 5	Average
	Response: average of six solenoids for flow (cfm).					
c (current)	.59	.54	.64	.60	.60	.594
r (redesigned)	.61	.59	.60	.65	.54	.598
Block average	.600	.565	.620	.625	.570	

Overall average .596

Response: average of six solenoids for flow (cfm) adjusted for block effect (average − block average + overall average).

Design	Block 1	Block 2	Block 3	Block 4	Block 5	Average
c (current)	.586	.571	.616	.571	.626	.594
r (redesigned)	.606	.621	.576	.621	.566	.598

Response: standard deviation of six solenoids for flow (cfm).

Design	Block 1	Block 2	Block 3	Block 4	Block 5	Average
c (current)	.067	.068	.040	.052	.041	.054
r (redesigned)	.033	.036	.030	.029	.024	.030
Block average	.050	.052	.035	.040	.033	

Overall average .042

Response: standard deviation of six solenoids for flow (cfm) adjusted for block effect (average − block average + overall average).

Design	Block 1	Block 2	Block 3	Block 4	Block 5	Average
c (current)	.059	.058	.047	.054	.050	.054
r (redesigned)	.025	.026	.037	.031	.033	.030

Factors for Design of the Production Process

After selecting the concepts for the production process (there are cases when a new concept for production process may not be necessary), the next step is to select factors for designing the production process. The QFD relation diagram should be used again with the control factors and noise factors coming from the production process.

The aim of parameter design should be to set control factors that will improve process capability by reducing the effect of noise factors (highest leverage for unit-to-unit noise). Examples from previous chapters will illustrate the strategy of parameter design as it applies to designing the production process.

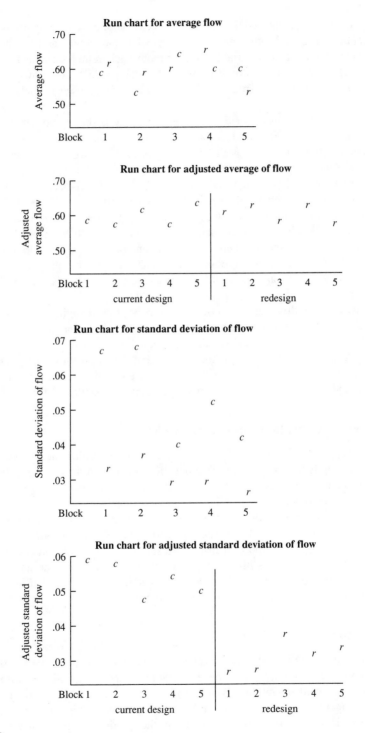

FIGURE 10.18
Four run charts for paired comparison block design

Example 5.5 showed that the control factor, application rate, gave us the smallest percentage of loose labels (smaller is better) when the target was 14 oz/hr. This factor interacted with a chunk variable (one or more background variables) used to define the block. Interaction was used to desensitize the product. A follow-up study would be a factorial design to study the background variables used to make up the chunk variable.

Example 6.2 has a shade of a dyed material as the quality characteristic. The target is at shade 220. The factorial design showed that running the process at high oxidation temperature makes the process less sensitive to variation in the material. The third factor, oven pressure, can be used to adjust the shade to 220.

Example 9.2 has fill weight as a quality characteristic. A number of factors that affected the average fill weight had been identified. A factorial design with a centerpoint was run with fill weight variation to determine presence of nonlinear effects. The results shown in Chapter 9 (Figure 9.9) indicate a nonlinear response with line speed. If speed is kept below 350 cans per hour, the variation in fill weight will be less. Also, temperature and consistency interact. Keeping temperature above 200° minimizes the variation in fill weight due to varying consistency of the incoming ingredients.

Output of this stage in process design is process definition. One now needs standardization—things or methods that have already been found to be good (standard parts, units, or modules, and standard procedures).

Development of production capability and development of the product must be done in parallel. This ensures smooth and efficient transition into factory operations, for product and production design must be managed together.

Process Specifications and Control Plan

Having applied the strategy of parameter design to designing the production process, the next step is to validate that the production process produces good product. Specifications and tolerances help to define acceptable outcomes.

As was emphasized in Chapter 2, there are two different approaches to the management of variability resulting from the production process. Figure 2.1 illustrated these two approaches. The aim should be to reduce the dependence on inspection to achieve quality. The purpose of inspection should be for improvement of process and reduction of costs. Mass inspection is costly and ineffective. The application of the model for improvement of quality and the methods of control charting and planned experimentation provide the basis for the control plan.

Pilot Run

Production capability must be developed to produce the parts and components, to assemble the product, and to have operating systems that are necessary for production and field operations. Building prototypes confirms the product. Often parts and components are handmade. Pilot runs with production tooling confirms the process. During a pilot run, parts from the production process are studied for the first time. A pilot run also provides the opportunity to predict future production capability. The methods of control charting and determining the capability of a process (Chapter 2) are useful here.

What are the important control factors? What are the noise factors that the product might encounter during production? Pilot runs provide the opportunity to establish the relationships between the control factors, noise factors, and quality characteristics as identified by the QFD relation diagram for the production process. Planned experimentation will increase the knowledge of these relationships. This will help establish the best operating conditions for the manufacturing process *before* production starts.

This concludes the activities related to improving quality during the design of a new product. Redesign of an existing product uses the same four major processes stated in Figure 10.2. Since current knowledge is higher, time until start of production is decreased.

10.6 DESIGN OF A NEW SERVICE

What about designing a new process, not for manufacturing a product, but for a new service? All the activities (Sections 10.2 through 10.5) involved in improvement of product quality apply. The product is service. What are the needs of the customer? What are the quality characteristics and their targets? What are the control factors and noise factors? These questions will be answered by applying the activities of Sections 10.2 through 10.5 to a new educational service—a new course in planned experimentation.

> **Example 10.3 Designing a New Course.** A new course in planned experimentation is being designed. What are the needs of the customer and the corresponding quality characteristics that guide the design of this new course?
>
> Figure 10.19 gives the quality characteristic diagram. This course should create an enjoyable learning experience for the students. It is also intended to enable them to apply the concepts to their jobs or hobbies. Fifteen quality characteristics were identified as important measures for this class.

What are the control factors in designing this new course? What are the noise factors? The QFD relation diagram is given in Figure 10.20. Eleven control factors are identified based on concepts developed from various sources of adult education research. The control factor *learning teams* relates to many of the quality characteristics. The five noise factors may affect the learning.

How can the noise factors be desensitized or eliminated? Best settings of control factors are based on experimenting with related classes. The method of control for noise factors has proved successful in other classes.

What is the best setting (team size and makeup) for the control factor *learning teams*? A planned experiment for a prototype class can help answer this question. Figure 10.21 gives an example of the planning form for such an experiment.

Creating any new service involves designing a process to provide that service. One difference between a process that provides a product and a process that produces a product pertains to inspection.

Needs of the customer: Enjoy learning new concepts and be able to apply concepts to job or hobby.

Quality characteristic[1]			
Primary[2]	**Secondary[3]**	**Tertiary[3]**	**#**
enjoy instruction	good teacher	pleasant person	1
		good instructor	2
	good class	good textbook	3
		good teaching process	4
enjoy learning	have fun in class	enjoy class	5
		enjoy environment	6
	clear instruction	good teaching aids	7
		good examples	8
	interesting	motivated	9
		involved	10
recognize applications	problem clear	good charter	11
	formulate problem	document current knowledge	12
		develop overall plan	13
know how to use methods successfully	use improvement cycle	use methods	14
		improvements made	15

[1] Do not include design factors in this list. (*Test*: You should *not* be able to set the levels of these quality characteristics.)

[2] Express in the language of the customer.

[3] To add more detail, subdivide into two or more quality characteristics.

FIGURE 10.19
Quality characteristic diagram for designing a course in planned experimentation

QFD Relation diagram for quality characteristics and factors for design

- ☐ design of product: system
- ☐ design of product: component of the system
- ☒ design of process: New Class

	Target	Class Size	Class Time	Content	Review & Follow Up	Facility	Visual Aids	Learning Teams	Computer Simulation	Experiential Learning	Materials	Instructor	Learning Problem	Knowledge of Student	Distractions of Student	Work Environment	Traffic
		control factors											noise factors				
1 Pleasant person													⊙				
2 Good instruction					○	○		○					⊙				
3 Good textbook					⊙							⊙			⊙		
4 Good teaching process							⊙	⊙	⊙	○			⊙		⊙		
5 Enjoy class		○				⊙				⊙		⊙	○		○		○
6 Enjoy environment		△	⊙				⊙		⊙				△		○	⊙	
7 Good teaching aids									⊙		⊙						
8 Good examples									⊙	⊙							
9 Motivated		△	△						⊙	⊙			○	○		○	
10 Involved		⊙	○			○			⊙	△		⊙	○	○		○	△
11 Good charter					⊙	○						○		⊙		⊙	○
12 Document current knowledge					⊙	○						○		⊙			
13 Develop overall plan					⊙	○						○	○				
14 Use methods						⊙						○	○	⊙		○	⊙
15 Improvements made						⊙							○		⊙		⊙
best setting	25	10 3-hr. classes	Textbook	3 months after class	Easy to get to	VCR, Overhead proj.	5 per Team	Manufacturing process	60% Class time	Desktop publishing	Single instructor					
method of control													Learning teams	Prerequisites	Learning teams	Hobbies or work problem	Location of class

- ⊙ = high
- ○ = medium
- △ = low

What planned experiments need to be run to test this diagram?

FIGURE 10.20

QFD relation diagram for designing a course in planned experimentation

313

FIGURE 10.21
Documentation of an experiment to study learning teams

1. Objective: Select size and makeup of a learning team for a planned experiment class.

2. Background information: Previous classroom experience with teams and a study of the literature on cooperative learning teams.

3. Experimental variables:

A. Response variables	Measurement technique
1. Concepts learned	Test
2. Enjoyed class	Interview
3. Involved	Observation

B. Factors under study	Levels	
1. Size of team	5	8
2. Makeup of team	Matched	Diverse

C. Background variables	Method of control
1. Learning disability	Measure
2. Knowledge of student	Chunk-type block variable
3. Distraction of student	Chunk-type block variable

Method of control: Treat the 2nd and 3rd background variables as noise factors. Create four blocks made up of different treatment combinations of the background variables.

4. Replication: Four replications (four blocks as noise factors) will be made using the four combinations of the two factors under study.

5. Methods of randomization: Table of random permutations of four was used to assign the four treatment combinations.

6. Design matrix: 2^2 design in four blocks.

Inspection can prevent bad product from reaching the customer, but this is often not possible for a service. It is too late; the service to the customer has already been provided. Since this is the case in designing a class, achieving quality by design is even more important.

10.7 SUMMARY

This chapter illustrated the importance of designing quality into new products and processes. The development started with Deming's conception of production as a system and the emphasis on the design and redesign stage in matching products and services to a need.

Four major processes in the design and redesign stage were identified. The key tasks of these processes included defining quality, setting targets, and designing products or processes that are close to the targets under a wide range of conditions.

By integrating the concepts of quality function deployment (QFD), the ideas developed by Taguchi, and the methods of planned experimentation into the model to improve quality, a strategy or road map for *quality by design* is created. Application of this strategy can:

- Accelerate the evolution of new product cycles.
- Reduce development costs.
- Improve the transition from R&D to manufacturing.
- Make the product or process robust against noise factors by selecting the proper level of control factors. This results in higher acceptance of the product and less warranty claims.

Finally, applying this strategy to the design and redesign stage of a product or process provides the greatest leverage for improvement of quality.

REFERENCES

Box, George (1988): "Signal-to-Noise Ratios, Performance Criteria and Transformations," *Technometrics*, Vol. 30, No. 1, February.

Clark, Kim B. (1989): "Development Game Belongs to the Japanese," *Autoweek*, May 1.

Hauser, John R. and Don Clausing (1988): "The House of Quality," *Harvard Business Review*, May–June.

Kackar, R. (1985): "Off-Line Quality Control, Parameter Design and the Taguchi Method," *Journal of Quality Technology*, Vol. 17, No. 4, October.

Ross, Phillip J. (1988): *Taguchi Techniques for Quality Engineering*, McGraw-Hill, New York.

Sullivan, Larry P. (1986): "Quality Function Deployment," *Quality Progress*, June.

Taguchi, Genichi, (1986): *Introduction to Quality Engineering, Designing Quality into Products and Processes*, Unipub, White Plains, NY.

Taguchi, Genichi, (1987): *System of Experimental Design, Volumes 1 and 2*, Unipub, White Plains, NY.

EXERCISES

10.1. Study the effect of poor quality product design and production process design for products in an organization.

10.2. Compare the processes in an organization with those in the evolution of a new product from Figure 10.2.

10.3. Identify the three sources of product performance variation (external, internal, and unit-to-unit) for the major products in an organization.

10.4. Complete a QFD relationship diagram for a new or existing product.

10.5. Determine how many planned experiments were carried out by product design engineers or manufacturing engineers in an organization in the last two years.

10.6. *The paper box factory.*
Design a paper box that stores change (quantity equal to the change in most people's pockets). Materials and technologies include

two pair of scissors

two small staplers

four crayons

a ream of 8 1/2″ × 11″ blank paper.

Appoint a product-planning team, a product design team, a production process design team, a manufacturing team, and a customer service team.

Use the process from Figure 10.2 to develop your product and production processes. What are the customer needs? (The box needs a removable lid) What are the quality characteristics? What is your measurement process? What are the product control factors? Noise factors?

Manufacturing should produce 50 boxes. Is the process stable? Capable?

Can the product be improved? How?

Can the process be improved? How?

CHAPTER
11

CASE
STUDIES

The aim of this chapter is to apply the model for improving quality and the philosophy and methodology presented in this book to provide a system for experimentation.

The first ten chapters are ordered to facilitate learning. A real project in any organization, however, would require a different order. Figure 11.1 illustrates this and serves as a quick reference to the chapters in this book where the concepts are treated more fully.

This chapter presents two applications of the methods presented in the previous ten chapters. One is a manufacturing case study. The other is a case study for redesigning a product. Other applications outside of manufacturing, R&D, and engineering could include administrative and other support functions for products or services.

11.1 SEQUENTIAL EXPERIMENTATION

Deming (1980) states that most problems in industry are analytic. The purpose of an analytic study (Chapter 3) is to improve performance of a product or process. The problem is prediction of future performance.

Degree of belief in our prediction is increased as knowledge of the product or process is increased. The model for improving quality provides a *sequential building of knowledge* with prediction as the aim. Figure 11.2 illustrates the iterative nature of the model for building of knowledge.

The two case studies presented in the next two sections of this chapter illustrate the sequential building of knowledge. The first case study involves a redesign of an existing milling process for the manufacture of metallic bricks. Eight cycles are completed to increase the degree of belief in the prediction that the changes made to the process will improve the quality of bricks in the future. As new knowledge is gained, appropriate changes are made to the process.

FIGURE 11.1
Concepts for improvement of quality

Activity	Analytic study	Model for improving quality	
		Current knowledge	Improvement cycle
Redesign an existing process	Yes, design analysis (Ch. 3)	supplier/customer flowchart cause and effect (Ch. 1)	Control chart (Ch. 2) Planned experiment (Ch. 4–9)
Design a new process	Yes, design analysis (Ch. 3)	quality characteristic factors for design, QFD (Ch. 10)	Planned experiment (Ch. 4–9) Control chart (Ch. 2)
Redesign an existing product	Yes, design analysis (Ch. 3)	quality characteristic factors for design, QFD (Ch. 10)	Planned experiment (Ch. 4–9) Control chart (Ch. 2)
Design a new product	Yes, design analysis (Ch. 3)	quality characteristic factors for design, QFD (Ch. 10)	Planned experiment (Ch. 4–9) Control chart (Ch. 2)

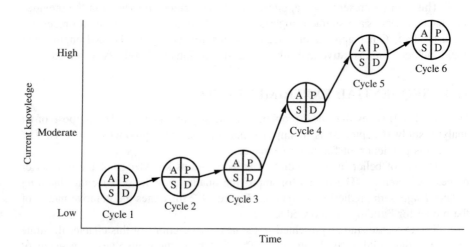

FIGURE 11.2
Sequential building of knowledge

The second case study involves a redesign of a wallpaper product. Seven cycles are completed to increase the degree of belief that a new wallpaper product will be an improvement over the current product.

11.2 CASE STUDY 1: IMPROVING A MILLING PROCESS

This case study is based on work in a small factory, Mid-State Brick Company. The factory contains multiple machining processes to make a simple product, a metallic brick. Figure 11.3 illustrates the production flow for the factory. The size and finish of the brick are important quality characteristics. Figure 11.4 contains the drawings and specifications for the brick.

Historically, the Mid-State factory had shipped all the bricks that it produced. Minimal checks or analyses of the quality of the bricks had been done. Recent complaints from customers about product quality forced an assessment of quality in the plant. Initial checks at final inspection showed many parts out of specifications for multiple dimensions and finish. The width of the brick and the finish on the sides were particular problems. A quality improvement team for the mill process was organized to address these issues.

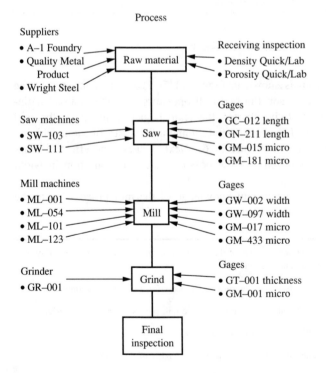

FIGURE 11.3
Process flow for Mid-State Brick Company

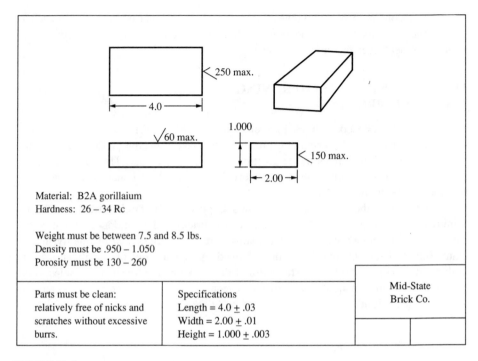

Material: B2A gorillaium
Hardness: 26 – 34 Rc

Weight must be between 7.5 and 8.5 lbs.
Density must be .950 – 1.050
Porosity must be 130 – 260

| Parts must be clean: relatively free of nicks and scratches without excessive burrs. | Specifications
Length = 4.0 ± .03
Width = 2.00 ± .01
Height = 1.000 ± .003 | Mid-State
Brick Co. |

FIGURE 11.4
Engineering drawing for the brick

The charter for the team is shown in Figure 11.5. The team consisted of the factory engineer, the mill supervisor, the two mill operators, and the contract maintenance person. After review of the charter, the team began focusing on the mill process. A schematic diagram of the mill operation and a flowchart of the machining process for each mill were prepared (Figure 11.6). Next, a cause-and-effect diagram (Figure 11.7) was prepared to summarize the possible causes of variation in width and microfinish.

Team discussions about their current knowledge of the mill process led to the following plan for the improvement effort:

Cycle 1   Document the current performance of the mill process and determine the capability of the process for width and microfinish.

             Determine the capability of the measurement process for width and microfinish.

Cycle 2   Identify the important sources of variation in the mill process using nested designs.

Cycle 3   Evaluate the different suppliers of mill cutters.

Other Cycles Conduct other studies as required to improve the quality of the bricks.

Last Cycle   Document the performance of the process using control charts.

FIGURE 11.5
Charter of quality improvement team for mill

General Description:
 Improve the quality of the bricks from the mill process. The focus should be on the width of the brick and the finish of the milled surface.

Expected Results:
 1. A capable process: All bricks meet specifications for width and microfinish on the milled surface.
 2. A better understanding of the factors that affect quality in the mill process.
 3. Procedures for continuous monitoring and improvement of the mill process.
 4. Understanding of improvement methods to apply to the other processes in the factory.

Boundaries:
 All four mills in the process should be addressed by the team. Studies can be done on one mill and the results transferred to the other mills. The operators should be the primary resources for measurement and other data collection.

Mill Machine Flow Chart

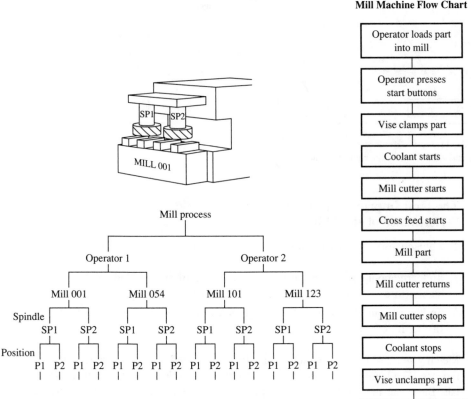

FIGURE 11.6
Flowcharts of the mill process

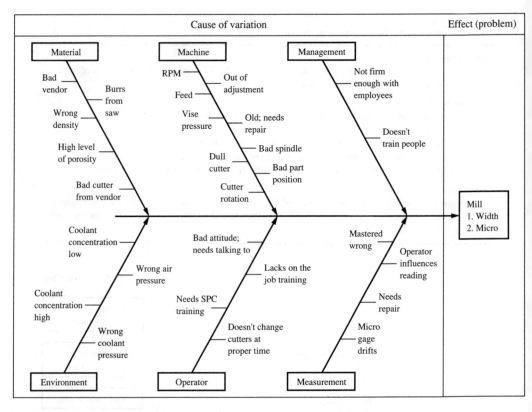

FIGURE 11.7
Cause and effect diagram for width and microfinish

The First Improvement Cycle: Current Performance of the Mills

The objective of the first improvement cycle was twofold:

1. To learn about the current performance of the mill process and, if the process was stable, to determine the process capability for the width and microfinish.
2. To learn about the quality of the gages used to measure the width and microfinish of the milled surface.

Meeting these objectives requires the information needed to describe the current mill process. The team felt that the process might be unstable for both width and microfinish. Based on the results at final inspection, the team felt the process would not be capable for either quality characteristic.

Control charts were planned for the process. None of the team members had any ideas about the important sources of variation, so it was difficult to develop a rational subgrouping strategy. The team decided to select one brick from each mill during every hour of operation for the next two days. The bricks from each mill would be measured for width and microfinish by the mill operator. Subgroups of four bricks would be formed from the bricks from each mill. The supervisor would gather the data and develop the control charts.

To document the quality of the measurement processes, the supervisor selected one part from the milled inventory each hour and measured it twice using both the width gage (GW-002) and the microfinish gage (GM-015). In addition, a certified master brick (width = 2.00 inches and microfinish = 150 μ) was measured using the gages each hour.

The team would meet at the end of the second day to review the control charts and complete the data analysis. Figure 11.8 contains the control chart for the width of the milled surface. No special causes were indicated by either the X-bar or R chart. The averages tended to cluster around the centerline, which indicated that an important source of variation might be included in the range. It appeared there were some important differences in the mills. The supervisor did not keep track of which part came from which mill, so no further analysis could be done with the data.

Although the clustering of the averages indicated that a systematic source of variation was probably present in the ranges, a capability analysis was done to give an indication of the quality of the parts being produced. The capability analysis of the process for width indicated that as many as one-half the parts were either too narrow or too wide. The capability of the process ranged from -35 to $+43$ thousandths of an inch from nominal. The tolerances for the width were ± 10 thousandths of an inch.

Figure 11.9 contains the control chart for the microfinish. The process was also stable for this quality characteristic. But the capability analysis indicated that only a small percentage of the bricks would have a microfinish less than the upper specification of 150 μ. The current process average of 166.6 μ would have to be reduced well below 150 μ in order to make all good bricks. The plant engineer thought the variation for microfinish was reasonably good but was very surprised to see how high the average was.

Figure 11.10 contains the control chart for precision (range chart for the two measurements of the same part) and bias (individual chart for the master part). Both charts were stable. The bias of the width measurement was $+5$ thousandth of an inch. The X-bar chart for the measured part (Figure 11.8) averaged about $+4$ thousandth, so this bias in the measurement indicated the bricks being produced actually averaged about -1 thousandth of an inch below the nominal value (2.00 inches).

The precision of the width measurement process was evaluated using the range of the two measurements of the same brick. The average range was 2.4 thousandth. Table 11.1 compares this variation to the variation of the process (from Figure 11.8). The variation in the measurement process represents less than 3 percent of the total variation in the process. The actual variation from brick to brick represents most of the variation. Thus the GW-002 gage provides adequate precision to learn about the variation in the process. But the standard deviation of the measurement process ($\sigma = 2.13$) represents a significant portion of the tolerance for width (± 10 thousandth inch). The capability of the measurement process would be about ± 6 thousandth inch.

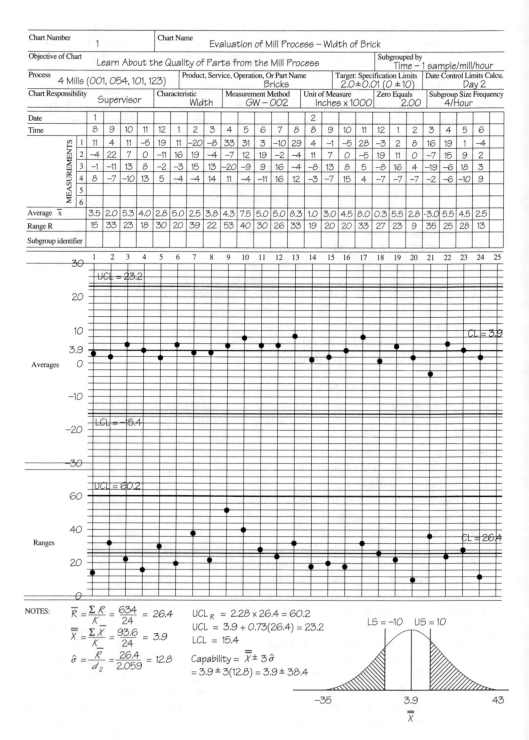

FIGURE 11.8
Initial width control chart for mill process

Chart Number	2	Chart Name	Evaluation of Mill Process – Microfinish of Milled surface of brick		

Objective of Chart	Learn About the Quality of Parts from the Mill Process		Subgrouped by	Time – 1 sample/mill/hour
Process	4 Mills (001, 054, 101, 123)	Product, Service, Operation, Or Part Name: Bricks	Target: Specification Limits <150	Date Control Limits Calcu. Day 2
Chart Responsibility	Supervisor	Characteristic: Microfinish	Measurement Method: GM – 107	Unit of Measure: Microinches	Zero Equals: 0	Subgroup Size Frequency 4/Hour

Date		1													2										
Time		8	9	10	11	12	1	2	3	4	5	6	7	8	8	9	10	11	12	1	2	3	4	5	6
MEASUREMENTS	1	167	159	160	170	162	169	163	160	164	174	155	160	170	171	166	170	162	167	170	164	166	181	176	178
	2	164	162	164	166	173	168	172	180	160	160	162	166	171	158	186	169	155	170	169	165	174	164	165	158
	3	168	167	168	182	168	173	162	159	170	177	168	163	168	160	159	160	163	154	175	165	158	167	173	160
	4	170	160	165	167	165	170	166	175	159	161	174	165	159	163	165	166	165	176	164	162	170	174	167	163
	5																								
	6																								
Average $\bar{x}$		167.3	162	164.3	171.3	167	170	165.8	167.3	163.3	168	165.5	163.5	167	163	171.5	166.3	161.5	166.8	169.5	164	167	171.5	170.3	164.8
Range R		6	8	8	16	11	5	10	21	11	17	16	6	11	13	21	10	11	22	11	3	16	17	11	20
Subgroup identifier																									

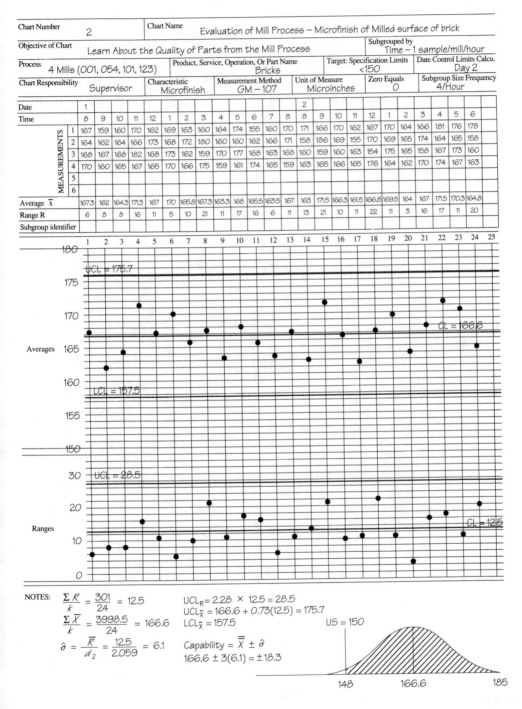

NOTES:

$$\frac{\Sigma R}{k} = \frac{301}{24} = 12.5$$

$$\frac{\Sigma \overline{X}}{k} = \frac{3998.5}{24} = 166.6$$

$$\hat{\sigma} = \frac{\overline{R}}{d_2} = \frac{12.5}{2.059} = 6.1$$

$UCL_R = 2.28 \times 12.5 = 28.5$

$UCL_{\overline{x}} = 166.6 + 0.73(12.5) = 175.7$

$LCL_{\overline{x}} = 157.5$

$US = 150$

Capability $= \overline{\overline{X}} \pm \hat{\sigma}$

$166.6 \pm 3(6.1) = \pm 18.3$

148   166.6   185

FIGURE 11.9
Initial microfinish control chart for mill

325

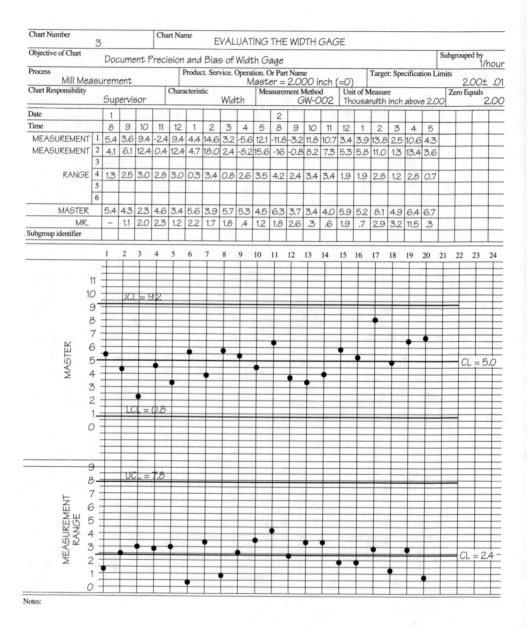

Chart Number	3			Chart Name			EVALUATING THE WIDTH GAGE																
Objective of Chart	Document Precision and Bias of Width Gage																			Subgrouped by 1/hour			
Process	Mill Measurement			Product. Service. Operation. Or Part Name Master = 2.000 inch (=0)														Target: Specification Limits 2.00± .01						
Chart Responsibility	Supervisor			Characteristic Width				Measurement Method GW-002				Unit of Measure Thousandth inch above 2.00						Zero Equals 2.00						
Date	1									2														
Time	8	9	10	11	12	1	2	3	4	5	8	9	10	11	12	1	2	3	4	5				
MEASUREMENT 1	5.4	3.6	9.4	-2.4	9.4	4.4	14.6	3.2	-5.6	12.1	-11.8	-3.2	11.8	10.7	3.4	3.9	13.8	2.5	10.6	4.3				
MEASUREMENT 2	4.1	6.1	12.4	0.4	12.4	4.7	18.0	2.4	-8.2	15.6	-16	-0.8	8.2	7.3	5.3	5.8	11.0	1.3	13.4	3.6				
3																								
RANGE 4	1.3	2.5	3.0	2.8	3.0	0.3	3.4	0.8	2.6	3.5	4.2	2.4	3.4	3.4	1.9	1.9	2.8	1.2	2.8	0.7				
5																								
6																								
MASTER	5.4	4.3	2.3	4.6	3.4	5.6	3.9	5.7	5.3	4.5	6.3	3.7	3.4	4.0	5.9	5.2	8.1	4.9	6.4	6.7				
MR.	−	1.1	2.0	2.3	1.2	2.2	1.7	1.8	.4	1.2	1.8	2.6	.3	.6	1.9	.7	2.9	3.2	11.5	.3				
Subgroup identifier																								

Notes:

FIGURE 11.10

Control chart for width measurement process

TABLE 11.1
Analysis of precision of width measurement

Variation for process

$\overline{R} = 26.4 \qquad \sigma_p = \overline{R}/d_2 = 26.40/2.059 = 12.8$

Variation of measurement process

$\overline{R}_m = 2.4 \qquad \sigma_m = \overline{R}_m/d_2 = 2.4/1.128 = 2.13$

Variation of product

$\sigma_{\text{product}} = \sqrt{(\sigma_{\text{process}})^2 - (\sigma_{\text{measurement}})^2}$

$\sigma_{\text{product}} = \sqrt{(12.8)^2 - (2.13)^2} = 12.62$

Summary of variation (units = thousandth inch)

Source of variation	Standard deviation (σ)	Variance component (σ^2)	Percent of variation
Product	12.62	159.3	97.3
Measurement	2.13	4.5	2.7
Total (process)	12.8	163.8	100.0

A similar analysis was done for the microfinish measurement process (GM-002). The bias was less than 1 μ, and the standard deviation for measurement was about 0.5 μ, which represents less than 1 percent of the process variation. Thus the microfinish gage was considered an unimportant source of variation for this process.

Based on the information obtained in this improvement cycle, the team decided to:

1. Calibrate the width gage to remove the bias. The mill operators will run the master brick twice a day on both the width and micro gages and plot the results on the master control charts developed in this cycle.

2. Notify management that most of the bricks being produced have a rougher finish than the microfinish specification allows and that about half the bricks are outside the width specifications.

3. Have the maintenance man investigate alternative width gages with better precision for future use.

Next, the team wanted to plan a study to learn about the important sources of variation in the width of the milled bricks.

The Second Improvement Cycle: Sources of Variation

The objective of the second improvement cycle was to determine the important sources of variation in the mill process. Figure 11.6 indicated that the mill process consists of the following components:

1. Two operators.
2. Two mills within each operator.
3. Two spindles within each mill.
4. Two positions within each spindle.

The team felt that the operators and the spindles within the mills would be the most important sources of variation for width. For microfinish, the team thought that the positions within each spindle would contribute most of the variation. A nested study was planned to evaluate the four sources of variation in the process relative to the part-to-part variation within a spindle.

The experiment was accomplished the next afternoon by selecting three parts from each position within each spindle for each mill and measuring the parts for width and microfinish. Figure 11.11 contains dot-frequency diagrams for both the microfinish and width data. For microfinish, none of the factors studied appears to be an important source of variation. Most of the variation is attributable to the variation from part to part within a position. None of the parts in the study were less than the upper specification of 150 μ. The problems with microfinish seem to be common throughout the mill process.

The dot-frequency diagram for width contains information about the variation in the width measurements. Differences between the mills are the biggest source of variation. These differences inflated the range on the control chart during cycle 1. Mills 054 and 123 are producing parts greater than the nominal width, while Mill 101 is making all parts slightly below nominal width. The positions for spindle 1, Mill 123 need to be aligned. The variation of parts from spindle 2, Mill 001 and spindle 1, Mill 054 seem to be greater than the variation of parts from the other spindles. The dot-frequency diagram was used to develop a repair/adjustment plan for the mills.

The actions taken in this cycle were primarily designed to reduce the variation in the width measurements. The contract maintenance man scheduled the following maintenance activities:

MILL	ACTIVITY
101	Adjust up .006 inches
001	Repair spindle 2
054	Repair spindle 1
	Adjust down .014 inch
123	Rebuild spindle 1
	Adjust down .012 inch

These activities were expected to take two days to complete.

While the maintenance activities were being completed, the team wanted to run studies on Mill 101 to learn about factors that cause the microfinish to be so high. Also, plans were discussed to develop control charts that could be used to determine the need for future maintenance.

The Third Improvement Cycle: Evaluating Mill Cutter Vendors

The Mid-State Brick Company currently bought mill cutters from five different vendors. When a cutter broke, it was replaced with whatever brand was available. The operators thought that the type of cutter could impact both the microfinish and the

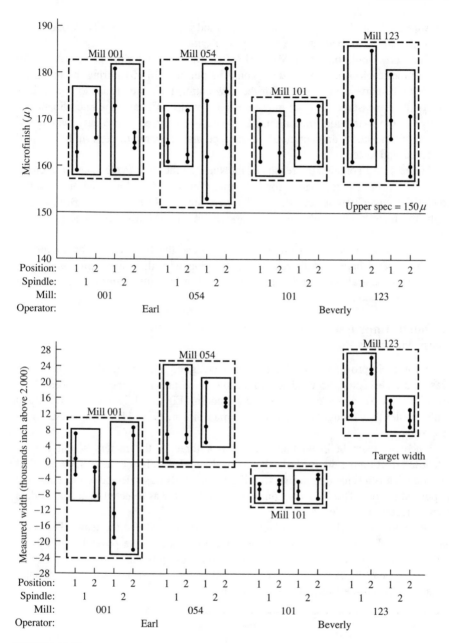

FIGURE 11.11
Dot-frequency diagrams for microfinish and width

width variation, but they disagreed on which brand gave the best results. A one-factor study was planned to evaluate the five types of cutters.

The study would be run on Mill 101 (after adjustment) during the next morning. One at a time, each of the five cutters would be put on Mill 101 during the morning. After about 10 parts were milled with the new cutter, three subgroups of four bricks each would be measured for both microfinish and width. The average microfinish and the range of the four width measurements would be analyzed to evaluate the cutters. At the end of the study, another cutter of the type that appeared to perform best would be installed and tested.

Figure 11.12 contains the results of the study in the third cycle. The Brite cutter had the best overall performance for both microfinish and width variation. When a second Brite cutter was tested, the results were similar. Even with the Brite cutter, though, the microfinish results averaged above 160 μ. The average range for width with the Brite cutter was about .004 inch.

The primary action from this cycle was to use the Brite cutter for all further tests. The supervisor agreed to schedule a meeting with the sales representative of the Brite cutter and also obtain information for all of the cutters from the purchasing group.

The Fourth Improvement Cycle: Screening Process Variables

For the fourth improvement cycle, the team designed a study to determine which process variables could be used to reduce the microfinish. Using the process flow chart (Figure 11.6) and the cause-and-effect diagram (Figure 11.7), the team identified seven variables they thought could affect the microfinish. Table 11.2 summarizes these potential factors.

Since the team believed that only two or three of the factors would have a substantial effect on the microfinish, they considered themselves to have a low level of knowledge. Thus a screening design (fractional factorial) was selected to eliminate the unimportant factors. The 2^{7-4} fractional factorial design was selected to accommodate the seven factors in Table 11.2.

Selecting the levels for each factor was a difficult task for the team. The plant engineer wanted to choose the levels near the extremes of the possible settings of the factors. Both of the operators were concerned about possible damage to the mills from running near the extremes for some of the factors. After discussion and debate, the team agreed on the following levels for the factors:

		Levels	
Factor		−	+
1 Mill rpm	(R)	300	700
2 Feed rate	(F)	4	10
3 Vise pressure	(V)	200	600
4 Coolant pressure	(P)	10	80
5 Coolant concentration	(C)	5	25
6 Air pressure	(A)	0	.01
7 Movement	(M)	Conventional	Climb

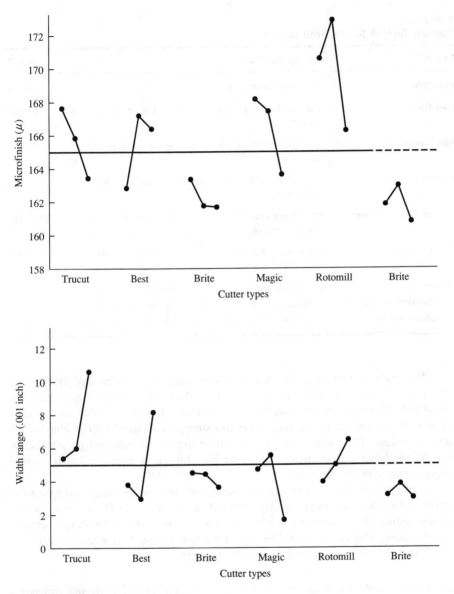

FIGURE 11.12
Results of one-factor study on cutter types

The study would be run on the Mill 101 using the Brite blade. After setting up the mill at the conditions in the design matrix, five parts would be run. Then two subgroups of four bricks each would be run and measured for microfinish and width. Although the focus of the study was on microfinish, the team wanted to continue to evaluate the width variation. The average microfinish and the range of the width measurements would be the response variables for the study.

TABLE 11.2
Possible factors for the mill process

Factor	Description of factors	Range
Mill rpm	The revolutions per minute that the mill cutter turns.	(10–900)
Feed Rate:	The rate that the mill cutter travels through the part in inches per minute.	(.00001–20)
Vise Pressure:	The amount of force that is clamping the part in pounds per square inch.	(1.5–950)
Coolant Pressure:	The pressure of the coolant flow on the part in pounds per square inch.	(2–100)
Coolant Concentration:	The amount of coolant concentration/amount of water. This is stated in percent.	(.00001–35)
Air Pressure:	The amount of air pressure blowing on the part while it is being milled.	(0–.125)
Movement:	The movement of the cutter into the part.	———
Conventional cutting	Pushes the part into the vise during the cutting process.	———
Climb cutting	Tries to pull the part out of the vise while it is cutting.	———

    The study was to be completed on the morning of the fifth day. During the fifth test, six cutters were broken. Figure 11.13 shows the results of the study for microfinish. Three factors were found to be important: mill rpm, feed rate, and coolant concentration. From the tests completed in this study, these three factors formed a full factorial design. The analysis of the cube in Figure 11.13 indicated no interaction between the factors. Best microfinish could be achieved at high rpm, high coolant concentration, and low feed rate.

    Figure 11.14 shows the results for width. Feed rate and possibly mill rpm were found to affect the width variation. The effect of changes in feed rate was greater at the low rpm setting. To minimize the width variation, low feed rates and high rpm settings were desirable. This is the same direction as the best conditions for microfinish.

    The following actions were taken after this cycle:

1. Set the coolant concentration at 25 percent. The coolant was controlled through a central system that affected all of the mills.
2. Set the feed rate at five inches per minute for each mill.
3. Set the mill rpm at 700 for each mill.

After maintenance was completed on the mills, the team planned to determine the capability for the mill process. Plans were also made to evaluate a new width gage that the supervisor had obtained.

Factor levels

Factor		Level	
		−	+
1 Mill rpm	(R)	300	700
2 Feed Rate	(F)	4	10
3 Vise Pressure	(V)	200	600
4 Coolant Pressure	(P)	10	80
5 Coolant Concentration	(C)	5	25
6 Air Pressure	(A)	0	.01
7 Movement	(M)	Conventional	Climb

Design Matrix for A 2^{7-4} pattern

Test	Run order	1 24 35 67	2 14 36 57	3 15 26 47	4 12 56 37	5 13 46 27	6 23 45 17	7 34 25 16	Average of $\bar{x}_s$ for sub-groups
1	8	−	−	−	+	+	+	−	133.6
2	2	+	−	−	−	−	+	+	149.6
3	5	−	+	−	−	+	−	+	154.0
4	3	+	+	−	+	−	−	−	158.1
5	1	−	−	+	+	−	−	+	160.2
6	7	+	−	+	−	+	−	−	127.3
7	4	−	+	+	−	−	+	−	178.7
8	6	+	+	+	+	+	+	+	133.0

Effects: −14.6 12.3 1.0 −6.2 −24.7 −1.2 −0.2

Run chart for microfinish (μ)

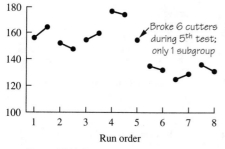

Broke 6 cutters during 5th test; only 1 subgroup

Note: All high coolant concentrations run last

Dot diagram of effects

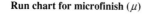

Response plots for important effects

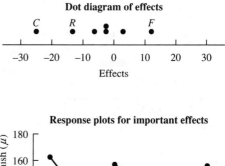

162.6 137.0   156.6 142.0   142.6 155.9

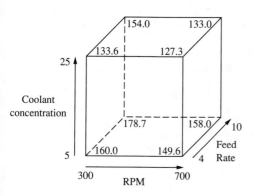

FIGURE 11.13
Fractional factorial study for microfinish

Factor levels

Factor		Level	
		−	+
1 Mill rpm	(R)	300	700
2 Feed Rate	(F)	4	10
3 Vise Pressure	(V)	200	600
4 Coolant Pressure	(P)	10	80
5 Coolant Concentration	(C)	5	25
6 Air Pressure	(A)	0	.01
7 Movement	(M)	Conventional	Climb

Design Matrix for A 2^{7-4} pattern

Test order	Run	1 24 35 67	2 14 36 57	3 15 26 47	4 12 56 37	5 13 46 27	6 23 45 17	7 34 25 16	Average of Rs for 2 sub-groups
1	8	−	−	−	+	+	+	−	3.9
2	2	+	−	−	−	−	+	+	2.9
3	5	−	+	−	−	+	−	+	6.1
4	3	+	+	−	+	−	−	−	5.4
5	1	−	−	+	+	−	−	+	3.2
6	7	+	−	+	−	+	−	−	4.6
7	4	−	+	+	−	−	+	−	6.0
8	6	+	+	+	+	+	+	+	4.1
Effects:		−0.9	2.1	−0.4	−0.5	0.0	−0.3	−0.6	

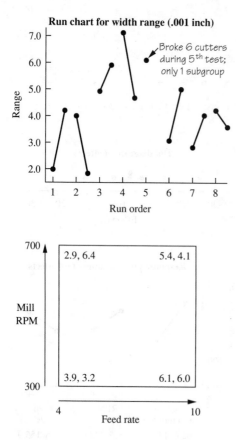

Run chart for width range (.001 inch)

Broke 6 cutters during 5th test; only 1 subgroup

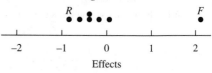

Dot diagram of effects

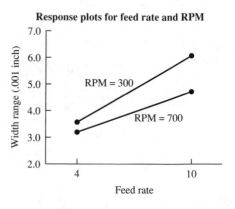

Response plots for feed rate and RPM

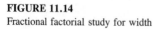

FIGURE 11.14
Fractional factorial study for width

The Fifth Improvement Cycle: Evaluate Effect of Improvements

Control charts were planned to evaluate the status of the mill process after the maintenance and other process changes had been made. The subgroups were set up to allow an assessment of each of the mills. During the next day, every half hour one of the mill operators selected a subgroup of four bricks from one of the mills. The operators coordinated the sampling so that the four mills would be rotated. The supervisor conducted a study of the new width gage following the design used in cycle 1.

Figure 11.15 shows the control charts for the width measurements. Both the average chart and the range chart indicated special causes. The measurements from Mill 001 were consistently high. This mill had been repaired after the third cycle, and was probably in need of adjustment. The other mills appeared to be averaging near the nominal width. There was an indication of a special cause on the range chart for Mill 123. All five of the ranges for this mill were above the centerline. Spindle 1 on this mill had been rebuilt after the third cycle.

Figure 11.16 shows the control charts for the microfinish measurements. The process was stable during day 6. A capability analysis shows great improvement—a very small percentage of the bricks are expected to be greater than the specification of 150 μ.

The study of the width gage (GW-097) showed no bias and better precision ($\sigma = 1.3$ thousandth inch) than the previous gage. Since the range chart for the width was not in control, this could not be compared to the process variation.

The following actions were taken after this cycle:

1. Adjust Mill 001 down by 5.5 thousandth of an inch.
2. Adopt the new width gage for all further measurements.

Plans for the next cycle were to run a nested design on Mill 123 to isolate the variation problem and then schedule appropriate maintenance to correct the problems. Also, the team planned to conduct a study to optimize the important factors identified during the fourth cycle.

The Sixth Improvement Cycle: Evaluating Important Factors

A nested design was planned to evaluate Mill 123 during the first hour of day 7. Three parts were selected from each position on each of the two spindles.

A 2^3 factorial design was used to optimize the factors identified as important in the fractional factorial design in the fourth cycle. Levels were selected around the best conditions from the previous study.

Figure 11.17 shows the results of the nested design on width for Mill 123. Figures 11.18 and 11.19 contain the summaries for the factorial studies for width and microfinish.

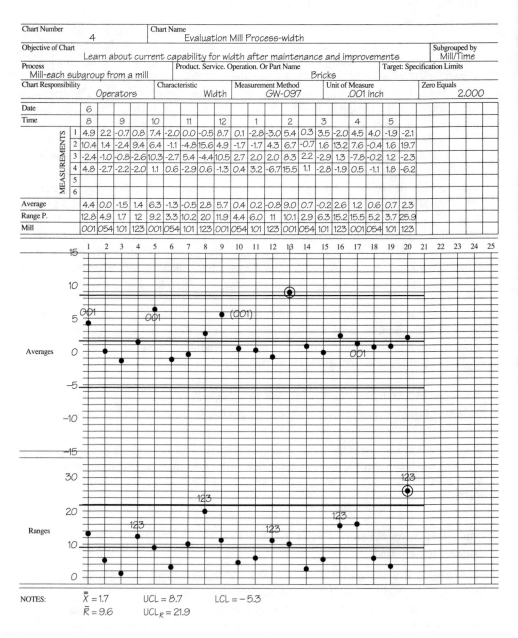

Chart Number		Chart Name														
4		Evaluation Mill Process-width														

Objective of Chart: Learn about current capability for width after maintenance and improvements — Subgrouped by Mill/Time

Process: Mill-each subgroup from a mill — Product. Service. Operation. Or Part Name: Bricks — Target: Specification Limits

Chart Responsibility: Operators — Characteristic: Width — Measurement Method: GW-097 — Unit of Measure: .001 inch — Zero Equals: 2.000

Date	6																			
Time	8		9		10		11		12		1		2		3		4		5	

MEASUREMENTS

1	4.9	2.2	-0.7	0.8	7.4	-2.0	0.0	-0.5	8.7	0.1	-2.8	-3.0	5.4	0.3	3.5	-2.0	4.5	4.0	-1.9	-2.1
2	10.4	1.4	-2.4	9.4	6.4	-1.1	-4.8	15.6	4.9	-1.7	-1.7	4.3	6.7	-0.7	1.6	13.2	7.6	-0.4	1.6	19.7
3	-2.4	-1.0	-0.8	-2.6	10.3	-2.7	5.4	-4.4	10.5	2.7	2.0	2.0	8.3	2.2	-2.9	1.3	-7.8	-0.2	1.2	-2.3
4	4.8	-2.7	-2.2	-2.0	1.1	0.6	-2.9	0.6	-1.3	0.4	3.2	-6.7	15.5	1.1	-2.8	-1.9	0.5	-1.1	1.8	-6.2
5																				
6																				

Average	4.4	0.0	-1.5	1.4	6.3	-1.3	-0.5	2.8	5.7	0.4	0.2	-0.8	9.0	0.7	-0.2	2.6	1.2	0.6	0.7	2.3
Range P.	12.8	4.9	1.7	12	9.2	3.3	10.2	20	11.9	4.4	6.0	11	10.1	2.9	6.3	15.2	15.5	5.2	3.7	25.9
Mill	001	054	101	123	001	054	101	123	001	054	101	123	001	054	101	123	001	054	101	123

NOTES: $\bar{\bar{X}} = 1.7$ UCL $= 8.7$ LCL $= -5.3$

$\bar{R} = 9.6$ UCL$_R = 21.9$

FIGURE 11.15
Control chart for width after improvements

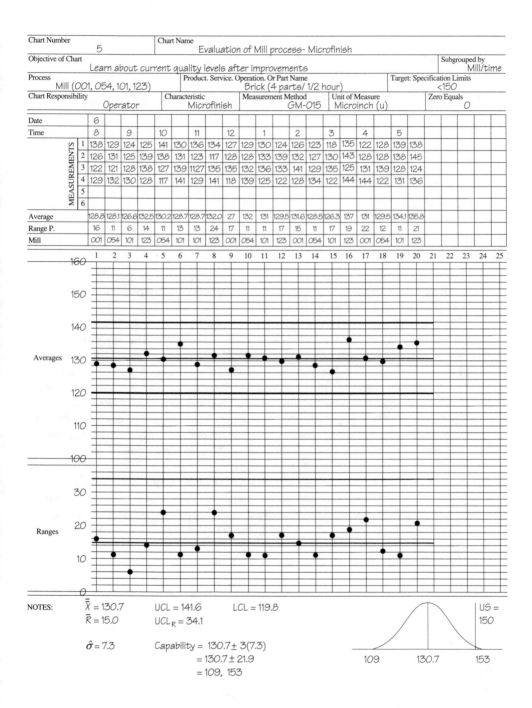

FIGURE 11.16
Control chart for microfinish after improvements

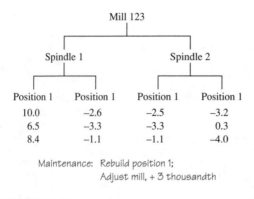

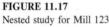

Maintenance: Rebuild position 1;
Adjust mill, + 3 thousandth

FIGURE 11.17
Nested study for Mill 123

The results of the nested design for Mill 123 indicated that position 1 on spindle 1 was out of line with the rest of the mill. After rebuilding this position, the entire mill was adjusted up 3 thousandth of an inch.

The factorial design was run on Mills 101 and 054. For each test condition, a subgroup of four parts was selected from each of the mills. The analysis of the variation in the measurements of width (Figure 11.18) indicated that Mill 101 had more variation than Mill 054. Maintenance was scheduled to check Mill 101 for needed repairs. The average range for the two mills was used as the response variable. The analysis of the effects indicated that the feed rate and the rpm coolant concentration interaction were the most important effects. Low coolant levels were desirable when running at high rpm, while high coolant levels were preferred when running at low rpm. The width variation increased as the feed rate was increased from 2 to 8 inches per minute.

The analysis of the microfinish confirmed the effects found in the earlier fractional factorial design for coolant concentration and feed rate. The increase in coolant above 25 percent showed a continuing decrease in microfinish. The effect of feed rate was the same as the previous study. If kept in the 500 to 800 range, rpm did not have an important effect on microfinish.

The following actions were taken as a result of this cycle:

1. Increase the coolant concentration to 35 percent. This should further improve the microfinish.
2. Set the rpm to 500. This should help the width variation while running at the high coolant level.

The team next planned to study the effect of costs at different feed rate levels. Since the feed rate directly affected the productivity of the mill process, the cost to the mill should be considered when setting the feed rate.

Three factor levels for width

Factor	Level −	Level +
1 Mill rpm	500	800
2 Feed Rate	2	8
3 Coolant Concentration (C)	20	30

Design Matrix: three factor factorial study of width

Width range

Test	Run order	1	2	3	12	13	23	123	Mill 101	Mill 054	Average
1	4	−	−	−	+	+	+	−	5.3	1.2	3.3
2	2	+	−	−	−	−	+	+	2.7	1.5	2.1
3	3	−	+	−	−	+	−	+	5.8	2.6	4.2
4	1	+	+	−	+	−	−	−	4.5	1.5	3.0
5	7	−	−	+	+	+	−	+	3.4	2.9	3.2
6	5	+	−	+	−	+	−	−	4.6	3.2	3.9
7	8	−	+	+	−	−	+	−	4.8	2.3	3.6
8	6	+	+	+	+	+	+	+	6.8	4.8	5.8
Effects:		.13	1.03	.98	.38	1.3	.13	.38			

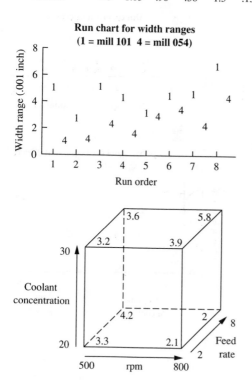

Run chart for width ranges
(1 = mill 101 4 = mill 054)

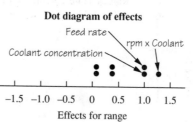

Dot diagram of effects

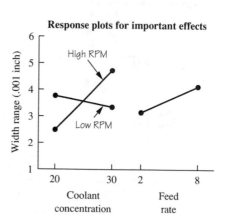

Response plots for important effects

FIGURE 11.18
Factorial design for width

339

Three factor levels for width

		Level	
	Factor	−	+
1	Mill rpm	500	800
2	Feed Rate	2	8
5	Coolant Concentration (C)	20	30

Design Matrix: three factor factorial study of width

Microfinish average (μ)

Test	Run order	1	2	3	12	13	23	123	Mill 101	Mill 054	Average
1	4	−	−	−	+	+	+	−	132.3	130.5	131.4
2	2	+	−	−	−	−	+	+	130.8	134.0	132.4
3	3	−	+	−	−	+	−	+	142.9	140.2	141.6
4	1	+	+	−	+	−	−	−	133.0	137.0	135.0
5	7	−	−	+	+	−	−	+	123.7	128.1	125.9
6	5	+	−	+	−	+	−	−	118.8	121.5	120.2
7	8	−	+	+	−	−	+	−	132.2	131.8	132.0
8	6	+	+	+	+	+	+	+	131.4	129.9	130.7
Effects:		−3.2	7.4	−8.0	−.8	−.4	1.0	3.1			

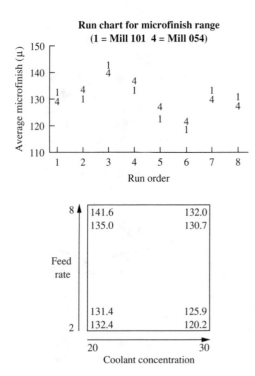

Run chart for microfinish range
(1 = Mill 101 4 = Mill 054)

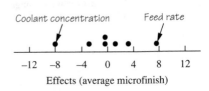

Dot diagram of effects

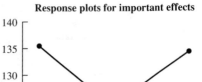

Response plots for important effects

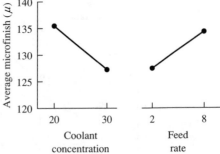

FIGURE 11.19
Factorial design for microfinish

The Seventh Improvement Cycle: Determining Optimum Levels

The supervisor asked the plant accountant to develop cost estimates in the mill process for running at feed rates from 2 to 10 inches per minute. The team planned a one-factor study to evaluate the variation of the width measurements and the microfinish average. One subgroup would be taken from each mill at feed rates of 2, 4, 6, 8, and 10 inches per minute.

The results of the study are presented in Table 11.3. The analysis of the data are shown in Figure 11.20.

Based on this analysis, the team decided to set the feed rate at 5 inches per minute. This should give a cost to mill of about $2.50 per brick and produce bricks that meet the quality specifications for both microfinish and width. The team felt that a final cycle could be run to demonstrate the performance of the process and test a control chart plan for continuous monitoring of the mill process.

The Eighth Improvement Cycle: Confirmation Of Improvements

After setting the feed rate to 5 inches per minute, process conditions were checked for each of the mills. Control charts were planned to be developed on the next day for both microfinish and width. Subgroups of four bricks would be selected from one of the mills every half hour. One brick would be selected from each of the four positions on the mill. The position and mill would be recorded on the chart for each brick. Thus, if a special cause was found on the chart, the particular problem spindle or position could be identified. The operators would each take responsibility for one of the charts.

TABLE 11.3
One-factor study to evaluate effect of feed rate

| | Width variation (range for subgroup of 4; unit = thousandth inch) | | | | | |
Feed rate (in/min)	Mill 001	Mill 054	Mill 101	Mill 123	Average	Cost per brick
2	5.6	4.2	1.7	2.8	3.60	$5.59
3						4.00
4	3.9	1.6	1.3	3.7	2.63	3.14
6	3.6	6.0	1.7	5.8	4.28	2.29
8	6.6	5.9	2.0	5.6	5.03	1.90
10	3.3	4.6	3.9	7.5	4.83	1.66

| | Microfinish (average for subgroup of 4; unit = microinches) | | | | | |
Feed rate (in/min)	Mill 001	Mill 054	Mill 101	Mill 123	Average	Cost per brick
2	118.9	125.6	128.6	120.1	123.3	$5.59
3						4.00
4	126.7	129.8	130.1	128.7	128.8	3.14
6	127.2	130.1	128.4	131.4	129.3	2.29
8	136.2	135.1	132.8	138.3	135.6	1.90
10	134.6	134.9	137.2	140.4	136.8	1.66

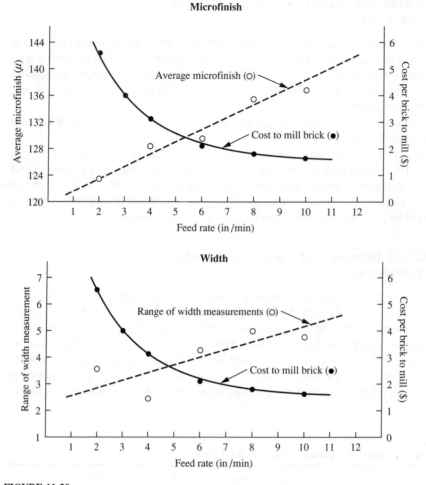

FIGURE 11.20
Analysis of cost and quality for feed rate

Figures 11.21 (microfinish) and 11.22 (width) show the control charts developed during day 9. Both of the charts indicate that the process is stable. Capability analyses for each of the quality characteristics indicated that all bricks milled are expected to meet the specifications.

Final Actions of the Team

The team reviewed their charter (Figure 11.5) and felt they had successfully accomplished all of the expected results. The operators would continue to use the control charts developed during the last (eighth) improvement cycle to monitor the mill process. Since the process was operating so well, the frequency of subgroups was reduced to one per hour. This would allow a check on each mill twice during a normal eight-hour shift.

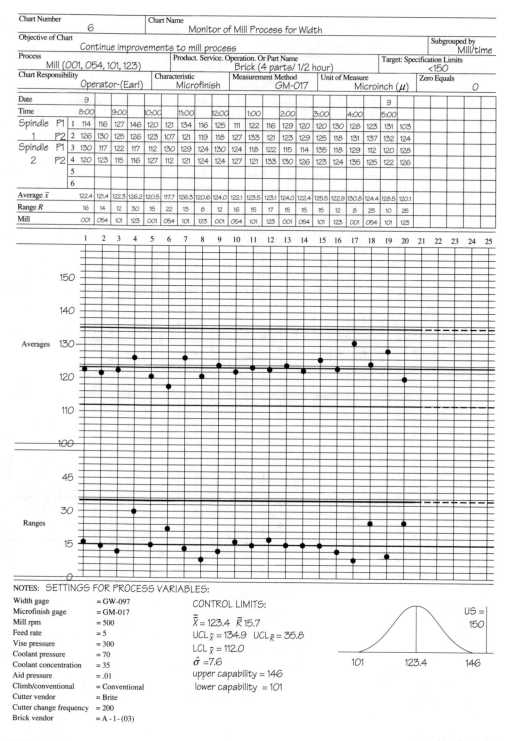

Chart Number		Chart Name				
6		Monitor of Mill Process for Width				

Objective of Chart: Continue improvements to mill process — Subgrouped by: Mill/time

Process: Mill (001, 054, 101, 123) — Product. Service. Operation. Or Part Name: Brick (4 parts/ 1/2 hour) — Target: Specification Limits <150

Chart Responsibility: Operator-(Earl) — Characteristic: Microfinish — Measurement Method: GM-017 — Unit of Measure: Microinch (μ) — Zero Equals: 0

Date		9																	9						
Time		8:00		9:00		10:00		11:00		12:00		1:00		2:00		3:00		4:00		5:00					
Spindle 1 P1	1	114	116	127	146	120	121	134	116	125	111	122	116	129	120	120	130	128	123	131	103				
P2	2	126	130	125	126	123	107	121	119	118	127	133	121	123	129	125	118	131	137	132	124				
Spindle 2 P1	3	130	117	122	117	112	130	129	124	130	124	118	122	115	114	135	118	129	112	120	128				
P2	4	120	123	115	116	127	112	121	124	124	127	121	133	130	126	123	124	135	125	122	126				
	5																								
	6																								
Average x̄		122.4	121.4	122.3	126.2	120.5	117.7	126.3	120.6	124.0	122.1	123.5	123.1	124.0	122.4	125.5	122.9	130.8	124.4	128.5	120.1				
Range R		16	14	12	30	15	22	13	8	12	16	15	17	15	15	15	12	8	25	10	25				
Mill		001	054	101	123	001	054	101	123	001	054	101	123	001	054	101	123	001	054	101	123				

NOTES: SETTINGS FOR PROCESS VARIABLES:

Width gage	= GW-097
Microfinish gage	= GM-017
Mill rpm	= 500
Feed rate	= 5
Vise pressure	= 300
Coolant pressure	= 70
Coolant concentration	= 35
Aid pressure	= .01
Climb/conventional	= Conventional
Cutter vendor	= Brite
Cutter change frequency	= 200
Brick vendor	= A - 1 - (03)

CONTROL LIMITS:

$\bar{\bar{X}} = 123.4$ $\bar{R}\ 15.7$

$UCL_{\bar{X}} = 134.9$ $UCL_{\bar{R}} = 35.8$

$LCL_{\bar{X}} = 112.0$

$\hat{\sigma} = 7.6$

upper capability = 146

lower capability = 101

US = 150

101   123.4   146

FIGURE 11.21

Control chart for microfinish

343

Chart Number 7

Chart Name Monitor of Mill Process for Width

Objective of Chart Continue Learning About Causes of Width Variation

Subgrouped by Mill/time

Process Mill (001, 054, 101, 123)

Product. Service. Operation. Or Part Name Brick (4 parts/ 1/2 hour)

Target: Specification Limits 2.00 ± .01(0 ± 10 thous.)

Chart Responsibility Operator-(Beverly)

Characteristic Width of milled surf.

Measurement Method GW-097

Unit of Measure Thousandth inch

Zero Equals 2.000

Date		9																		9					
Time		8:00	9:00	10:00	11:00	12:00	1:00	2:00	3:00	4:00	5:00														
SPINDLE 1 P1	1	-4.1	-0.1	-.7	.3	-2.5	-.7	.8	-1.5	-2.6	2.2	-.5	1.2	.3	-1.5	1.8	1.6	-5.0	-2.6	1.9	1.8				
P2	2	-0.5	0.1	-.3	-2.0	-4.1	-1.8	-.6	.9	-3.0	-2.0	0	-1.0	-1.4	1.7	-.8	.9	3.7	1.5	1.1	-1.6				
SPINDLE 2 P1	3	0.3	-0.7	-1.4	.4	2.8	.7	1.6	1.7	3.9	-3.0	-4.3	-4.1	1.3	-2.2	1.1	-.9	1.8	-3.3	-2.3	1.6				
P2	4	1.4	0.9	-2.6	-1.8	4.1	2.0	3.2	.5	-.8	3.0	.4	-1.4	-3.0	-1.7	1.3	-1.3	-.9	1.0	.9	0				
	5																								
	6																								
Average x̄		-.72	0.0	-1.2	-.8	.1	0	1.2	.4	-.6	.1	-1.1	-1.3	-.7	-.9	.9	.1	-.1	-.9	.4	-.1				
Range R.		5.4	1.6	2.3	2.4	8.2	3.8	3.8	3.2	7.0	6.0	4.7	5.3	4.3	4.0	2.6	2.9	10.7	4.8	4.3	3.4				
Mill		001	054	101	123	001	054	1.1*	123	001	054	101	123	001	054	123	123	001	101	101	123	001	054	101	123

(Averages control chart: UCL ≈ 3.5, LCL ≈ -3.5, centerline ≈ 0. Ranges control chart: UCL ≈ 10.3, centerline ≈ 4.5.)

NOTES: SETTINGS FOR PROCESS VARIABLES:

Width gage	= GW-097
Microfinish gage	= GM-017
Mill rpm	= 500
Feed rate	= 5
Vise pressure	= 300
Coolant pressure	= 70
Coolant concentration	= 35
Aid pressure	= .01
Climb/conventional	= Conventional
Cutter vendor	= Brite
Cutter change frequency	= 200
Brick vendor	= A - 1 -(03)

CONTROL LIMITS:

$\bar{\bar{x}} = 0.26 \quad \bar{R} = 4.53$

$UCL_{\bar{x}} = 3.04 \quad UCL_{\bar{R}} = 10.3$

$LCL_{\bar{x}} = 3.56$

$\hat{\sigma} = -6.9$

upper capability $= -.26 + 3(2.2)$
$= 6.3$

lower capability $= -.26 - 3(2.2)$
$= -6.9$

LS = -10   US = +10

-6.9 -0.3 6.3

FIGURE 11.22
Control chart for width

The team was looking forward to sharing their experiences in improving the mill process with the other operations in the plant. A meeting was scheduled to present their work to the other employees of the Mid-State Brick company. After this meeting, two new quality improvement teams would be formed to work on the saw and grinder processes.

11.3   CASE STUDY 2: REDESIGN A WALLPAPER PRODUCT

A company that designs and manufactures wall coverings for home and commercial use has decided to replace an existing wallpaper product with a more up-to-date line. Although the current product has been very successful over the past three years, marketing had defined some new colors and patterns that they believed would improve sales and customer satisfaction. A product development team composed of marketing, R&D, engineering, and manufacturing personnel was formed to redesign the current wallpaper product. The charter for the team is shown in Figure 11.23.

The team documented all current knowledge concerning the old product. Data were summarized from all sources of customer feedback—including final users, builders, installers, retailers, and all internal customers from production and engineering.

The overall task for the team was to design a new product to accommodate the new color and patterns that was close to the targets of the quality characteristics under a wide range of conditions. To do this, the team completed the four processes of designing and redesigning products. These processes are:

1. Define quality.
2. Generate product concepts.
3. Design product.
4. Design production process.

FIGURE 11.23
Charter of product development team for a redesign of a wallpaper product

---

General description:
    Redesign the current wallpaper product.

Expected results:
    1. Improved appearance of the pattern and color choices.
    2. Other quality characteristics will be at least as good as the old product.
    3. Easy to install.
    4. It will wear well and be easy to clean.
    5. Manufacturing costs should not be increased.

Boundaries:
    1. Stay within the R&D budget allocated for the project.
    2. Be able to use existing manufacturing process.
    3. Product must be available to retailers by next spring.

The first task is to define quality by identifying the quality characteristics for the new product. The resulting quality characteristic diagram is given in Figure 11.24.

The next task was to generate product concepts. The team listed all products that related to the charter. They included products from competitors as well as their own. R&D had developed some new formulations based on a different chemistry that resulted in brighter colors and more distinctive patterns.

/usr3/macdir(single 2)

Needs of the customer:	Decorative walls		
	Quality characteristic [1]		
Primary [2]	**Secondary** [3]	**Tertiary** [3]	**#**
looks good	consistent with fashion trends		1
	attractive		2
	lack of visible seams		3
easy to clean	time to clean		4
	effort to clean		5
wears well	resistance to scratches		6
	stain resistance		7
	gloss retention		8

[1] Do not include design factors in this list. (*Test*: You should *not* be able to set the levels of these quality characterisitics.)
[2] Express in the language of the customer.
[3] To add more detail, subdivide into two or more quality characteristics.

FIGURE 11.24
Quality characteristic diagram for redesigning wallpaper

The concepts tested related to composition and thickness of the three layers of the wallpaper. Wallpaper is illustrated in the drawing below.

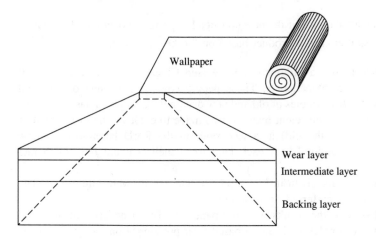

The team made a subjective assessment of product concepts. The summary table is given in Figure 11.25.

Attributes	Product concept		New concepts		
	Current product	Best competitor	1	2	3
Product form	Wallpaper	Wallpaper	Handmade prototypes		
Ease in manufacturing	yes	yes	yes	yes	yes
Needs addressed:					
1. Consistent with trends	3	4	5	5	5
2. Attractive	2	3	5	4	5
3. Lack of visible seams	5	4	4	3	4
4. Time to clean	4	3	5	4	3
5. Effort to clean	3	2	4	3	3
6. Resistance to scratches	4	3	5	4	4
7. Stain resistance	3	3	5	4	4
8. Gloss retention	3	3	5	5	4

Scale: 5 = greatly exceeds needs
4 = exceeds needs
3 = meets needs
2 = partially meets needs
1 = fails to meets needs

FIGURE 11.25
Assessment of product concepts

The First Improvement Cycle: Preference Test

The team decided to run a study with a twofold objective:

1. Determine if the appearance of the new concept 1 is preferred over the old product.
2. Evaluate the installation of a product based on the new concept.

The plan was to use a judgment sample (see Chapter 3 for a discussion of judgment samples) of eight panelists. These people covered the range of potential customers. Five wallpaper samples of old and new design were prepared based on color and pattern combinations. Sufficient material was made to evaluate the installation of the new product (against the old) at a test room in the R&D laboratory. Sample presentation was done randomly for each panelist. Panelists could not collaborate with each other.

The resultant data are given in Figure 11.26. The results were impressive. The new concept was clearly preferred over the old.

The team discussed the results with the panelists. The panelists that preferred the old to the new on samples 1, 2, and 3 based their preference on the new samples having too wide a gap at the seams. This resulted in the seam being visible. A return trip to the lab indicated that the first three samples of the new design did have visible seams. The problem is illustrated below.

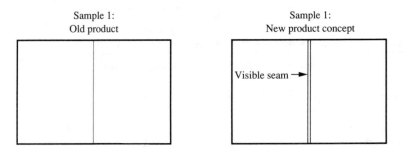

Sample 1:
Old product

Sample 1:
New product concept

Visible seam

Previous testing in R&D had not identified a seam problem. They believed humidity during installation could be a factor contributing to the seam problem. They have no objective measurement for the gap of seams.

The team decided any further testing of new product for causes of visible seams would require a gage. Engineering agreed and developed a gage.

The Second Improvement Cycle: Test of New Gage

The next activity of the team was to evaluate the variability of the new gage with respect to the gap of seams. A three-factor nested experiment was selected to evaluate the following components:

1. Samples.
2. Position within each sample.
3. Measurement within each position.

Panelist	Samples (color/pattern)				
	1	2	3	4	5
1	N	N	N	N	N
2	N	N	O	N	N
3	O	N	N	N	N
4	N	O	N	N	N
5	N	N	N	N	N
6	N	N	N	N	N
7	O	O	O	N	N
8	N	N	N	N	N

N = preferred new product
O = preferred old product

FIGURE 11.26
Overall preference test for new vs. old product

The planning form is given in Figure 11.27. The five samples are the samples of the new product from the previous preference study. Three positions per sample will be measured to determine the unevenness of seams. Three measurements will be made per position by different operators. The design matrix is given in Figure 11.28.

Figure 11.29 shows a run chart of the data. No obvious trends or other special causes are seen. The dot-frequency diagram is given in Figure 11.30.

The dot-frequency diagram indicates that most of the variation was attributable to sample differences and, to a lesser extent, to position within sample. The gage results were very repeatable. Variation between positions within sample is easily seen. The first three samples had the greatest problem with seams, which confirmed the previous preference study. (Note: the gage measurement scale is .1 mils and $0 =$ deviation from target.)

The team decided to accept the gage for additional product development work and to take measurements from four different positions within a sample and calculate averages and standard deviations for any future studies. The team then proceeded to the process of designing the product based on the selection of product concept.

Documentation of current knowledge of the new product is displayed as a QFD relation diagram in Figure 11.31. (*Note:* This diagram is analogous to the cause-and-effect diagram used to define current knowledge of an existing process.) Eight control factors represent design parameters that can be set to specific levels. Those thought to have a strong relationship to the quality characteristics can be adjusted to move the quality characteristics closer to target.

FIGURE 11.27
Form for documentation of a planned experiment for cycle 2

1. Objective:
Evaluate the variability of the new gage with respect to seams.

2. Background information:
This measurement system is new.

3. Experimental variables:

A. Response variables	Measurement technique

1. Visible seams (.1 mils)	New gage.

B. Factors under study	Levels

1. Samples (new product)	Five samples from last study.
2. Position within sample	Three positions per sample made along the seam.
3. Measurement within position	Three measurements made with three operators.

C. Background variables	Method of control

1. Installer	One person.
2. Type of adhesive	Use standard adhesive.
3. Calibration	Checked at the beginning and end of study.

4. Replication:
The study has to be completed in one day. Only one replication of the experimental pattern will be done.

5. Methods of randomization:
The five samples and three positions per sample were randomly ordered.

6. Design matrix: (see Figure 11.28)

7. Data collection forms: (not shown here)

8. Planned methods of statistical analysis:
Run chart and dot-frequency diagram.

9. Estimated cost, schedule, and other resource considerations:
Study can be completed in one day using the laboratory testing room.

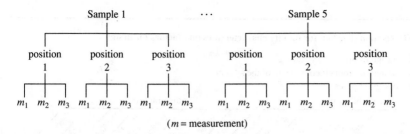

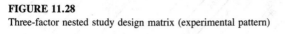

(*m* = measurement)

FIGURE 11.28
Three-factor nested study design matrix (experimental pattern)

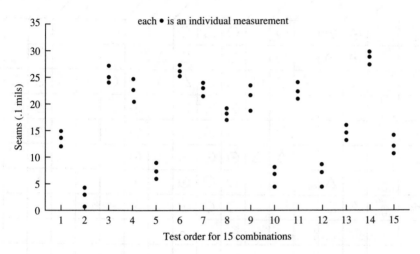

FIGURE 11.29
Run chart for three-factor nested study

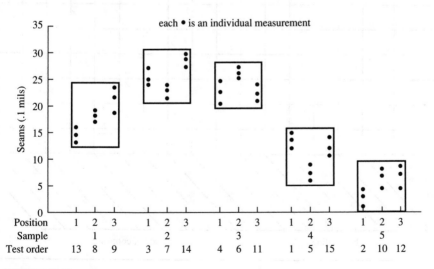

FIGURE 11.30
Dot-frequency diagram for three-factor nested study

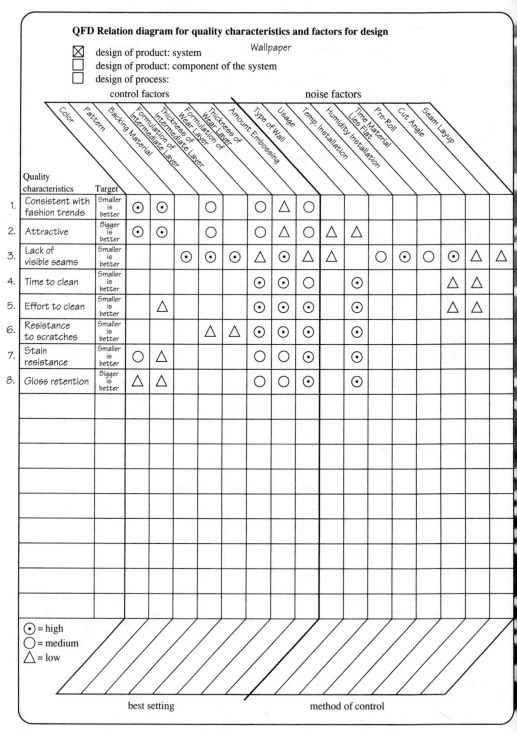

What planned experiments need to be run to test this diagram?

FIGURE 11.31
QFD Relation diagram for redesigning wallpaper

Of the eight noise factors identified in Figure 11.31 as affecting the quality characteristics, most relate to visible seams. The team developed an overall plan to test the relationships in the QFD relation diagram, but in this case study only the work on visible seams and appearance will be described. The identified noise and control factors are illustrated below.

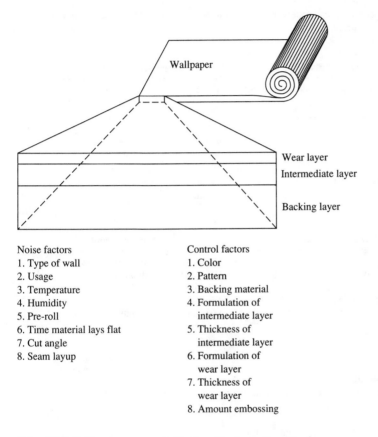

Noise factors
1. Type of wall
2. Usage
3. Temperature
4. Humidity
5. Pre-roll
6. Time material lays flat
7. Cut angle
8. Seam layup

Control factors
1. Color
2. Pattern
3. Backing material
4. Formulation of intermediate layer
5. Thickness of intermediate layer
6. Formulation of wear layer
7. Thickness of wear layer
8. Amount embossing

The Third Improvement Cycle: Factors for Design

The objective of the next cycle is to plan an experiment to select factors for design that have an effect on the quality characteristics.

The plan for the cycle is to vary the formulation of intermediate and wear layers to simulate the current and new product. The belief is that the humidity during installation may be interacting with the new product and causing the seam problem. The planning form is given in Figure 11.32. The design matrix is given in Figure 11.33.

The run charts for the averages and standard deviations of visible seams at each of the 16 combinations are contained in Figure 11.34. No obvious patterns were present. Three of the points on the average chart obviously indicate special causes. Further analysis will disclose that those points were due to changing the factors under study.

FIGURE 11.32
Documentation of the wallpaper experiment for cycle 3

1. Objective:
Determine the effect control factors and noise factors have on a redesign for wallpaper.

2. Background information:
The redesign has a new color and pattern that has a beautiful appearance. There is a need to reduce the magnitude of visible seams.

3. Experimental variables:

A. | Response variables | Measurement technique |
|---|---|
| 1. Visible seams (average, standard deviation) | new gage (.1 mils) |

B. | Factors under study | Levels | |
|---|---|---|
| 1. Formulation | old | new |
| 2. Type of wall | wood | concrete |
| 3. Temperature | 50° | 90° |
| 4. Humidity | 40% | 80% |
| 5. Time material lays flat | 2 min. | 4 min. |
| 6. Pre-roll | 0 min. | 10 min. |
| 7. Cut angle | 45° | 90° |
| 8. Seam layup | low | high |

C. | Background variables | Method of control |
|---|---|
| 1. Installer | One person, standard instructions |
| 2. Type of adhesive | Use standard for all applications |
| 3. Thickness of intermediate layer | 4.0 mils |
| 4. Thickness of wear layer | 1.0 mils |
| 5. Backing material | Material A (same as old) |

4. Replication:
Measurements for visible seams at four positions per sheet.

5. Methods of randomization:
Order of the 16 runs was randomized using a random permutation table.

6. Design matrix:
2^{8-4} fractional factorial design (see Figure 11.33).

7. Data collection forms: (see margin in Figure 11.33)

8. Planned methods of statistical analysis:
Compute average and standard deviation of four readings per sheet.

9. Estimated cost, schedule, and other resources:
Evaluated in the test room with temperature and humidity controls. Six days are required to complete.

Test	Run order	1 F	2	3 T	4 H	5	6 P	7	8	12 37 56 48	13 27 46 58	14 36 57 28	15 26 47 38	16 25 34 78	17 23 45 68	24 35 67 18	Average	Standard deviation
1	8	−	−	−	+	+	+	−	+	+	+	−	−	−	+	−	0	4
2	15	+	−	−	−	−	+	+	+	−	−	−	−	+	+	+	11	11
3	16	−	+	−	−	+	−	+	+	−	+	+	−	+	−	−	4	5
4	6	+	+	−	+	−	−	−	+	+	−	+	−	−	−	+	4	5
5	3	−	−	+	+	−	−	+	+	+	−	−	+	+	−	−	9	12
6	12	+	−	+	−	+	−	−	+	−	+	−	+	−	−	+	42	18
7	9	−	+	+	−	−	+	−	+	−	−	+	+	−	+	−	5	5
8	7	+	+	+	+	+	+	+	+	+	+	+	+	+	+	+	77	14
9	1	+	+	+	−	−	−	+	−	+	+	−	−	−	+	−	18	11
10	4	−	+	+	+	+	−	−	−	−	−	−	−	+	+	+	21	10
11	10	+	−	+	+	−	+	−	−	−	+	+	−	+	−	−	80	25
12	13	−	−	+	−	+	+	+	−	+	−	+	−	−	−	+	12	9
13	11	+	+	−	−	+	+	−	−	+	−	−	+	+	−	−	9	5
14	2	−	+	−	+	−	+	+	−	−	+	−	+	−	−	+	2	4
15	14	+	−	−	+	+	−	+	−	−	−	+	+	−	+	−	1	8
16	5	−	−	−	−	−	−	−	−	+	+	+	+	+	+	+	1	2

Divisor = 8

Average   23.5 −2.0 29 11.5 4.5 12.0 −3.5 2.0 −4.5 19.0 10.4 −0.5 16 −3.5 5.5

Standard deviation   5.8 −3.8 7.5 2.0 −0.2 0.8 0.1 0 −3.0 2.2 −0.2 −1.5 2.5 −2.8 −0.2

FIGURE 11.33
2^{8-4} design matrix for the wallpaper study

The effects of the factors computed from the design matrix were included in Figure 11.33. The dot diagrams are contained in Figure 11.35.

One control factor, formulation, and three noise factors, pre-roll, humidity, and temperature, were found to be important. Temperature was a real surprise. It was thought that humidity was the problem.

Analysis of the two cubes from Figure 11.36 (see also Figure 11.37) as a full factorial design indicates a possible interaction between formulation and temperature for averages. These interactions accounted for the three high points on the run chart for averages in Figure 11.34.

The prediction of the team was that the new concept will have installation problems with seams. Claims could be high! The plan will be to run another cycle with a follow-up study using a new run of material and to learn more about the interactions.

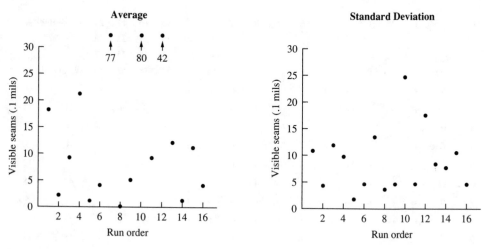

FIGURE 11.34
Run charts for the wallpaper study

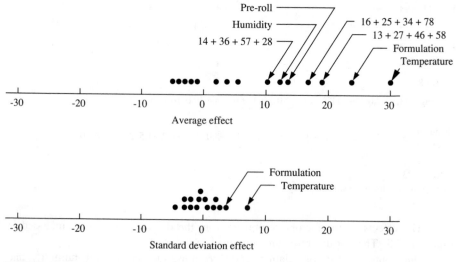

FIGURE 11.35
Dot diagrams for the wallpaper study

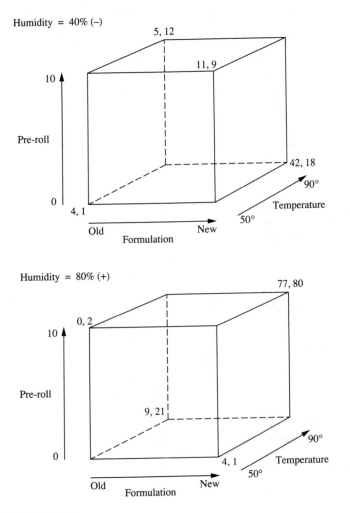

FIGURE 11.36
Cube for the wallpaper study (2^{8-4} design, any three factors form a full factorial design)

The Fourth Improvement Cycle: Follow up Study

The objective of this study was to use a new run of material and determine which interactions persist. A three-factor experiment with formulation (old and new), pre-roll, and temperature/humidity as a chunk variable with the low level at both low temperature and low humidity and the high level at both high temperature and high humidity. It was believed that these combinations will create the most extreme conditions as noise factors in the field.

The planning form for the study is given in Figure 11.38. Visible seams and appearance are the response variables. Appearance will be scored subjectively by an appearance team. A 2^3 factorial design, given in Figure 11.39, was chosen so that all interactions can be studied.

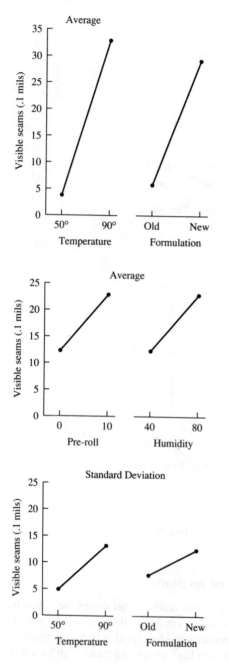

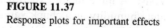

FIGURE 11.37

Response plots for important effects

The run charts for the averages and standard deviations of visible seams at each of the eight combinations are contained in Figure 11.40. No obvious patterns were present. The effects of the factors computed from the design matrix are included in Figure 11.39.

The dot diagrams for the average and standard deviation of visible seams are contained in Figure 11.41. Results are similar to the previous experiment and the

FIGURE 11.38
Documentation of the wallpaper experiment for cycle 4

1. Objective:
Determine the effect of control factors and noise factors on a new run of material for wallpaper.

2. Background information:
The redesign has a new color and pattern that has a beautiful appearance. There is a need to reduce the magnitude of visible seams. Previous study indicated three noise factors (temperature, pre-roll, and humidity) may contribute to the seam problem.

3. Experimental variables:

A. Response variables	Measurement technique
1. Visible seams (average, standard deviation)	new gage (.1 mils)
2. Appearance	subjective scoring

B. Factors under study	Levels	
	old	new
1. Formulation	old	new
3. Temperature/Humidity	50°/40%	90°/80%
3. Pre-roll	0 min.	10 min.

C. Background variables	Method of control
1. Installer	One person, standard instructions
2. Type of adhesive	Use standard for all applications
3. Thickness of intermediate layer.	4.0 mils
4. Thickness of wear layer	1.0 mils
5. Backing material	Material *A* (same as old)
6. Type of wall	Wood
7. Time material lays flat	4 hours
8. Cut angle	90°
9. Seam layup	Low

4. Replication:
Four measurements for visible seams per sheet.

5. Methods of randomization:
Order of the eight runs was randomized using a random permutation table.

6. Design matrix: 2^3 factorial design

7. Data collection forms: (see margin in Figure 11.39)

8. Planned methods of statistical analysis:
Run charts, dot diagram, cube, and response plots.

9. Estimated cost, schedule, and other resources:
Evaluated in the test room with temperature and humidity controls. One day is required to complete. Appearance scoring done by appearance team.

dot diagram in Figure 11.35. Formulation interacted with the temperature/humidity combination to affect the average of visible seams.

Pre-roll showed a larger effect on standard deviations of visible seams. Pre-roll had an effect for the average in cycle 3 that was not confirmed in this study. The

Test	Run order	1 F	2 T/H	3 P	12	13	23	123	Response Average	Standard Deviation
1	3	−	−	−	+	+	+	−	5	8
2	4	+	−	−	−	−	+	+	5	8
3	1	−	+	−	−	+	−	+	10	14
4	7	+	+	−	+	−	−	−	32	16
5	6	−	−	+	+	−	−	+	3	2
6	2	+	−	+	−	+	−	−	10	6
7	5	−	+	+	−	−	+	−	13	8
8	8	+	+	+	+	+	+	+	40	15

Divisor = 4

Average effect	14.0	18.0	3.5	10.5	3.0	2.0	−0.5
Standard Deviation	3.2	7.2	− 3.8	1.2	2.2	0.2	0.2

FIGURE 11.39
2^3 design matrix for the wallpaper study

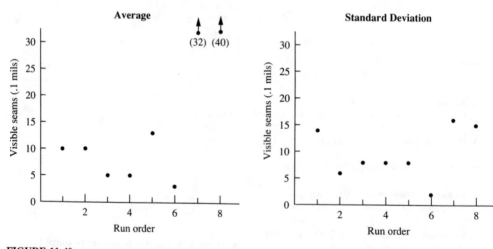

FIGURE 11.40
Run charts for the wallpaper study

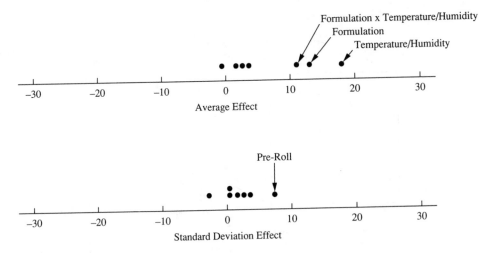

FIGURE 11.41
Dot diagram for the wallpaper study

standard-deviation response plot showed less visible seams with a 10-minute pre-roll waiting period. The team attributed this to the new run of material. This factor will be watched in future studies.

Analysis of the cube in Figure 11.42*a* revealed the interaction of formulation and temperature/humidity. Response plots for the important factors are given in Figure 11.42*b*.

The result of cycle 4 was a confirmation of cycle 3. The team still had a problem with seams. Cycle 4, however, showed that taking a pre-roll step of allowing 10 minutes for the roll to lie flat prior to installation helped some with the seams. The team therefore decided that pre-roll could be controlled as a noise factor by including this finding in the instructions given to the installer.

Because the appearance team had preferred the new product to the old in all other quality characteristics, the product development team decided to study the effect of some of the control factors on variations in the noise factors identified during installation of the wallpaper.

The Fifth Improvement Cycle: Back to Design Factors

The objective of the next cycle was to determine the effect that installation factors had on seams and appearance in a new prototype formulation. The hope was to reduce the effect of noise factors by changing the levels of three control factors:

Factors	Old prototype	New prototype
Wear layer	1.0 mils	2.0 mils
Intermediate layer	formula 1	formula 2
Backing layer	material *A*	material *B*

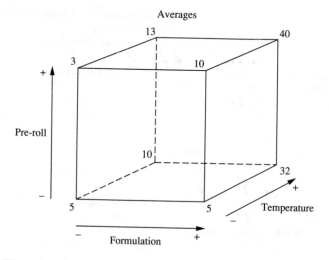

FIGURE 11.42a
Cube plot for the wallpaper experiment

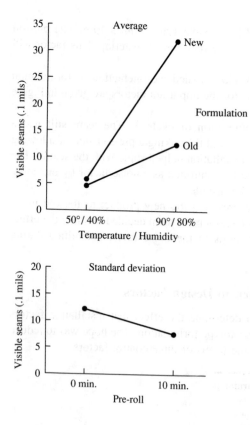

FIGURE 11.42b
Response plots for the wallpaper experiment

Hopefully, the new settings for these three control factors would desensitize the product to the noise factor temperature/humidity and the need for a 10-minute pre-roll.

Figure 11.43 contains the planning form for the study. Three control factors and two installation (noise) factors are studied. Four measurements were made on seams for each sheet, and averages and standard deviations were calculated. A 2^{5-1} design was chosen because there was no confounding of the two-factor interactions.

At completion of the experiment, prototypes were tested for appearance. It was found that appearance of the new design remained good throughout the 16 runs. Laboratory testing for scratches, stains, and gloss retention confirmed the robustness of the new prototype.

Figure 11.44 contains the response plots for standard deviations of visible seams that summarize the important results of the experiment.

The first plot shows a strong interaction between thickness of wear level, a control factor, and temperature/humidity, a noise factor. The effect of this noise factor on visible seams is less with the 2.0-mil wear layer. The thinner wear level (1.0 mils) was affected by the high temperature/humidity combination.

The second plot shows a strong interaction between the other two control factors. Backing material and the formulation of the intermediate layer interacted with respect to seams. Because both backing materials are needed for different applications, material A is chosen with formula 1 and material B is chosen with formula 2.

The third plot demonstrates a strong relationship between an installation factor and visible seams. By laying out the roll for 10 minutes before installation, the seams will have less variation. This adds to the installation time.

Based on these analyses and the team's current knowledge, the following conclusions are:

Control factors	Best setting
Thickness of wear layer	2.0 mils
Backing material A	formula 1
Backing material B	formula 2

Noise factors	Method of control
Pre-roll	Have installation instructions include a 10-minute pre-roll step.
Temperature/humidity	Desensitized by setting control factor, *thickness of wear layer*, at 2.0 mils.

Updates to the QFD relation diagram are given in Figure 11.45

The team next had to design the production process. No new equipment was needed for the production process. Backing material B was new and required a new supplier. The increase in wear thickness to 2.0 mils increased costs. This increase was offset by the lower cost of backing material B.

Training based on the change to the production process was conducted for all those involved. Short production runs indicated no new production problems. The changes made in the redesign of the product did not result in any difficulties for the production process.

FIGURE 11.43
Documentation of the wallpaper experiment for cycle 5

1. Objective:
Determine the effect of installation factors on a redesign for wallpaper by changing three control factors.

2. Background information:
The redesign has a new color and pattern that has a beautiful appearance. There has been a problem with seams. Probable cause is high temperature or high humidity at the installation site.

3. Experimental variables:

A.	Response variables	Measurement technique
	1. Visible seams (average, standard deviation)	new gage (.1 mils)
	2. Appearance	subjective scoring

B.	Factors under study	Levels	
	1. Backing material	Material A	Material B
	2. Formulation of intermediate layer	Formula 1	Formula 2
	3. Thickness of wear layer	1.0 mils	2.0 mils
	4. Temperature/humidity	50°/40%	90°/80%
	5. Pre-roll	0 min.	10 min.

C.	Background variables	Method of control
	1. Installer	One person, standard instructions
	2. Type of adhesive	Use standard for all applications
	3. Wall	Use coarse material
	4. Thickness of intermediate layer	4.0 mils
	5. Formulation of wear layer	Same as old design
	6. Time material lays flat	4 hours
	7. Cut angle	90°
	8. Seam layup	low

4. Replication:
Four measurements for visible seams per sheet.

5. Methods of randomization:
Order of the 16 runs was randomized using a random permutation table.

6. Design matrix: 2^{5-1} fractional factorial design.

7. Data collection forms: (not shown here)

8. Planned methods of statistical analysis:
Run charts, dot diagram, cube, and response plots.

9. Estimated cost, schedule, and other resources:
Evaluated in the test room with temperature and humidity controls. Four days are required to complete. Appearance scoring done by appearance team.

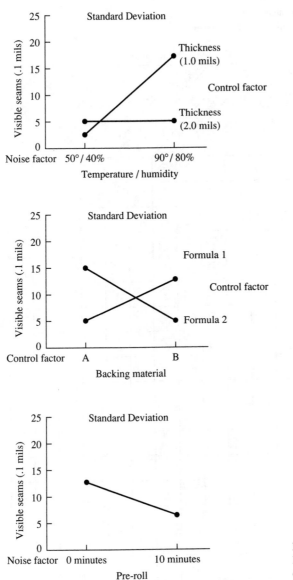

FIGURE 11.44
Response plots for the wallpaper experiment

The team had decided to proceed with a pilot run of samples for each of the five basic color/pattern product types evaluated in cycle 1 (see Figure 11.26). Formulation for each type of color/pattern was as follows:

Product type (color/pattern)	1	2	3	4	5
Backing material	A	A	A	B	B
Formula	1	1	1	2	2

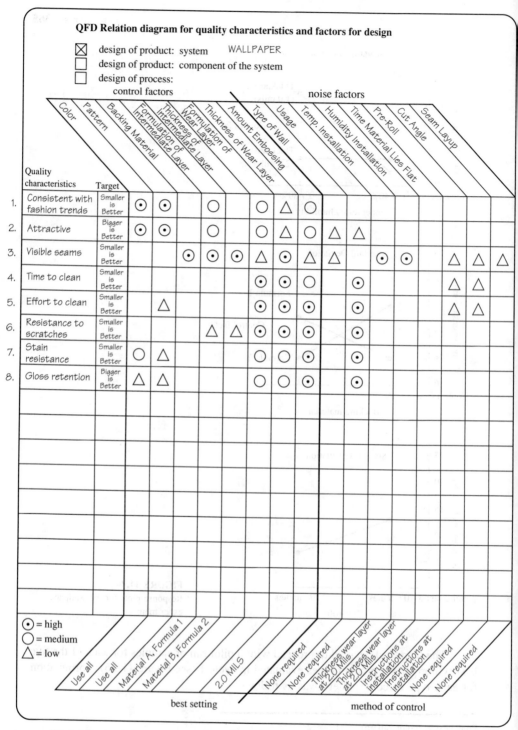

What planned experiments need to be run to test this diagram? 2^{8-4} cycle 3, 2^3 cycle 4, 2^{5-1} cycle 5

FIGURE 11.45
Updated QFD relation diagram for redesigning wallpaper

Backing material *A* was chosen for the first three product types because of the successful use of that material for other current product types similar to them.

The Sixth Improvement Cycle: Field Test

The objective of this study was to field test the pilot run of wallpaper. All five product types from the pilot run were evaluated along with the old product. A field test was chosen over the temperature and humidity environment of the R&D lab.

This cycle must increase understanding of performance of the new design with respect to potential problems. Including the old product with the pilot product of the new design under widely varying conditions should increase the team's degree of belief about its future predictions of product performance.

Field sites were used to create a blocking factor. The blocks chosen were judged likely to create extreme conditions. Figure 11.46 gives the planning form for the study.

Important results of the study are given in Figure 11.47. A run chart for standard deviation of visible seams is given in Figure 11.48.

The run chart in Figure 11.48 confirmed the results of the fifth cycle. Standard deviations for visible seams are 14 or less. The pilot run of five product types compared favorably with the old product types. The team predicted that claims against the redesigned product will not exceed claims against the old product.

For all five product types, appearance exceeded the old product at all four field sites. The action of the team was a decision to start production of the redesigned product.

The Seventh Improvement Cycle: Capability

During cycle 6, production continued to run the pilot for four days to determine capability of the processes. The objective of the seventh cycle was to learn about the current performance of the production processes before production of the redesigned product.

Figure 11.49 contains control charts for thickness of the wear layer. Five samples of wallpaper were taken every two hours and measured for thickness. Control limits were calculated at the end of 20 subgroups. A special cause occurred at startup on July 21. A check of viscosity indicated too high a reading. Viscosity was brought down and the pilot continued.

Process capability is a prediction of individual measurements of thickness of wear layer. The action from cycle 5 was to increase wear layer from 1.0 to 2.0 mils. This would help desensitize the possible problem of high temperature and high humidity that could occur during installation. The capability from this cycle was from 1.25 to 2.87 mils. The average was 2.06 mils. This was slightly higher than the target of 2.0 mils. The thickness of the wear layer from samples involved in the field test of cycle 6 ranged from 1.4 to 2.5 mils.

Based on the knowledge gained from this cycle, the following actions were taken:

FIGURE 11.46

Documentation of the wallpaper experiment for cycle 6

1. Objective:

Determine the performance of a redesign of a wallpaper product under various field conditions. This information will be used to determine if the current product should be replaced by the new design.

2. Background information:

Previous experiments using fractional factorial and factorial designs lead to changes in the current design. Noise factors during installation are affecting the variation in visible seams.

3. Experimental variables:

A.

Response variables	Measurement technique
1. Visible seams (average, standard deviation)	gage (.1 mils)
2. Appearance	subjective scoring

B.

Factors under study	Levels
1. Product types	1 2 3 4 5 6 = old product

C.

Background variables	Method of control
1. Type of adhesive	Use standard for all applications
2. Time material lays flat	2 hours
3. Cut angle	Standard at 90°
4. Seam layup	Set at low
5. Pre-roll	Instructions (10 min.)
6. Thickness of wear layer	Measure thickness

Method of control for other background variables: Create four blocks (field sites) made up of different treatment combinations of the background variables.

Background factor	Block 1	Block 2	Block 3	Block 4
1. Field site	N.E.	N.W.	S.E.	S.W.
2. Humidity	middle	low	high	low
3. Temperature	middle	low	high	high
4. Wall	wood	old	cement	old

4. Replication:

Four measurements for visible seams per sheet.

5. Methods of randomization:

Order of the six levels of product types within block is randomized using a random permutation table.

6. Design matrix: (attach copy) Not shown.

7. Data collection forms: (attach copy) See Figure 11.47.

8. Planned methods of statistical analysis:

Compute the average and standard deviation of the four measurements. Plot run charts for the six sample types adjusted for block effect.

Response: standard deviation of visible seams (run order in parentheses)

Product	Block 1	Block 2	Block 3	Block 4
1	(3) 2	(2) 4	(6) 12	(3) 5
2	(4) 7	(4) 6	(1) 9	(1) 7
3	(6) 2	(5) 12	(3) 12	(5) 1
4	(1) 9	(1) 6	(2) 14	(2) 10
5	(2) 7	(3) 4	(4) 10	(4) 8
6 = old	(5) 9	(6) 10	(5) 12	(6) 11

FIGURE 11.47
Test results for wallpaper study

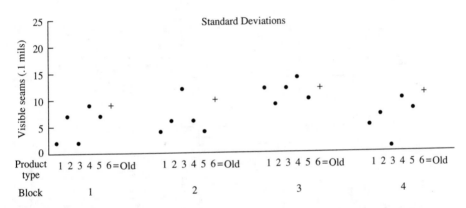

FIGURE 11.48
Run chart for block design of wallpaper study, cycle 6

1. Target for thickness of wear layer should continue at 2.0 mils (increases raise production costs).
2. Process engineering should do a study to establish the relationship of viscosity to thickness of wear layer.
3. Monitor viscosity with control charts. Take action on special causes. Identify common causes of variation.
4. Discuss results of pilot with product development team.

Final Actions of the Product Development Team

The team reviewed their charter (Figure 11.23). Based on their current knowledge, the expected results can be achieved. The team was able to stay within the boundaries of the charter.

The team recommended proceeding with production of the redesigned wallpaper. Recommendation was approved and production started the next week.

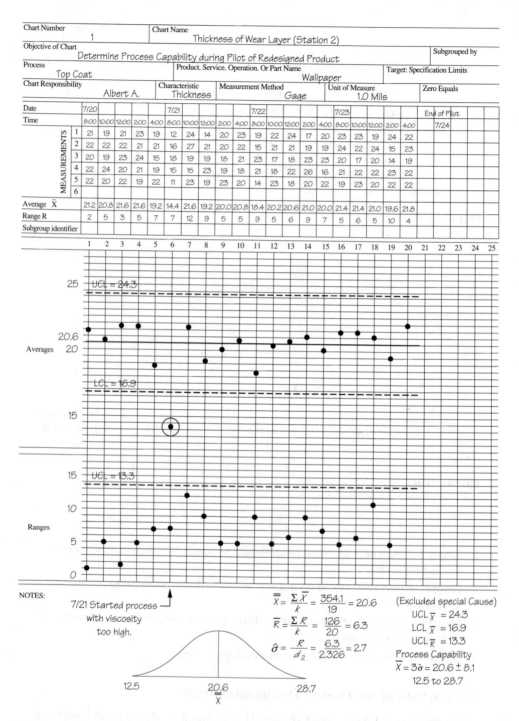

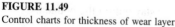

FIGURE 11.49
Control charts for thickness of wear layer

REFERENCES

Deming, W. Edwards (1980): Dedication to 50th Anniversary Commemorative Reissue of the 1931 publication of *The Economic Control of Quality of Manufactured Product*, American Society for Quality Control.

EXERCISES

11.1. *Study each case study*. Recalculate and plot all charts and graphs.
Discuss possible alternative approaches to the design and analysis of data.
Discuss why both cases were analytic studies. Were there any enumerative aspects of the two case studies?

11.2. *Study case studies involved with improving quality*. Compare and discuss approaches used to those presented in this book.

11.3. *Create your own case study for improving quality from an area of interest to you.*

APPENDIX

FORMS

A.1 WORKSHEETS FOR CURRENT KNOWLEDGE AND THE IMPROVEMENT CYCLE

This appendix contains two worksheets that have been used extensively in the application of the model for improving quality. Each worksheet is designed so that information can be documented in the enclosed area, with questions relating to that information included below the area. The flow of each worksheet is from top to bottom, following the description of the model developed in Section 1.3, and each worksheet is four pages in length.

Improvement of Quality

CURRENT KNOWLEDGE

DATE _____

CHARTER OF THE TEAM

GENERAL DESCRIPTION

EXPECTED RESULTS

BOUNDARIES FOR THE ACTIVITIES

- Does the charter give guidance to the team?

TEAM LEADER

TEAM MEMBERS

- Were representatives of all groups affected by the above charter considered for team membership?
- Does the team consider this effort to be important?

by

Version 3.3

Associates in Process Improvement

LIST THE PROCESSES RELATED TO THE CHARTER AND CHOOSE THE INITIAL PROCESS FOR STUDY

• Are the right people on the team for the process selected?

DESCRIBE THE INITIAL PROCESS FOR STUDY

| Suppliers | Inputs | | Outcomes | Customers |

Process name:

Process owner:

Key stages in the process:

QUALITY-CHARACTERISTICS

Selected Inputs	Quality-characteristic	Quality-characteristic of the process	Selected Outcomes	Quality-characteristic

Questions to consider when planning an improvement cycle:
• Do customers need to be surveyed to help define quality-characteristics?
• Is a Pareto analysis needed to identify the most important quality-characteristics?
• Did you get input from managers and workers in the process?
• Is the quality of the measurement process for these characteristics documented?
• Are historical data available?
• Have control charts been developed for the quality-characteristics?

OPERATIONAL DEFINITIONS FOR SELECTED QUALITY-CHARACTERISTICS

FLOWCHART (OR PICTURE) THE PROCESS

- Is this the actual process in use today?
- Can you identify opportunities for improvement (For example: complexity, waste, layers of inspection)?
- Are there obvious improvements to the process that should be made?
- Is there need to run a cycle to standardize procedures for the process?

IF APPROPRIATE, DEVELOP CAUSE AND EFFECT DIAGRAMS FOR THE IMPORTANT QUALITY-CHARACTERISTICS OR PROBLEMS IN THE PROCESS

• Do common causes or special causes dominate?

STATE SOME POTENTIAL IMPROVEMENT CYCLES (CONSIDER YOUR ANSWERS TO THE QUESTIONS ASKED PREVIOUSLY)

• What should be accomplished by the initial cycles?   • What other processes need to be studied?
• Are resources available to carry out these cycles?

SELF ASSESSMENT OF THE TEAM

• Are the right people on the team?   • Does the team still agree on the importance of this effort?
• Have ground rules and administrative guidelines for the team been established and followed?
• Is the owner of the process involved in the team's activities?

Improvement of Quality

IMPROVEMENT CYCLE

DATE_____

TEAM_____

CYCLE# _____

A	P
S	D

PLAN

OBJECTIVE OF CYCLE

• What additional knowledge is necessary to take action?

QUESTIONS TO BE ANSWERED FROM THE DATA OBTAINED IN THIS CYCLE

PREDICTIONS

• Are historical data available to answer the questions above?

• Does the team agree on these predictions?

Version 3.3

by

Associates in Process Improvement

DEVELOP A PLAN TO ANSWER THE QUESTIONS (WHO, WHAT, WHERE, WHEN)

- Did your plan consider the following methods:
  - Data Collection Forms
  - Pareto Diagrams
  - Control Charts
  - Frequency Plots
  - Planned Experimentation
  - Survey Methods
  - Simulation/Modeling
  - Scatter Diagrams
  - Run Charts
  - Engineering Analysis
- Did you assign responsibilities for collection and analysis of the data?
- Is training needed?
- Is the plan consistent with the charter?
- Can the plan be carried out on a small scale?
- Have you considered people outside the team who will be affected by this plan?

DO

OBSERVATIONS IN CARRYING OUT THE PLAN

- Identify the things observed that were not part of the plan.
- Document what went wrong during the data collection.
- Begin analysis of the data as it is collected.
- Evaluate the data for changes over time. (control chart or run chart).

STUDY

ANALYSIS OF DATA

COMPARE THE ANALYSIS OF THE DATA TO THE CURRENT KNOWLEDGE

- Do the results of this cycle agree with predictions made in the planning phase?
- Under what conditions could the conclusions from this cycle be different?
- What are the implications of the unplanned observations and problems during the collection of the data?
- Do the data help answer the questions posed in the plan?

SUMMARIZE THE NEW KNOWLEDGE GAINED IN THIS CYCLE

- Revise the current knowledge to reflect this learning (update flowcharts and cause and effect diagrams).
- Will this new knowledge apply elsewhere?

A	P
S	D

ACT

WHAT CHANGES ARE TO BE MADE TO THE PROCESS?

LIST OTHER ORGANIZATIONS AND PEOPLE THAT WILL BE AFFECTED BY THE CHANGES

- Is the cause system sufficiently understood?
- Has an appropriate action or change been developed or selected?
- Have the changes been tested on a small scale?
- Are there forces in the organization that will help or hinder the changes?
- Have the responsibilities to implement and evaluate the changes been communicated?
- Will the actions or changes improve performance in the future?

OBJECTIVE OF NEXT CYCLE

A.2 TABLES OF RANDOM NUMBERS

Randomization is the objective assignment of factor level combinations to experimental units. Tables of random numbers have been developed for use in making this random assignment. This appendix contains two different types of random number tables: a general random number table and a table of random permutations. When random permutations are used, each number of a series is selected without replacement until all numbers in the series have been selected.

Table A.2-1 contains random numbers from 00 to 99. Multiple columns can be combined to obtain larger numbers if needed. For example, the fist digit from the second column can be combined with the first column to obtain random numbers from 000 to 999.

Tables A.2-2 and A.2-3 contain random permutations. The numbers 1 through 16 are permuted in Table A.2-2; each column contains three permutations. The numbers 1 through 8 are permuted in Table A.2-2; each column contains five random permutations.

There are a number of ways to select numbers from the tables. An informal procedure would be to drop a pencil on the page of interest and start selecting numbers from the point of the pencil. A more formal approach would be to randomly select a row and column number from Table A.2-2 and use these numbers to designate the starting point or the permutation. One could also start in one corner of the table and mark off the numbers or permutations as they are used.

TABLE A.2–1
Random numbers

	1	2	3	4	5	6	7	8	9	10	11	12	13	14	15	16	17	18	19	20
1	09	40	31	49	80	12	45	78	42	07	65	81	25	95	58	23	93	62	57	20
2	58	98	25	75	13	53	43	18	26	58	09	54	53	22	76	67	30	12	09	50
3	11	23	37	28	19	74	47	58	81	11	97	11	64	98	47	84	22	33	03	35
4	72	02	97	35	03	33	22	84	06	53	56	58	41	17	92	89	79	76	02	12
5	15	27	99	71	42	18	80	76	64	35	54	84	01	69	15	57	31	05	42	76
6	80	01	44	18	33	58	85	93	39	72	55	12	93	40	78	84	99	41	75	09
7	33	49	64	04	33	17	56	52	80	39	63	98	69	83	60	96	16	01	06	35
8	45	33	54	88	44	52	79	41	11	43	02	16	04	21	27	78	87	22	09	75
9	06	63	02	80	67	17	76	40	16	38	47	51	04	64	32	43	26	42	09	05
10	23	32	09	37	73	26	99	45	06	67	22	65	79	04	82	31	54	26	39	42
11	26	78	51	43	82	85	95	22	87	36	20	79	55	70	11	09	70	11	70	98
12	92	79	45	85	62	10	77	45	40	12	37	31	12	53	99	94	79	35	37	72
13	14	76	60	56	39	71	24	05	21	23	69	33	95	94	30	01	21	28	20	39
14	61	32	61	34	33	45	32	65	33	87	25	09	46	21	96	29	83	20	71	61
15	47	53	35	97	69	05	01	71	98	75	81	66	59	92	19	07	03	74	94	58
16	94	54	30	03	66	54	58	28	13	04	65	79	76	51	29	14	36	65	99	48
17	82	87	13	22	35	58	11	14	33	11	74	84	83	09	66	88	83	67	47	84
18	36	67	26	13	15	96	08	00	46	10	33	36	72	23	10	05	56	82	06	22
19	97	65	95	60	61	28	66	43	75	66	39	22	30	96	82	72	34	03	93	37
20	37	14	98	95	57	75	99	03	30	00	47	89	30	74	72	78	82	06	11	93
21	74	91	65	50	67	44	71	81	43	39	19	36	40	71	24	05	36	42	64	06
22	50	04	94	15	85	04	68	19	60	64	84	68	61	60	83	43	00	81	42	32
23	92	89	70	59	85	33	67	62	66	50	17	85	42	68	81	12	42	23	32	67
24	53	66	53	52	16	44	33	42	10	79	14	80	67	76	98	88	75	85	00	98
25	94	62	73	33	36	87	39	96	78	05	20	90	02	31	67	80	28	16	91	75
26	68	57	06	61	56	04	84	05	43	45	74	74	75	66	33	87	47	31	72	81
27	06	84	59	23	81	29	05	48	37	23	45	71	68	33	49	68	74	09	83	93
28	85	79	41	68	14	79	11	28	89	05	52	31	49	68	45	06	64	39	12	91
29	16	39	54	23	35	94	86	85	64	32	22	01	90	81	02	73	39	39	75	83
30	90	03	87	19	44	10	79	38	89	16	90	65	36	77	50	19	87	33	81	21
31	18	68	85	69	43	94	89	88	45	61	52	75	23	68	43	82	70	19	94	10
32	78	92	03	53	60	27	38	11	92	17	90	28	93	55	88	85	42	09	35	34
33	59	54	68	32	48	10	95	67	39	22	59	88	30	94	01	92	56	47	38	29
34	09	79	59	43	18	43	76	68	49	33	98	05	32	19	64	02	36	76	65	06
35	54	37	70	85	29	42	78	70	59	00	06	43	56	68	41	38	74	77	85	90
36	14	92	32	82	02	15	56	65	78	32	45	61	92	40	54	29	76	50	33	65
37	12	67	52	09	54	74	44	95	06	83	21	99	29	61	02	98	59	45	72	19
38	03	61	29	07	74	38	76	38	29	65	43	02	18	75	19	45	69	30	27	81
39	89	90	89	65	63	99	97	65	55	91	25	25	65	22	01	25	42	31	69	43
40	44	46	80	83	08	27	66	78	03	56	82	42	45	13	03	67	75	09	49	26
41	95	60	34	69	60	77	02	97	22	82	38	35	96	74	57	16	47	98	94	38
42	62	51	42	73	96	38	71	11	43	19	69	79	75	39	17	66	26	22	17	99
43	24	43	17	93	11	81	58	47	73	19	28	37	50	09	12	39	30	67	76	73
44	22	05	09	58	26	18	43	53	13	75	25	76	42	92	05	31	57	15	69	51
45	21	52	96	85	32	49	16	44	79	39	07	43	87	24	34	59	38	72	18	30

TABLE A.2–2
Random permutations of the numbers 1–16

	1	2	3	4	5	6	7	8	9	10	11	12	13	14	15	16
1	9	4	6	9	8	2	6	12	4	2	12	9	16	13	5	10
2	8	11	13	1	14	1	2	5	2	1	5	2	15	7	4	4
3	11	15	16	2	5	8	7	13	3	8	13	4	13	1	11	3
4	2	5	15	8	13	11	13	15	7	11	15	16	12	16	3	8
5	15	9	7	3	9	4	11	3	6	4	3	5	10	9	1	14
6	10	1	3	16	12	15	8	2	16	15	2	13	11	11	8	6
7	3	13	10	14	15	12	1	10	9	12	10	6	2	2	12	1
8	5	7	11	13	6	13	5	11	8	13	11	10	1	12	2	12
9	6	2	4	11	2	3	3	4	13	3	4	11	9	6	10	16
10	13	10	2	12	4	16	16	9	14	16	9	3	3	10	15	5
11	16	14	12	7	16	6	12	1	5	6	1	1	4	15	13	7
12	12	6	5	6	3	7	10	7	10	7	7	12	14	3	14	11
13	14	3	8	4	10	10	9	16	1	10	16	8	5	4	9	9
14	1	16	9	10	11	14	4	14	15	14	14	7	8	5	16	13
15	7	8	1	5	1	9	15	6	12	9	6	14	7	8	7	15
16	4	12	14	15	7	5	14	8	11	5	8	15	6	14	6	2
17	16	5	10	16	2	8	5	14	3	6	9	4	14	13	2	2
18	13	4	4	13	16	3	1	9	9	16	12	11	15	15	3	15
19	3	11	3	3	11	7	9	1	12	13	3	3	4	3	4	8
20	9	3	8	9	15	11	10	5	11	12	14	15	7	4	10	16
21	5	1	14	5	5	14	13	10	7	15	7	5	5	5	16	4
22	10	8	6	10	6	16	14	15	5	11	10	8	10	9	7	9
23	6	12	1	6	4	4	8	16	13	14	1	9	11	1	9	13
24	1	2	12	1	10	13	12	3	10	8	2	7	13	11	1	5
25	2	10	16	2	1	1	15	2	2	10	5	6	12	2	12	14
26	15	15	5	15	12	12	2	13	14	2	4	1	9	16	11	10
27	8	13	7	8	13	5	7	8	6	4	11	12	1	12	13	6
28	4	14	11	4	9	6	4	4	1	9	8	13	6	7	5	3
29	11	9	9	11	3	15	16	11	4	1	6	14	3	14	14	7
30	14	16	13	14	8	2	11	7	16	7	13	10	16	10	6	11
31	12	7	15	12	14	10	6	12	15	5	16	16	2	6	8	1
32	7	6	2	7	7	9	3	6	8	3	15	2	8	8	15	12
33	4	3	7	1	14	6	7	1	1	9	5	6	1	8	13	7
34	8	6	2	10	1	7	5	4	15	11	14	15	16	13	16	13
35	16	9	14	14	13	2	12	7	9	14	12	10	7	2	4	4
36	7	1	5	11	9	5	4	13	5	16	8	16	11	3	6	2
37	11	7	10	8	8	9	1	11	16	6	2	7	9	12	8	9
38	1	8	8	4	6	10	9	2	6	3	6	12	3	15	3	12
39	2	15	1	12	16	1	14	15	2	8	9	9	14	5	2	14
40	9	5	4	16	4	12	13	8	7	13	7	11	5	16	7	3
41	14	11	11	3	2	15	15	6	10	12	11	5	10	10	10	11
42	15	10	15	7	5	11	8	14	14	7	10	14	15	11	12	16
43	6	12	3	15	11	4	16	5	12	1	4	2	12	14	14	10
44	10	2	12	13	15	8	10	3	11	2	3	3	13	4	9	8
45	3	4	16	2	3	14	3	12	8	5	1	1	8	7	1	6
46	12	14	13	9	7	16	2	10	1	15	16	8	6	9	15	15
47	5	16	6	5	10	3	11	9	4	4	13	4	2	6	11	5
48	13	13	9	6	12	13	6	16	3	10	15	13	4	1	5	1

TABLE A.2–3
Random permutations of the numbers 1–8

	1	2	3	4	5	6	7	8	9	10	11	12	13	14	15	16	17	18	19	20
1	2	3	1	3	3	6	7	4	3	2	1	1	7	7	5	4	3	3	8	5
2	5	5	8	1	8	2	8	6	6	4	8	6	2	4	4	5	7	8	7	6
3	7	2	6	7	4	3	6	3	7	6	2	3	8	3	1	7	5	7	3	2
4	1	7	5	8	7	8	3	7	8	8	4	4	4	6	7	1	6	4	2	1
5	8	4	2	2	6	4	1	8	1	7	5	7	3	1	6	6	1	2	4	7
6	4	1	4	5	5	7	5	2	4	5	3	2	1	2	8	2	4	5	6	4
7	6	8	7	4	1	1	4	5	5	3	7	8	5	8	3	8	2	6	5	8
8	3	6	3	6	2	5	2	1	2	1	6	5	6	5	2	3	8	1	1	3
9	5	4	5	4	5	2	5	6	8	2	8	8	8	6	4	6	6	3	8	6
10	4	3	8	5	4	6	3	5	3	4	5	6	6	7	6	1	5	7	7	2
11	8	7	2	2	3	1	1	4	7	1	1	2	7	1	2	4	4	5	3	3
12	6	8	4	1	8	8	2	7	5	7	6	1	4	4	7	5	2	4	1	7
13	2	5	3	6	7	3	8	3	1	8	7	4	1	5	1	2	8	8	2	5
14	1	2	7	8	6	7	4	1	2	6	3	7	2	8	5	7	1	2	5	8
15	3	1	6	3	1	5	6	8	6	5	4	5	3	2	3	8	3	6	6	1
16	7	6	1	7	2	4	7	2	4	3	2	3	5	3	8	3	7	1	4	4
17	8	6	6	4	4	8	3	4	2	4	3	5	4	8	6	7	6	1	8	2
18	3	3	2	6	6	2	8	2	6	3	8	7	8	1	1	8	5	6	7	1
19	6	7	3	1	1	4	6	5	1	6	7	6	7	5	4	6	7	5	5	6
20	5	8	8	3	7	5	5	8	7	7	1	3	6	4	5	5	3	4	2	8
21	7	1	4	5	8	1	2	3	5	8	6	2	3	2	2	1	8	2	4	5
22	2	2	7	2	5	6	1	6	3	2	2	8	5	6	7	2	4	7	3	3
23	1	5	5	7	3	7	7	7	4	5	5	1	2	3	3	3	2	3	1	7
24	4	4	1	8	2	3	4	1	8	1	4	4	1	7	8	4	1	8	6	4
25	7	2	1	8	8	7	8	3	1	5	8	2	1	6	7	8	5	1	4	5
26	6	5	4	5	4	6	6	2	8	6	6	8	4	5	3	7	3	8	3	8
27	4	8	8	2	3	5	3	5	6	8	7	6	2	7	1	4	1	4	2	1
28	2	1	2	4	6	3	5	7	3	3	3	3	8	8	4	3	6	2	8	6
29	8	7	7	3	7	1	2	6	4	1	2	1	3	1	6	5	8	3	6	3
30	5	3	6	6	1	2	7	4	2	7	1	4	6	3	2	2	4	6	7	2
31	3	6	3	1	5	8	1	8	5	4	4	5	7	4	8	1	2	5	1	4
32	1	4	5	7	2	4	4	1	7	2	5	7	5	2	5	6	7	7	5	7
33	3	8	5	5	2	2	6	3	1	6	6	6	3	2	6	1	6	1	3	7
34	1	4	8	7	6	6	2	7	3	5	2	5	5	1	7	3	5	7	4	8
35	6	5	4	1	1	7	5	6	8	8	5	3	6	7	3	6	8	2	8	1
36	4	1	1	6	4	5	1	1	5	7	8	1	7	4	4	8	3	5	5	3
37	5	3	3	3	5	3	4	2	4	3	4	2	2	8	5	7	7	6	2	2
38	8	2	2	2	8	4	8	4	6	1	1	8	1	5	2	5	1	4	6	4
39	2	7	4	8	7	1	7	5	2	2	7	7	4	3	8	6	4	3	7	6
40	7	6	7	4	3	8	3	8	7	4	3	4	8	6	1	2	2	8	1	5

A.3   FORM FOR PLANNING A CONTROL CHART

1. OBJECTIVE OF THE CHART:

2. SAMPLING, MEASUREMENT, AND SUBGROUPING
   Variable(s) to be charted:
   Method of measurement:
   Magnitude of measurement variation:
   Point of sampling:
   Strategy for subgrouping:
   Frequency of subgroups:

3. MOST LIKELY SPECIAL CAUSES:

4. NOTES REQUIRED:

Note	Responsibility

5. REACTION PLAN FOR OUT-OF-CONTROL POINTS (attach copy):

6. ADMINISTRATION:

Task	Responsibility

   Making measurements:
   Recording data on charts:
   Computing statistics:
   Plotting statistics:
   Extending/changing control limits:
   Filing:

7. SCHEDULE FOR ANALYSIS:

A.4 FORM FOR DOCUMENTATION OF A PLANNED EXPERIMENT

1. **OBJECTIVE:**

2. **BACKGROUND INFORMATION:**

3. **EXPERIMENTAL VARIABLES:**

A. Response variables | Measurement technique
   1.
   2.
   3.

B. Factors under study | Levels
   1.
   2.
   3.
   4.
   5.
   6.
   7.

C. Background variables | Method of control
   1.
   2.
   3.

4. **REPLICATION:**

5. **METHODS OF RANDOMIZATION:**

6. **DESIGN MATRIX:** (attach copy)

7. **DATA COLLECTION FORMS:** (attach copies)

8. **PLANNED METHODS OF STATISTICAL ANALYSIS:**

9. **ESTIMATED COST, SCHEDULE, AND OTHER RESOURCE CONSIDERATIONS:**

A.5   DESIGN MATRICES FOR LOW CURRENT KNOWLEDGE

One half of a 2^3 Factorial Design
(any two factors form a full factorial pattern)

2^{3-1} Design Matix

Test	Run Order	1 23	2 13	3 12	Response
1		−	−	+	
2		+	−	−	
3		−	+	−	
4		+	+	+	

Divisor = 2

Effect

One Sixteenth of a 2^7 Factorial Design
(any two factors form a full factorial pattern)

2^{7-4} Design Matrix

Test	Run Order	1 24 35 67	2 14 36 57	3 15 26 47	4 12 56 37	5 13 46 27	6 23 45 17	7 34 25 16	Response
1		−	−	−	+	+	+	−	
2		+	−	−	−	−	+	+	
3		−	+	−	−	+	−	+	
4		+	+	−	+	−	−	−	
5		−	−	+	+	−	−	+	
6		+	−	+	−	+	−	−	
7		−	+	+	−	−	+	−	
8		+	+	+	+	+	+	+	

Divisor = 4

Effect

Design matrix for a 2^{15-11} pattern (any two factors form a full factorial pattern)

Test	Run order	1	2	3	4	5	6	7	8	9	10	11	12	13	14	15
		. . . . Two-factor interactions														
1		−	−	−	−	+	+	+	+	+	+	−	−	−	−	+
2		+	−	−	−	−	−	−	+	+	+	+	+	+	−	−
3		−	+	−	−	−	+	+	−	−	+	+	+	−	+	−
4		+	+	−	−	+	−	−	−	−	+	−	−	+	+	+
5		−	−	+	−	+	−	+	−	+	−	+	−	+	+	−
6		+	−	+	−	−	+	−	−	+	−	−	+	−	+	+
7		−	+	+	−	−	−	+	+	−	−	−	+	+	−	+
8		+	+	+	−	+	+	−	+	−	−	+	−	−	−	−
9		−	−	−	+	+	+	−	+	−	−	−	+	+	+	−
10		+	−	−	+	−	−	+	+	−	−	+	−	−	+	+
11		−	+	−	+	−	+	−	−	+	−	+	−	+	−	+
12		+	+	−	+	+	−	+	−	+	−	−	+	−	−	−
13		−	−	+	+	+	−	−	−	−	+	+	+	−	−	+
14		+	−	+	+	−	+	+	−	−	+	−	−	+	−	−
15		−	+	+	+	−	−	−	+	+	+	−	−	−	+	−
16		+	+	+	+	+	+	+	+	+	+	+	+	+	+	+

Divisor = 8

Effect

A.6 DESIGN MATRICES FOR MODERATE CURRENT KNOWLEDGE

Some balanced incomplete block designs

				Block number									
t	b	k	r	1	2	3	4	5	6	7	8	9	10
3	2	3	2	A	A	B							
				B	C	C							
4	2	6	3	A	A	A	B	B	C				
				B	C	D	C	D	D				
4	3	4	3	A	A	A	B						
				B	B	C	C						
				C	D	D	D						
5	2	10	4	A	A	A	A	B	B	B	C	C	D
				B	C	D	E	C	D	E	D	D	E
5	3	10	6	A	A	A	A	A	A	B	B	B	C
				B	B	B	C	C	D	C	C	D	D
				C	D	E	D	E	E	D	E	E	E
5	4	5	4	A	A	A	A	B					
				B	B	C	B	C					
				C	C	D	D	D					
				D	E	E	E	E					
6	3	10	5	A	A	A	A	A	B	B	B	C	D
				B	B	C	C	D	C	C	D	E	E
				E	F	D	F	E	D	E	F	F	F
6	4	15	10	A	A	A	A	A	A	A	A		
				B	B	B	B	B	B	C	C		
				C	C	C	D	D	E	D	D		
				D	E	F	E	F	F	E	F		
				A	A	B	B	B	B	C			
				C	D	C	C	C	D	D			
				E	E	D	D	E	E	E			
				F	F	E	F	F	F	F			
6	5	6	5	A	A	A	A	A	B				
				B	B	B	B	C	C				
				C	C	C	D	D	D				
				D	D	E	E	E	E				
				E	F	F	F	F	F				

t = Number of factor levels or combinations
b = Block size (number of experimental units per block)
k = Number of blocks required for balanced design
r = Number of replications of each factor level required

A, B, C, D, E, and F represent factor levels or factor combinations.

One Half of a 2^4 Factorial Design
(any three factors form a full factorial pattern)

2^{4-1} Design Matrix

Test	Run order	1	2	3	4	14 / 23	24 / 13	34 / 12	Response
1		−	−	−	−	+	+	+	
2		+	−	−	+	+	−	−	
3		−	+	−	+	−	+	−	
4		+	+	−	−	−	−	+	
5		−	−	+	+	−	−	+	
6		+	−	+	−	−	+	−	
7		−	+	+	−	+	−	−	
8		+	+	+	+	+	+	+	

Divisor = 4
Effect

One Half of a 2^5 Factorial Design (any four factors form a full factorial pattern)

2^{5-1} Design Matrix

Test	Run Order	1	2	3	4	5	12	13	14	15	23	24	25	34	35	45	Response
1		−	−	−	−	+	+	+	+	−	+	+	−	+	−	−	
2		+	−	−	−	−	−	−	−	−	+	+	+	+	+	+	
3		−	+	−	−	−	−	+	+	+	−	−	−	+	+	+	
4		+	+	−	−	+	+	−	−	+	−	−	+	+	−	−	
5		−	−	+	−	−	+	−	+	+	−	+	+	−	−	+	
6		+	−	+	−	+	−	+	−	+	−	+	−	−	+	−	
7		−	+	+	−	+	−	−	+	−	+	−	+	−	+	−	
8		+	+	+	−	−	+	+	−	−	+	−	−	−	−	+	
9		−	−	−	+	−	+	+	−	+	+	−	+	−	+	−	
10		+	−	−	+	+	−	−	+	+	+	−	−	−	−	+	
11		−	+	−	+	+	−	+	−	−	−	+	+	−	−	+	
12		+	+	−	+	−	+	−	+	−	−	+	−	−	+	−	
13		−	−	+	+	+	+	−	−	−	−	−	−	+	+	+	
14		+	−	+	+	−	−	+	+	−	−	+	+	+	−	−	
15		−	+	+	+	−	−	−	−	+	+	+	−	+	−	−	
16		+	+	+	+	+	+	+	+	+	+	+	+	+	+	+	

Divisor = 8
Effect

One Sixteenth of a 2^8 Factorial Design
(any three factors form a full factorial pattern)

										2^{8-4} Design Matrix							
Test	Run Order	1	2	3	4	5	6	7	8	12 37 56 48	13 27 46 58	14 36 57 28	15 26 47 38	16 25 34 78	17 23 45 68	24 35 67 18	Response
1		−	−	−	+	+	+	−	+	+	+	−	−	−	+	−	
2		+	−	−	−	−	+	+	+	−	−	−	−	+	+	+	
3		−	+	−	−	+	−	+	+	−	+	+	−	+	−	−	
4		+	+	−	+	−	−	−	+	+	−	−	−	+	−	+	
5		−	−	+	+	−	−	+	+	+	−	−	+	+	−	−	
6		+	−	+	−	+	−	−	+	−	+	−	+	−	−	+	
7		−	+	+	−	−	+	−	+	−	−	+	+	−	+	−	
8		+	+	+	+	+	+	+	+	+	+	+	+	+	+	+	
9		+	+	+	−	−	−	+	−	+	+	−	−	−	+	−	
10		−	+	+	+	+	−	−	−	−	−	−	−	+	+	+	
11		+	−	+	+	−	+	−	−	−	+	+	−	+	−	−	
12		−	−	+	−	+	+	+	−	+	−	+	−	−	−	+	
13		+	+	−	−	+	+	−	−	+	−	−	+	+	−	−	
14		−	+	−	+	−	+	+	−	−	+	−	+	−	−	+	
15		+	−	−	+	+	−	+	−	−	−	+	+	−	+	−	
16		−	−	−	−	−	−	−	−	+	+	+	+	+	+	+	

Divisor = 8
Effect

Design matrix for a 2^{16-11} pattern (any three factors form a full factorial design)

Test Order	1	2	3	4	5	6	7	8	9	10	11	12	13	14	15	16	..........Two-factor interactions..........
1	−	−	−	−	+	+	+	+	+	+	−	−	−	−	+	+	− − − − + + + + + + + − − − − +
2	+	−	−	−	−	−	−	+	+	+	+	+	+	+	−	−	+ + − − − − − − + + + + + + − −
3	−	+	−	−	−	+	+	−	−	+	+	+	−	+	−	+	− + − + − − + + − − + + + − + −
4	+	+	−	−	+	−	−	−	−	−	+	−	−	+	+	+	+ + − − + − − − + − − + − + + +
5	−	−	+	−	+	−	+	−	+	−	+	−	+	+	−	+	− − + − + − + − + − + − + + −
6	+	−	+	−	−	+	−	−	+	−	−	+	−	+	+	+	+ − + − − + − − + − − + − + +
7	−	+	+	−	−	−	+	+	−	−	−	+	+	−	+	+	− + + − − − − + + − − + + − +
8	+	+	+	−	+	+	−	+	−	−	+	−	−	−	−	+	+ + + − + + − + + − + − − − −
9	−	−	−	+	+	+	−	+	−	−	−	+	+	+	−	+	− − − + + + − + − − − + + + −
10	+	−	−	+	−	−	+	+	−	−	+	−	−	+	+	+	+ − − + − − + + − − + + − + +
11	−	+	−	+	−	+	−	−	+	−	+	−	+	−	+	+	− + − + − + − − + − + − + + +
12	+	+	−	+	+	−	+	−	+	−	−	+	−	−	−	+	+ + − + + − + − + − + − + − −
13	−	−	+	+	+	−	−	−	−	+	+	+	−	+	+	−	− − + + + − − + + + − − + + −
14	+	−	+	+	−	+	+	−	−	+	−	−	+	−	+	+	+ − + + − + + − + + − + + − −
15	−	+	+	+	−	−	−	+	+	+	−	−	−	+	−	+	− + + + − − − + + + − − − + −
16	+	+	+	+	+	+	+	+	+	+	+	+	+	+	+	+	+ + + + + + + + + + + + + + +
17	+	+	+	+	−	−	−	−	−	+	+	+	+	−	−	−	− + + + + + + − − − − + +
18	−	+	+	+	+	+	+	−	−	−	−	−	+	+	−	+	− − − − + + + + + + − −
19	+	−	+	+	+	−	−	+	+	−	−	−	+	−	+	−	+ − − − + + − + + + + − + −
20	−	−	+	+	−	+	+	+	+	−	+	+	−	−	−	+	+ − + + − − + + − + + − + + +
21	+	+	−	+	−	+	−	+	−	+	−	+	−	−	+	−	+ − + − + − + − + + + − + + +
22	−	+	−	+	+	−	+	+	−	+	+	−	+	−	−	+	+ − + − + − + − + − + + + +
23	+	−	−	+	+	+	−	−	+	+	+	−	−	+	−	−	+ + − − − + + − − + + − + + − +
24	−	−	−	+	−	−	+	−	+	+	−	+	+	+	−	+	+ + − + + − + − − + − − − −
25	+	+	+	−	−	−	−	+	−	+	+	+	−	−	−	+	− − − − + + + − + − − − + + + −
26	−	+	+	−	+	+	−	−	+	+	−	+	+	−	−	−	+ − − + − − + + − − + +
27	+	−	+	−	+	−	+	+	−	+	−	+	−	+	−	−	+ − + − + − − + − + − + − +
28	−	−	+	−	−	+	−	+	−	+	+	−	+	+	+	−	+ + − + + − + − − + − − − −
29	+	+	−	−	−	−	+	+	+	+	−	−	−	+	+	−	− − − + + + − − + + + − − +
30	−	+	−	−	+	−	+	+	−	+	+	−	+	+	−	+	− + + − + + − + − − + − − −
31	+	−	−	−	+	+	+	−	−	−	+	+	+	−	+	−	− + + + − − + + + − − − + −
32	−	−	−	−	−	−	−	−	−	−	−	−	−	+	+	+	+ + + + + + + + + + + + +

Divisor = 16
Effect

2^3 Factorial Design in two-blocks
(any three factors form a full factorial pattern)

						1B	2B	3B	
Test	Run order	1	2	3	B	23	13	12	Response
Block 1									
1		−	−	+	+	−	−	+	
2		+	−	−	+	+	−	−	
3		−	+	−	+	−	+	−	
4		+	+	+	+	+	+	+	
Block 2									
5		+	+	−	−	−	−	+	
6		−	+	+	−	+	−	−	
7		+	−	+	−	−	+	−	
8		−	−	−	−	+	+	+	
Divisor = 4									
Effect									

B = Blocking variable.

2^4 design in two blocks of eight (any four factors form a full factorial pattern)

							1	1	1	2	2	2	3	3	4	Response
Test	Run order	1	2	3	4	B	2	3	4	B	3	4	B	4	B	B
Block 1																
1		−	−	−	−	+	+	+	+	−	+	+	−	+	−	−
2		+	+	−	−	+	+	−	−	+	−	−	+	+	−	−
3		+	−	+	−	+	−	+	−	+	−	+	−	−	+	−
4		−	+	+	−	+	−	−	+	−	+	−	+	−	+	−
5		+	−	−	+	+	−	−	+	+	+	−	−	−	−	+
6		−	+	−	+	+	−	+	−	−	−	+	+	−	−	+
7		−	−	+	+	+	+	−	−	−	−	−	−	+	+	+
8		+	+	+	+	+	+	+	+	+	+	+	+	+	+	+
Block 2																
9		−	+	−	−	−	−	+	+	+	−	−	−	+	+	+
10		−	−	+	−	−	+	−	+	+	−	+	+	−	−	+
11		+	+	+	−	−	+	+	−	−	+	−	−	−	−	+
12		−	−	−	+	−	+	+	−	+	+	+	−	−	+	−
13		+	+	−	+	−	+	−	+	+	−	−	+	+	−	−
14		+	−	+	+	−	−	+	+	−	−	+	+	−	−	−
15		−	+	+	+	−	−	−	−	+	+	+	−	+	−	−
16		+	−	−	−	−	−	−	−	−	+	+	+	+	+	+
Divisor = 8																
Effect																

B = Blocking variable.

Design matrix for a 2^{15-10} pattern in two blocks (any three factors form a full factorial pattern)

Test	Run order	1	2	3	4	5	6	7	8	9	10	11	12	13	14	15	b	.... Two-factor interactions ....
Block 1																		
1		−	−	−	−	+	+	+	+	+	+	−	−	+	−	+	+	+ − − + − + + − − + + − + − − +
2		+	−	−	−	+	−	−	+	+	+	+	+	+	−	−	+	− − + + + + − − + + − − − − + +
3		−	+	−	−	−	+	+	−	+	+	+	+	−	+	−	+	− + − + + − + − + − + − + − + +
4		+	+	−	−	−	−	−	−	+	+	−	−	−	+	+	+	− + + − + + − + − + − + + − − +
5		+	−	+	−	+	+	+	−	−	−	−	+	+	+	+	+	− + − + − − + − + + − + − + − +
6		−	−	+	−	−	+	+	+	−	−	+	−	−	−	+	+	+ + + + − − − − − − + + + + + +
7		+	+	+	−	−	−	−	+	+	+	−	−	+	+	−	+	+ + − − − − + + + + − − + − + +
8		−	+	+	−	+	−	−	−	+	+	+	−	−	−	+	+	+ − + − − + − + + − + − − − + +
9		−	−	−	+	+	−	−	−	−	−	+	+	−	+	+	+	+ − − + + − + − − + + − − + − +
10		+	−	−	+	−	−	−	+	−	−	−	−	+	−	+	+	+ + + + − − − − + + + + − − + +
11		−	+	−	+	+	+	+	+	−	−	−	−	−	−	−	+	+ + − − + + − − − − + + + + + +
12		+	+	−	+	+	−	−	−	−	−	+	+	+	−	+	+	− + + − + + − + + − + − + − − +
13		+	−	+	+	−	+	+	−	+	+	+	−	−	+	−	+	− − + + + + − − − − + + + + + +
14		−	−	+	+	+	+	+	+	+	+	−	−	+	−	−	+	+ − − + − + + − + − − + + − − +
15		−	+	+	+	−	−	−	−	+	+	−	−	−	+	−	+	+ + − − − − + + − − + + − + − +
16		+	+	+	+	+	+	+	+	+	+	+	+	+	+	+	+	+ + + + + + + + + + + + + + + +

Test	1	2	3	4	5	6	7	8	9	10	11	12	13	14	15	b	Two-factor interactions . . . .
Block 2																	
17	+	+	+	+	−	−	−	−	−	−	+	+	+	+	−	−	+ − − + − + + − − + + − + − − +
18	−	+	+	+	+	+	+	−	−	−	+	−	−	+	+	−	− − + + + + − − + + − − − − + +
19	+	+	+	+	+	+	+	−	+	−	−	+	+	−	+	−	− + − + − + − + − + − + − + − +
20	−	−	+	+	+	+	+	+	+	−	+	−	−	−	−	−	− + + − − + + − + − − + + − − +
21	+	+	−	+	−	−	−	+	+	+	−	+	+	−	+	−	− + + − + + − − + − + + − − + +
22	−	+	−	+	+	+	+	+	−	+	+	−	−	−	−	−	+ + + − − + − + − + − + − + − +
23	+	−	−	+	−	−	−	−	+	+	−	−	−	−	−	−	+ − + − − + − + − + − + − − + +
24	−	−	−	+	+	+	+	−	+	+	+	+	+	+	+	−	+ − − + + − − + − + − − + + − +
25	+	+	+	−	−	−	−	−	+	−	−	+	+	−	+	−	− + + − + − + − + − − + + − + +
26	−	+	+	−	+	+	+	−	+	−	+	−	−	−	−	−	+ + − + − + + − + − − + − + − +
27	+	+	+	−	+	+	+	+	−	−	−	+	+	+	+	−	+ + − + − + + − − + + − + − − +
28	−	−	+	−	−	−	−	+	−	−	+	−	−	+	−	−	+ + + − + − + − − + − + − + − +
29	+	+	−	−	−	−	−	+	−	+	−	+	+	+	+	−	+ + + − + − + − + − + − + − − +
30	−	+	−	−	+	+	+	+	+	+	+	−	−	+	−	−	− + + − − + + − − + − + − + − +
31	+	−	−	−	−	−	−	−	−	+	−	+	+	−	+	−	+ − − + − + + − + − − + + − − +
32	−	−	−	−	+	+	+	−	−	+	+	+	+	−	−	−	+ + + + − − − − + + + + − − − +

Divisor = 16

Effect

B = Blocking variable.

Design matrix for a 2^{7-3} pattern in two blocks
(any three factors form a full factorial pattern)

Test	Run Order	1	2	3	4	5	6	7	B	12 37 56 4B	13 27 46 5B	14 36 57 2B	15 26 47 3B	16 25 34 7B	17 23 45 6B	24 35 67 1B	Response
Block 1																	
1		−	−	−	+	+	+	−	+	+	+	−	−	−	+	−	
2		+	−	−	−	−	+	+	+	−	−	−	−	+	+	+	
3		−	+	−	−	+	−	+	+	−	+	+	−	+	−	−	
4		+	+	−	+	−	−	−	+	+	+	−	+	−	−	+	
5		−	−	+	+	−	−	+	+	+	−	−	+	+	−	−	
6		+	−	+	−	+	−	−	+	−	+	−	+	−	−	+	
7		−	+	+	−	−	+	−	+	−	−	+	+	−	+	−	
8		+	+	+	+	+	+	+	+	+	+	+	+	+	+	+	
Block 2																	
9		+	+	+	−	−	−	+	−	+	+	−	−	−	+	−	
10		−	+	+	+	+	−	−	−	−	−	−	−	+	+	+	
11		+	−	+	+	−	+	−	−	−	+	+	−	+	−	−	
12		−	−	+	−	+	+	+	−	+	−	+	−	−	−	+	
13		+	+	−	−	+	+	−	−	+	−	−	+	+	−	−	
14		−	+	−	+	−	+	+	−	−	+	−	+	−	−	+	
15		+	−	−	+	+	−	+	−	−	−	+	+	−	+	−	
16		−	−	−	−	−	−	−	−	+	+	+	+	+	+	+	

Divisor = 8

Effect

B = Blocking variable

Four Factor Design (3^{4-1})

Test	Run Order	A	B	C	D	Response
1		−	−	−	−	
2		+	+	0	−	
3		0	0	+	−	
4		0	+	−	−	
5		−	0	0	−	
6		+	−	+	−	
7		+	0	−	−	
8		0	−	0	−	
9		−	+	+	−	
10		+	+	−	0	
11		0	0	0	0	
12		−	−	+	0	
13		−	0	−	0	
14		+	−	0	0	
15		0	+	+	0	
16		0	−	−	0	
17		−	+	0	0	
18		+	0	+	0	
19		0	0	−	+	
20		−	−	0	+	
21		+	+	+	+	
22		+	−	−	+	
23		0	+	0	+	
24		−	0	+	+	
25		−	+	−	+	
26		+	0	0	+	
27		0	−	+	+	

Code	Interpretation
−	low level of factor
0	middle level of factor
+	high level of factor

Note: Most of the information on 2-factor interactions are clear. The following effects have some confounding:

$$AB \text{ with } CD$$
$$AC \text{ with } BD$$
$$AD \text{ with } BC.$$

If the interactions effects with factor A can be considered neglible, then the remaining interactions are clear.

Design matrices for three-level factorials

Code	Interpretation
−	Low level of factor
0	Middle level of factor
+	High level of factor

Five-factor design (3^{5-1})

Test	A	B	C	D	E
1	−	−	−	−	−
2	+	−	0	−	−
3	0	−	+	−	−
4	0	+	−	−	−
5	−	+	0	−	−
6	+	+	+	−	−
7	0	0	0	−	−
8	−	0	+	−	−
9	+	0	−	−	−
10	0	+	+	0	−
11	−	+	−	0	−
12	+	+	0	0	−
13	+	0	+	0	−
14	−	0	−	0	−
15	0	0	0	0	−
16	−	−	−	0	−
17	0	−	0	0	−
18	+	−	+	0	−
19	+	0	0	+	−
20	0	0	+	+	−
21	−	0	−	+	−
22	−	−	0	+	−
23	+	−	+	+	−
24	0	−	−	+	−
25	−	+	+	+	−
26	+	+	−	+	−
27	0	+	0	+	−

Test	A	B	C	D	E
28	0	0	−	−	0
29	−	0	0	−	0
30	+	0	+	−	0
31	+	−	−	−	0
32	0	−	0	−	0
33	−	−	+	−	0
34	+	+	0	−	0
35	0	+	+	−	0
36	−	+	−	−	0
37	+	−	+	0	0
38	0	−	−	0	0
39	−	−	0	0	0
40	−	+	+	0	0
41	+	+	−	0	0
42	0	+	0	0	0
43	−	0	−	0	0
44	+	0	0	0	0
45	0	0	+	0	0
46	−	+	0	+	0
47	+	+	+	+	0
48	0	+	−	+	0
49	0	0	0	+	0
50	−	0	+	+	0
51	+	0	−	+	0
52	0	−	+	+	0
53	−	−	−	+	0
54	+	−	0	+	0

Test	A	B	C	D
55	+	+	−	−
56	0	+	0	−
57	−	+	+	−
58	−	0	−	−
59	+	0	0	−
60	0	0	+	−
61	−	−	0	−
62	+	−	+	−
63	0	−	−	−
64	−	0	+	0
65	+	0	−	0
66	0	0	0	0
67	0	−	+	0
68	−	−	−	0
69	+	−	0	0
70	0	+	−	0
71	−	+	0	0
72	+	+	+	0
73	0	−	0	+
74	−	−	+	+
75	+	−	−	+
76	+	+	0	+
77	0	+	+	+
78	−	+	−	+
79	+	0	+	+
80	0	0	−	+
81	−	0	0	+

Note: All two-factor interactions are clear in this design

A.7 DESIGN MATRICES FOR HIGH CURRENT KNOWLEDGE

2^2 Factorial Design

Test	Run Order	1	2	12	Response
1		−	−	+	
2		+	−	−	
3		−	+	−	
4		+	+	+	

Divisor = 2
Effect

2^3 Factorial Design

		Design Matrix						
Test	Run Order	1	2	3	12	13	23	123	Response
1		−	−	−	+	+	+	−	
2		+	−	−	−	−	+	+	
3		−	+	−	−	+	−	+	
4		+	+	−	+	−	−	−	
5		−	−	+	+	−	−	+	
6		+	−	+	−	+	−	−	
7		−	+	+	−	−	+	−	
8		+	+	+	+	+	+	+	

Divisor = 4
Effect

Design matrix for a 2^4 factorial pattern

Test	Run Order	1	2	3	4	12	13	14	23	24	34	123	124	134	234	1234	Response
1		−	−	−	−	+	+	+	+	+	+	−	−	−	−	+	
2		+	−	−	−	−	−	−	+	+	+	+	+	+	−	−	
3		−	+	−	−	−	+	+	−	−	+	+	+	−	+	−	
4		+	+	−	−	+	−	−	−	−	+	−	−	+	+	+	
5		−	−	+	−	+	−	+	−	+	−	+	−	+	+	−	
6		+	−	+	−	−	+	−	−	+	−	−	+	−	+	+	
7		−	+	+	−	−	−	+	+	−	−	−	+	+	−	+	
8		+	+	+	−	+	+	−	+	−	−	+	−	−	−	−	
9		−	−	−	+	+	+	−	+	−	−	−	+	+	+	−	
10		+	−	−	+	−	−	+	+	−	−	+	−	−	+	+	
11		−	+	−	+	−	+	−	−	+	−	+	−	+	−	+	
12		+	+	−	+	+	−	+	−	+	−	−	+	−	−	−	
13		−	−	+	+	+	−	−	−	−	+	+	+	−	−	+	
14		+	−	+	+	−	+	+	−	−	+	−	−	+	−	−	
15		−	+	+	+	−	−	−	+	+	+	−	−	−	+	−	
16		+	+	+	+	+	+	+	+	+	+	+	+	+	+	+	

Divisor = 8

Effect

Two Factor Design (3^2)

Test	A	B
1	−	−
2	0	−
3	+	−
4	−	0
5	0	0
6	+	0
7	−	+
8	0	+
9	+	+

Three Factor Design (3^3)

Test	A	B	C
1	−	−	−
2	0	−	−
3	+	−	−
4	−	0	−
5	0	0	−
6	+	0	−
7	−	+	−
8	0	+	−
9	+	+	−
10	−	−	0
11	0	−	0
12	+	−	0
13	−	0	0
14	0	0	0
15	+	0	0
16	−	+	0
17	0	+	0
18	+	+	0
19	−	−	+
20	0	−	+
21	+	−	+
22	−	0	+
23	0	0	+
24	+	0	+
25	−	+	+
26	0	+	+
27	+	+	+

Code	Interpretation
−	Low level of factor
0	middle level of factor
+	High level of factor

GLOSSARY

accuracy degree of variation in individual measurements from the accepted standard value.

analytic study a study in which action will be taken on a cause-and-effect system to improve performance of a product or a process in the future.

attribute data (classification data or count data) (for classification data), the quality characteristic recorded in one of two classes. (for count data) the number of incidences of a particular type recorded.

background variable a variable that potentially can affect a response variable in an experiment but is not of interest as a factor; sometimes called a *noise variable* or *blocking variable*.

bias amount of deviation of the average of individual measurements from the accepted standard value.

blocks groups of experimental units treated in a similar way in an experimental design; usually defined by background variables.

capability of a process a prediction of the individual outcomes or measurements of a quality characteristic from a stable process.

cause system a particular combination of causes of variation that affect a quality characteristic.

cause-and-effect diagram a diagram that organizes potential causes into general categories such as methods, materials, machines, and people, and illustrates the common relationships with quality characteristics.

charter the first component of the model for improving quality, consisting of a general description, expected results, and boundaries for an improvement project. The general description provides an initial orientation for the team. The expected results provide more specific details to support the general description. The boundaries for the activities define the scope of the improvements.

chunk variable a variable developed by forming blocks of a certain combination of background variables.

common causes those causes that are inherent in the process over time, affect everyone working in the process, and affect all outcomes of the process.

composite design a design for evaluating nonlinear factor effects that is constructed by adding selected factor combinations to two-level factorial designs.

confounded effects the average effect of a factor, or a differential effect between factors (interactions), combined indistinguishably with the effects of other factor(s), block factor(s), or interaction(s).

control factors factors that can be assigned at specific levels, set by those designing the product or process (not directly changed by the customer).

control chart a statistical tool used to distinguish variation in a process due to common causes and variation due to special causes. Common types are the C chart, U chart, NP chart, P chart, X chart, X-bar and R chart, and X-bar and S chart.

current knowledge the second component of the model to improve quality, this a priori knowledge is essential to planning any change (via an improvement cycle, the third component) to a product or process. Development of current knowledge includes focusing on a process or product, identifying quality characteristics, and using other methods such as a flowchart or cause-and-effect diagram to document the knowledge of the team.

customer the person or group that receives or uses the outcome of a process.

degree of belief the extent to which an experimenter has made conclusions from an analytic study; cannot be quantified.

design matrix a simple listing of the combinations of factors to be run in a study.

dot diagram a plot, using a number line centered at zero, of the estimated effects from a factorial study. Effects clustered near zero on the diagram cannot be distinguished from variation due to nuisance variables and should not be considered important.

dot frequency diagram a graph used to analyze data from a nested experimental pattern. The graph is designed to visually partition the variation in the data among the nested factors.

effect the change in the response variable that occurs when a factor or background variable is changed from one level to another. The effect must be further described in the context in which it is used (a linear effect, an interaction effect, etc.).

enumerative study a study in which action will be taken on the universe.

evolutionary operation (EVOP) a strategy to run experiments on a process in operation. The EVOP studies were designed to be run by operators on a full-scale manufacturing process while they continued to produce output of satisfactory quality.

experiment   an analytic study to provide a basis for action.

experimental pattern   the arrangement of factor levels and experimental units in the design.

experimental unit   the smallest division of material in an experiment such that any two units may receive different combinations of factors. Examples of experimental units are a part, a batch, one pound of material, an individual person, or a 10-square-foot plot of ground.

external noise factors   noise factors relating to the environment in which the product is used or distributed.

factor   a variable that is deliberately varied or changed in a controlled manner in an experiment to observe its effect on the response variable; sometimes called an *independent variable* or *causal variable*.

flowchart   a display of the various stages in a process, using different types of symbols, to demonstrate the flow of product or service over time.

fractional factorial design   an experimental pattern in which only a particular fraction of the factor combinations required for the complete factorial experiment is selected to be run.

frame   a list of identifiable, tangible units, some or all of which belong to the universe, and any number of which may be selected and studied.

improvement cycle   an adaptation of the scientific method (consisting of four phases—plan, do, study, act) used to increase a team's knowledge about the product or process and to provide a systematic way of accomplishing change.

incomplete block design   an experimental pattern in which the number of experimental units in a block is less than the number of combinations of factors and levels.

interaction   a situation in which the effect that a factor has on the response may depend on the levels of some of the other factors.

internal noise factors   noise factors relating to product deterioration with age or use.

judgment samples   a selection of experimental units and conditions that is based on a judgement other than random selection.

level   a given value or specific setting of a quantitative factor, or a specific option of a qualitative factor, that is included in the experiment. The levels of a factor selected for study in the experiment may be fixed at certain values of interest, or they may be chosen from many possible values.

mixture designs   a special type of study in which the factors are ingredients that are mixed together. The response variables are thought to depend on the relative proportions of the components of the mixture rather than the absolute concentrations.

needs   objectives defined by the producer that will add value to a society. Customers are defined, and then products and services are designed to satisfy the customers.

nested or hierarchal design an experiment to examine the effect of two or more factors in which the same level of a factor cannot be used with all levels of other factors.

noise factors factors that can potentially affect the quality characteristic but cannot be controlled at the design phase.

nonlinear effect a change in a response variable that is not linearly related to the corresponding change in a factor.

nuisance variable an unknown background variable that can affect a response variable in an experiment, sometimes called a lurking variable or extraneous variable.

operational definition a definition that gives communicable meaning to a concept by specifying how the concept is applied within a particular set of circumstances.

paired-comparison experiments a single factor at two levels and one background variable.

parameter design a strategy intended to reduce the effect of all noise variables on quality characteristics for a product or process.

Pareto diagram a chart that orders data of occurrences by type, category, or other classification. Its purpose is not to identify the cause of some effect, but rather to point the right effect to study.

planned experimentation a collection of methods and a strategy to make a change to a product or process and observe the effect of that change on one or more quality characteristics with the purpose of helping experimenters gain the most information with the resources available.

planned grouping arrangement of experimental units in blocks.

precision of a measurement process degree of variation in individual measurements of the same item.

prediction a declaration of the value or state of some characteristic of a process or its outcome in the future.

process a set of causes and conditions that repeatedly come together to transform inputs into outcomes.

process improvement the continuous endeavor to learn about the cause system in a process and to use this knowledge to change the process to reduce variation and complexity and to improve customer satisfaction.

process owner the person at the lowest level in the hierarchy of the organization who has the authority to make fundamental changes to the process.

QFD relation diagram a matrix for relating the quality characteristics to factors that should be addressed in the design of a new product or process. This matrix helps define the current knowledge for a product and is analogous to the cause-and-effect diagram for current knowledge of a process.

quality characteristics a trait, preferably measurable, of an input or outcome of a process or a measure of performance of a process used to define quality.

quality characteristics diagram a matrix used to define quality by relating needs to quality characteristics.

random sample a sample of items selected using a random number table or similar device such that the selection of any particular unit depends purely on chance.

randomization the objective assignment of combinations of factor levels to experimental units.

randomized block design an experimental pattern in which the size of the block equals the number of combinations of factor and levels (either a single background variable or a chunk variable).

replication repetition of experiments, experimental units, measurements, treatments, etc. as part of the planned experiment.

response plots a plot illustrating the relationship between the response and important factors in an experiment.

response variable a variable observed or measured in an experiment, sometimes called a dependent variable. The response variable is the outcome of an experiment and is often a quality characteristic or a measure of performance of the process.

run chart plot of data in time order.

screening designs a set of fractional factorial designs used by the experimenter with a low level of knowledge to screen out unimportant factors.

sequential experimentation sequential building of knowledge using the improvement cycle (PDSA) with prediction as the aim.

special causes causes that are not in the process all the time or do not affect everyone, but arise because of specific circumstances.

stable process a process in which variation in outcomes arises only from common causes.

subgrouping organizing (classifying, stratifying, grouping, etc.) data from the process in a way that is likely to give the greatest chance for the data in each subgroup to be alike and the greatest chance for data in other subgroups to be different. The aim of rational subgrouping is to include only common causes of variation within a subgroup, with all special causes of variation occurring between subgroups.

supplier the person or group that provides an input to the process.

Taguchi loss function a quadratic function that relates loss of customer satisfaction with distance from a target of some quality characteristic.

tolerance design a strategy that helps set acceptable levels of variation for control factors (considering costs) of a product or process after parameter design has been performed.

unit-to-unit noise factors variations in the manufacturing process.

universe the entire group (e.g., people, material, invoices) possessing certain properties of interest.

unstable process a process in which variation is a result of both common and special causes.

variable data measured numerical values of the quality characteristic: a dimension, physical attribute, cost, or time.

variance components estimates of the variation due to each factor in a nested study.

INDEX